Instabilities and Nonequilibrium Structures III

Mathematics and Its Applications

Managing Editor:

M. HAZEWINKEL
Centre for Mathematics and Computer Science, Amsterdam, The Netherlands

Editorial Board:

F. CALOGERO, *Università degli Studi di Roma, Italy*
Yu. I. MANIN, *Steklov Institute of Mathematics, Moscow, U.S.S.R.*
M. NIVAT, *Université de Paris VII, Paris, France*
A. H. G. RINNOOY KAN, *Erasmus University, Rotterdam, The Netherlands*
G.-C. ROTA, *M.I.T., Cambridge, Mass., U.S.A.*

Volume 64

Instabilities and Nonequilibrium Structures III

edited by

E. Tirapegui

Facultad de Ciencias Físicas y Matemáticas,
Universidad de Chile,
Santiago, Chile

and

W. Zeller

Instituto de Física,
Universidad Católica de Valparaíso,
Valparaíso, Chile

KLUWER ACADEMIC PUBLISHERS

DORDRECHT / BOSTON / LONDON

Library of Congress Cataloging-in-Publication Data

Instabilities and nonequilibrium structures III / edited by E.
Tirapegui and W. Zeller.
p. cm. -- (Mathematics and its applications)
Includes index.
ISBN 0-7923-1153-1
1. Fluid dynamics--Congresses. 2. Stochastic processes--Congresses. 3. Stability--Congresses. I. Tirapegui, Enrique.
II. Zeller, W. III. Title: Instabilities and nonequilibrium structures three. IV. Series: Mathematics and its applications (Kluwer Academic Publishers)
QA911.I523 1991
532--dc20 91-18950

ISBN 0-7923-1153-1

Published by Kluwer Academic Publishers,
P.O. Box 17, 3300 AA Dordrecht, The Netherlands.

Kluwer Academic Publishers incorporates
the publishing programmes of
D. Reidel, Martinus Nijhoff, Dr W. Junk and MTP Press.

Sold and distributed in the U.S.A. and Canada
by Kluwer Academic Publishers,
101 Philip Drive, Norwell, MA 02061, U.S.A.

In all other countries, sold and distributed
by Kluwer Academic Publishers Group,
P.O. Box 322, 3300 AH Dordrecht, The Netherlands.

Printed on acid-free paper

Printed in the Netherlands

SERIES EDITOR'S PREFACE

'Et moi, ..., si j'avait su comment en revenir, je n'y serais point allé.'

Jules Verne

The series is divergent; therefore we may be able to do something with it.

O. Heaviside

One service mathematics has rendered the human race. It has put common sense back where it belongs, on the topmost shelf next to the dusty canister labelled 'discarded non-sense'.

Eric T. Bell

Mathematics is a tool for thought. A highly necessary tool in a world where both feedback and non-linearities abound. Similarly, all kinds of parts of mathematics serve as tools for other parts and for other sciences.

Applying a simple rewriting rule to the quote on the right above one finds such statements as: 'One service topology has rendered mathematical physics ...'; 'One service logic has rendered computer science ...'; 'One service category theory has rendered mathematics ...'. All arguably true. And all statements obtainable this way form part of the raison d'être of this series.

This series, *Mathematics and Its Applications*, started in 1977. Now that over one hundred volumes have appeared it seems opportune to reexamine its scope. At the time I wrote

> "Growing specialization and diversification have brought a host of monographs and textbooks on increasingly specialized topics. However, the 'tree' of knowledge of mathematics and related fields does not grow only by putting forth new branches. It also happens, quite often in fact, that branches which were thought to be completely disparate are suddenly seen to be related. Further, the kind and level of sophistication of mathematics applied in various sciences has changed drastically in recent years: measure theory is used (non-trivially) in regional and theoretical economics; algebraic geometry interacts with physics; the Minkowsky lemma, coding theory and the structure of water meet one another in packing and covering theory; quantum fields, crystal defects and mathematical programming profit from homotopy theory; Lie algebras are relevant to filtering; and prediction and electrical engineering can use Stein spaces. And in addition to this there are such new emerging subdisciplines as 'experimental mathematics', 'CFD', 'completely integrable systems', 'chaos, synergetics and large-scale order', which are almost impossible to fit into the existing classification schemes. They draw upon widely different sections of mathematics."

By and large, all this still applies today. It is still true that at first sight mathematics seems rather fragmented and that to find, see, and exploit the deeper underlying interrelations more effort is needed and so are books that can help mathematicians and scientists do so. Accordingly MIA will continue to try to make such books available.

If anything, the description I gave in 1977 is now an understatement. To the examples of interaction areas one should add string theory where Riemann surfaces, algebraic geometry, modular functions, knots, quantum field theory, Kac-Moody algebras, monstrous moonshine (and more) all come together. And to the examples of things which can be usefully applied let me add the topic 'finite geometry'; a combination of words which sounds like it might not even exist, let alone be applicable. And yet it is being applied: to statistics via designs, to radar/sonar detection arrays (via finite projective planes), and to bus connections of VLSI chips (via difference sets). There seems to be no part of (so-called pure) mathematics that is not in immediate danger of being applied. And, accordingly, the applied mathematician needs to be aware of much more. Besides analysis and numerics, the traditional workhorses, he may need all kinds of combinatorics, algebra, probability, and so on.

In addition, the applied scientist needs to cope increasingly with the nonlinear world and the extra mathematical sophistication that this requires. For that is where the rewards are. Linear models are honest and a bit sad and depressing: proportional efforts and results. It is in the nonlinear world that infinitesimal inputs may result in macroscopic outputs (or vice versa). To appreciate what I am hinting at: if electronics were linear we would have no fun with transistors and computers; we would have no TV; in fact you would not be reading these lines.

There is also no safety in ignoring such outlandish things as nonstandard analysis, superspace and anticommuting integration, *p*-adic and ultrametric space. All three have applications in both electrical engineering and physics. Once, complex numbers were equally outlandish, but they frequently proved the shortest path between 'real' results. Similarly, the first two topics named have already provided a number of 'wormhole' paths. There is no telling where all this is leading - fortunately.

Thus the original scope of the series, which for various (sound) reasons now comprises five subseries: white (Japan), yellow (China), red (USSR), blue (Eastern Europe), and green (everything else), still applies. It has been enlarged a bit to include books treating of the tools from one subdiscipline which are used in others. Thus the series still aims at books dealing with:

- a central concept which plays an important role in several different mathematical and/or scientific specialization areas;
- new applications of the results and ideas from one area of scientific endeavour into another;
- influences which the results, problems and concepts of one field of enquiry have, and have had, on the development of another.

How can one control instabilities and nonequilibrium structures, or, better, make good use of them. Nature does so quite effectively and on a large scale; so do all kinds of athletes; by and large technology does not. The first steps consist in understanding and describing such structures and that is a branch of mathematics that is in full swing right now as this volume will testify. I called it a branch of mathematics in the previous line but actually it is more like a web of interconnections linking very many concepts and ideas: commutative and noncommutative (quantum) probability, universality in chaos, topological solitons, cellular automata, fragmentation, changes in symmetry, bifurcations, pattern formation, renormalization, ... This list but a few of the characteristic phrases and all of them are discussed in the present volume which thus represents a rich source of material from the unstable nonequilibrium world (which is just like the one we live in; fortunately for us).

The shortest path between two truths in the real domain passes through the complex domain.

J. Hadamard

La physique ne nous donne pas seulement l'occasion de résoudre des problèmes ... elle nous fait pressentir la solution.

H. Poincaré

Never lend books, for no one ever returns them; the only books I have in my library are books that other folk have lent me.

Anatole France

The function of an expert is not to be more right than other people, but to be wrong for more sophisticated reasons.

David Butler

Bussum, June 1991

Michiel Hazewinkel

FOREWORD

We present here a selection of the lectures given at the Third International Workshop on Instabilities and Nonequilibrium Structures in Valparaíso, Chile, in December 1989. The Workshop was organized by Facultad de Ciencias Físicas y Matemáticas of Universidad de Chile and by Instituto de Física of Universidad Católica de Valparaíso where it took place.

This periodic meeting takes place every two years in Chile and aims to contribute to the efforts in Latin America towards the development of scientific research. This development is certainly a necessary condition for progress in our countries and we thank our lectures for their warm collaboration to fulfill this need. We are also very much indebted to the Chilean Academy of Sciences for sponsoring officially this Workshop.

We thank also our sponsors and supporters for their valuable help, and most especially UNESCO, National Science Foundation (USA), and Fundación Andes of Chile. The efforts of M.Alain Siberchicot, Scientific Advisor at the French Ambassy in Santiago, have been essential for our success and we acknowledge here the generous support of the Scientific Cooperation Program of France for Chile. Ms. Alda Bertoni and Cecilia Campos deserve a special mention for their remarkable work during the realization of the Workshop. We acknowledge also the help of Ms. Elsa Ñanco in the material preparation of this book. We are grateful to Professor Michiel Hazewinkel for including this book in his series and to Dr. David Larner of Kluwer for this continuous interest and support to this project.

E. Tirapegui
W. Zeller

LIST OF SPONSORS OF THE WORKSHOP

- Academia Chilena de Ciencias
- Academia de Ciencias de América Latina
- Facultad de Ciencias Físicas y Matemáticas de la Universidad de Chile
- Instituto de Física de la Universidad Católica de Valparaíso
- CONICYT (Chile)
- Ministere Francais des Affaires Etrangeres
- UNESCO
- International Centre for Theoretical Physics (Trieste)
- DFG (Germany)
- FNRS (Belgium)
- Fundacion Andes (Chile)
- Sociedad Chilena de Física
- Departamento Técnico de Investigación y de Relaciones Internacionales de la Universidad de Chile

TABLE OF CONTENTS

PREFACE

The articles presented here deal with mathematical and physical aspects of dynamical systems, stochastic effects, instabilities and pattern formation in extended systems.

Part I is devoted mainly to the mathematics of dynamical systems with a finite number of variables and several papers review recent developments on temporal chaos. Articles on statistical mechanics and cellular automata are also included here.

Part II collect works on the effect of noise on dynamical systems near bifurcation points. Approximation methods to calculate properties of the stochastic processes involved are presented and the approach based on functional integration is discussed for white and colored noise.

In Part III we have included papers which emphasize experimental or phenomenological aspects. Recent results on the study of extended systems are presented here and point to the development of a qualitative theory of the non linear partial differential equations which modelize these systems. Topics such as turbulence, spatio-temporal chaos and the role of topological defects receive special attention.

E. Tirapegui and W. Zeller (eds.), Instabilities and Nonequilibrium Structures III, 1.

PART I.

DYNAMICAL SYSTEMS AND STATISTICAL MECHANICS

FAST DYNAMO FOR SOME ANOSOV FLOWS

P.COLLET
Centre de Physique Théorique
UPRA 14 du CNRS
Ecole Polytechnique
F-91128 Palaiseau Cedex (France)

I Introduction.

The transport of passive quantities by hydrodynamical flows is an important problem in theoretical as well as applied Science. In this paper, we will concentrate on the study of the evolution of a passive vector transported by a flowing liquid. An important application (among several others) is the development of magnetic fields in astrophysical situations [1,2,3,4,5,6].

In the transport of a passive quantity by a fluid flow, one assumes that there is no back influence of the transported quantity (here the magnetic field) on the evolution of the flow. This is of course a reasonable assumption as long as the amplitude of the passive quantity remains small enough. We will also make another important simplification which is often considered as a starting point for the more general problem, namely we will assume that the fluid flow is time stationary and incompressible.

Let v denote the (stationary) velocity field of the fluid, and h the magnetic field. The evolution equations for the system are given by

$$\partial_t h = R^{-1}\Delta h - [v,h]\ , \tag{1}$$

$$\mathrm{div} h = 0\ . \tag{2}$$

In these equations, R is the magnetic Reynolds number, and $[v,h]$ denotes the Lie bracket of two vector fields which is given by

$$[v,h]^l = v^i \partial_i h^l - h^i \partial_i v^l\ .$$

In this equation, as in all this paper we have adopted the convention of summing over identical covariant and contravariant identical indices.

In astrophysical situations, the magnetic Reynolds number R is very large and we will denote by ϵ its (small) inverse.

Equations (1) and (2) can in principle be integrated to give the time evolution of the magnetic field starting form a given initial condition. This time evolution will of course depend on the flow v. In several situations it is known that the magnetic field will decay to zero at large time. Several such so called anti-dynamo theorems have been described in the literature [6]. A simple example of such results is for a regular flow on a two dimensional compact manifold: the magnetic field will ultimately decay to zero.

A major step in the theory of fast dynamos was made by Arnold et al. [7] when they suggested that if the vector field v has positive Liapounoff exponents, then one can suspect an exponential

E. Tirapegui and W. Zeller (eds.), Instabilities and Nonequilibrium Structures III, 5–11.

fast growth of the magnetic field (fast dynamo) at least for small values of ϵ. Recall that since the vector field v is conservative and regular, one needs to consider at least three dimensional situations to allow for a positive Liapounoff exponent. An example was constructed in [7] with an exponential growth rate indeed equal to the Liapounoff exponent. See also [13].

A simplified approach to the result of Arnold et al. can be obtained by considering equation (1) with $\epsilon = 0$. One gets

$$\partial_t h = -[v, h] \,. \tag{3}$$

This equation can be solved explicitly using the flow φ_t associated to the vector filed v (recall that it is a family of diffeomorphisms such that φ_0 is the identity and $\partial_t \varphi_t(x) = v(\varphi_t(x))$). It is easy to verify that the solution of equation (3) with initial condition $h(x,0)$ is given by

$$h(x,t) = D\varphi_t(\varphi_{-t}(x))\, h(\varphi_{-t}(x), 0) \,. \tag{4}$$

This formula is obvious in Lagrangian coordinates. If the vector field v has a positive Liapounoff exponent, the matrix $D\varphi_t$ will grow exponentially fast in time when applied to a generic vector, and we get a fast dynamo.

The above argument is however based on several weak points. The first one is that we cannot compare without care equations (1) and (3) because we dropped the term with highest derivative (degenerate perturbation). This is similar to the problem of the semi classical limit in Quantum Mechanics. The spectrum of the unperturbed problem is however probably very different here. In Quantum Mechanics, the unperturbed operator is a multiplication operator which has a rather regular continuous spectrum. Here the unperturbed spectrum is not well known (see [8] for related questions) but can be computed in some special cases where it appears to be an infinitely degenerate discrete spectrum. A second difficulty is that we have been rather careless with the use of the Liapounoff exponent. In particular one should use the Liapounoff exponent of the absolutely continuous invariant measure. Finally, if one considers compact manifolds, one should work using invariant differential operators. All these difficulties were solved for a special case in [7].

In this paper, we will consider a more general situation. Although we will study vector fields on three dimensional compact Riemaniann manifolds, we will use flat coordinates in order to simplify the notations. Let $\mathcal{L}_\epsilon$ denote the differential operator which appears on the right hand side of equation (1). Since we are on a compact manifold, this operator being elliptic has a discrete spectrum, which can only accumulate at the infinite left hand side of the complex plane. Let $\lambda_{\max}$ be the largest real part of the points in the spectrum of $\mathcal{L}_\epsilon$. This number will in general depend on ϵ (and v). We will say that v generates a fast dynamo if

$$\lambda = \liminf_{\epsilon \to 0} \lambda_{\max} > 0 \,.$$

We will call λ the dynamo rate. Note that there is an interchange of limits compared to the naive approach explained above. One first computes the point in the spectrum with maximal real part (this gives of course the behaviour of the magnetic field for large time) and then take the limit $\epsilon \to 0$. As we will see below these two limits do not commute in general because of the influence of the large deviations of the Liapounoff exponent.

II Stochastic Solution.

In this section we will explain how to solve equation (1) using probabilistic techniques. In order to use these techniques, we first have to get rid of the constraint (2).

Proposition 1. *Let h be a vector field satisfying equation (1) but not necessarily equation (2). Let u be the divergence of h. Then u satisfies the diffusion equation*

$$\partial_t u = \epsilon \Delta u - v^i \partial_i u .$$

This result follows at once from the fact that the velocity field v has divergence 0.

Corollary 2. *Any eigenvector of the operator $\mathcal{L}_\epsilon$ corresponding to an eigenvalue with positive real part has a zero divergence.*

Let h be such an eigenvector, u its divergence and λ the corresponding eigenvalue. The result follows at once from the identity

$$2\Re\lambda \int \overline{u} u \, d\mu = -\epsilon/2 \int \partial^i \overline{u} \partial_i u \, d\mu - \int v^i \partial_i (\overline{u} u) \, d\mu ,$$

where $d\mu$ is the Riemann measure. Since v has zero divergence, the last term is zero and the result follows.

As a consequence, since we are only interested with eventual eigenvalues with positive real part, it is enough to understand the spectrum of $\mathcal{L}_\epsilon$ without assuming the constraint (2).

We can now write the solution of the evolution equation (1) using a Feynman-Kac formula. Let X be the stochastic process solving the Ito stochastic differential equation

$$dX^i = -v^i(X)dt + \epsilon^{1/2} dw^i , \tag{5}$$

where the w^i for $i = 1, 2, 3$ are three independent Brownian motions, and the initial condition is

$$X(0) = x .$$

The map

$$x \to X(t) = X(x, t)$$

is a stochastic semigroup [9], and using stochastic calculus, one can now solve equation (1).

Theorem 3. *The solution of (1) with initial condition $h(x, 0)$ is given by the vector Feynman-Kac formula*

$$h(x,t) = \mathbf{E}\left(DX(x,t)^{-1} h(X(x,t), 0)\right) . \tag{6}$$

The analogy with equation (4) is obvious, we have reversed time ($-v$ in equation (5)) for the convenience of the following analysis. There is however a difference for the growth rate of the solution. Equation (6) suggests an amplitude for the magnetic field given by

$$\mathbf{E}\left\{\|DX(x,t)^{-1}\|\right\} ,$$

and we will see below that this is indeed the correct answer. The growth rate is given by

$$\lim_{t\to\infty} t^{-1} \log \mathbf{E}\left\{\|DX(x,t)^{-1}\|\right\} ,$$

while the (average) Liapounoff exponent is given by

$$\lim_{t\to\infty} t^{-1}\mathbf{E}\left\{\log\|DX(x,t)^{-1}\|\right\} .$$

These two quantities are in general not equal because of the large deviations of the Liapounoff exponent. The first one being larger than the second one. To see this effect more clearly, one can define the pressure $F(\beta)$ by

$$F(\beta) = \lim_{t\to\infty} t^{-1}\log\mathbf{E}\left\{\|DX(x,t)^{-1}\|^{\beta}\right\} .$$

Our growth rate is equal to $F(1)$ and the Liapounoff exponent is $F'(0)$. Since F is a convex function, these two quantities can be equal only if F is affine (we assume F differentiable). In that case, the Legendre transform of F (often denoted by $f(\alpha)$) is trivial, and there is no large deviation (multifractal) effect for the Liapounoff exponent [10,11]. This is the case for the model investigated in [12].

The above considerations are of course only true for positive and small values of ϵ. In particular the quantity $F(1)$ defined above is also a function of ϵ. It turns out that this quantity has a limit when ϵ tends to zero which is the pressure of the Liapounoff exponent for the Sinaï-Bowen-Ruelle measure μ of the flow which is here the Riemaniann volume measure. In other words we have the following theorem

Theorem 4. *Suspensions of two dimensional Anosov maps which are Euclidean in the time direction generate a fast dynamo with dynamo rate given by*

$$F(1) = \lim_{t\to\infty} t^{-1}\log\int\|D\varphi_{-t}(x)\|\,d\mu(x) .$$

As explained before this theorem extends the results of Arnold et al. to the situation where the Liapounoff exponent has non trivial large deviations. The Euclidean assumption means that in the direction of the suspension the Riemaniann metric and the vector field are trivial.

Similar results were also obtained in [5,13,14,15,16] using discrete time approximations. We will discuss this approximation in more details in the next section.

III Estimation of the Dynamo Rate.

Using the results of the previous section, we will now estimate the dynamo rate, *i.e.* the eigenvalue of the operator $\mathcal{L}_\epsilon$ with largest real part. The idea is now to introduce an operator describing the evolution of the magnetic field over a fixed time and analysing this operator in terms of the deterministic flow ($\epsilon = 0$).

A natural way to do this for suspensions of Anosov maps is to use the Poincare section. There are some technical difficulties for defining random Poincare sections in the general case, and we refer to [12] for a precise definition of the suspensions for which the following analysis can be applied. The Poincare section denoted by P_ϵ will be obtained by integrating the stochastic equation (5) over a finite amount of time. If ϵ is small, the result will be a small perturbation of the deterministic Poincare section, at least with large probability. Using the theory of large deviation, one can prove that the random Poincare map and the deterministic Poincare map differ by a small amount with probability larger than [17]

$$1 - e^{-O(1)/\epsilon} .$$

The estimate on the largest eigenvalue will be obtained by estimating the large n behaviour of quantities of the form

$$\mathbf{E}\left\{DP_\epsilon^{-n}(h \circ P_\epsilon^n(x))\right\} , \tag{7}$$

which is a discrete analog of (6). Note that the mappings P_ϵ are not autonomous because they depend of the stochastic noise which evolves with time. They however satisfy Markoff properties. In particular, in the above formula, P_ϵ^n is not the nth power of P_ϵ but a product of mappings with shifted noise dependence.

As explained above, we can control P_ϵ except on a set of small probability, and this implies that we can only derive some estimates in the above formula for $n \ll \epsilon^{-1}$. In order be sure that we estimate the largest eigenvalue, we have however to consider the limit of infinitely large n. These two seemingly opposite constraints can be reconciled if we have an estimate of the gap between the real part of the largest eigenvalue, and the next one (actually we need also estimates on spectral projections). If the gap is large enough, the largest eigenvalue will dominate the value of (7) even for moderately large values of n. Notice that this is a main difference with problems involving discrete time intervals with bounded noise. If we wait long enough, large deviations will almost surely appear and destroy any direct relation between the deterministic and stochastic systems. We need the more precise spectral information contained in the gap in order to be able to estimate the eigenvalue with largest real part.

We now explain how the estimate is done first by reducing the vector setting to the scalar case. We first recall that the deterministic Poincare map has an invariant field of unstable directions. Using the method of invariant cones, one can prove that the stochastic Poincare map has a (covariant) stochastic field of unstable directions provided large deviation events do not occur. There is also a field of stable directions which is contracted by the tangent mapping. We refer to [12] for a discussion of these notions, and in particular for the non autonomous aspects of the problem.

Considering equation (7), we conclude that for a generic vector field h the vector

$$DP_\epsilon^{-n}(h \circ P_\epsilon^n(x))$$

will be aligned with the (stochastic) stable direction at x and time n which we denote by $e(x,n)$. Using again the theory of invariant cones, one can show that there is a Hölder continuous sequence of strictly positive random functions ρ such that

$$DP_\epsilon^{-1}e(P_\epsilon(x),n) = \rho(x,n+1)e(x,n+1) .$$

Here we have explicitly written the non autonomous dependence of e and ρ but not of P_ϵ for simplicity. It follows that in order to estimate the largest eigenvalue and the gap of $\mathcal{L}_\epsilon$, we have to estimate the same quantities for the scalar operator Q given by

$$Qf(x) = \mathbf{E}\left\{\rho(x,1)f(P_\epsilon(x))\right\} ,$$

where f is a continuous function.

We now observe that this operator maps positive functions f into positive functions. We will explain below that it is even positivity improving, namely if f is a continuous function which is positive and not identically zero, its image is a strictly positive function. For these operators, one can apply a theorem of Birkhoff which can be stated as follows in the present situation.

Theorem 5. *Let K be a linear operator on continuous functions given by a strictly positive continuous kernel $K(x,y)$. Let k be the finite quantity*

$$k = \sup_{x,y,z,t} \frac{K(x,y)K(z,t)}{K(x,t)K(y,z)} .$$

Then the eigenvalue with maximal modulus of the operator K is real and positive, the corresponding eigenvector is a strictly positive function, and the gap with the rest of the spectrum can be bounded explicitly in terms of k .

We refer to [18] for a proof of this theorem, and we now explain why the hypothesis are satisfied in the present context . If one neglects the multiplication by the random function ρ, the kernel of the operator is simply the transition probability. Therefore, since the stochastic process is basically a diffusion, there is at positive time a non zero probability to jump at any point. Moreover, if we wait long enough the diffusion will eventually invade all the manifold more or less evenly. Therefore we expect the above number k to be of a reasonable size after some time has elapsed *i.e.* if we take for K some (large enough) power of Q. Note that in the present situation, the diffusion effect is much faster due to the drifting of the chaotic deterministic system. The invasion time can be shown to be of order $\log \epsilon^{-1}$ instead of the usual $\epsilon^{-1/2}$. This is indeed the time needed for a piece of unstable manifold of length $\mathcal{O}(1)\epsilon^{1/2}$ to be so chaotically stretched and folded that any point in the manifold is at a distance at most $\epsilon^{1/2}$ of the image. The proof is of course more difficult because the function ρ is not a constant. We refer to [12] for the precise estimates and a complete proof of theorem 4 along these lines.

References

[1] T.Dombre, U.Frisch, J.M.Greene, J.M.Henon, A.Mehr, A.M.Soward. J.Fluid Mech. **167**, 353 (1986).

[2] D.Galloway, U.Frisch. Geophys. and Astrophys. Fluid Dyn. **36**, 53 (1986).

[3] S.Childress, P.Collet, U.Frisch, A.D.Gilbert, A.D.Moffat, G.M.Zaslavsky. Report on the Workshop "Small-diffusivity dynamos and dynamical systems", Nice 1989.

[4] A.M.Soward. J. Fluid Mech. **180**, 267 (1987).

[5] J.Finn. Chaotic generation of magnetic fields - The fast dynamo problem. Preprint University of maryland 1989.

[6] H.K.Moffat. *Magnetic Field Generation in Electrically Conducting Fluids.* Cambridge University Press, Cambridge 1978.

[7] V.I.Arnold, Ya.B.Zeldovich, A.A.Ruzmaikin, D.D.Sokoloff. Sov. Phys. J. E. T. P. **54**, 1083 (1981).

[8] C.Chicone, R.C.Swanson, in *Global Theory of Dynamical Systems,* Z.Nitecki and C.Robinson editors, Lecture Notes in Mathematics 819, Springer, Berlin, Heidelberg, New York 1980.

[9] N.Ikeda, S.Watanabe. *Stochastic Differential Equations and Diffusion Processes.* North Holland, Amsterdam, Tokyo 1981

[10] Y.Kifer. Izv. Akad. Nauk. SSSR, **38**, 1083 (1974), and J. Analyse Math. **47**, 111 (1986).

[11] L.S.Young. Ergod. Th. & Dynam. Sys. **6**, 311 (1986).

[12] P.Collet. Fast dynamos generated by Euclidean suspensions of Anosov flows. In preparation.

[13] M.M.Vishik. Geophys. Fluid Dyn. to appear.

[14] B.J.Bayly, S.Childress. Phys. Rev. Lett. **59**, 1573 (1987), Geophys. and Astrophys. Fluid. Dyn. **44**, 211 (1988).

[15] J.M.Finn, E.Ott. Phys. Fluids, **31**, 2992 (1988) and Preprint University of Maryland 1989.

[16] A.Lesne. In preparation.

[17] M.Freidlin, I.Ventsel. *Random Perturbations of Dynamical Systems.* Springer, Berlin, Heidelberg, New York 1984.

[18] G.Birkhoff. *Lattice Theory.* Amer. Math. Soc., Providence 1967.

HOW HORSESHOES ARE CREATED

Jean-Marc GAMBAUDO* and Charles TRESSER**

* Laboratoire de Mathématique, U.A. 168, Université de Nice Parc Valrose 06034, NICE Cedex, FRANCE.

** Department of Mathematics, I.B.M. Research division, T.J. Watson Research Center, Yortown Heights, NY 10598, U.S.A.

E. Tirapegui and W. Zeller (eds.), Instabilities and Nonequilibrium Structures III, 13–25.

1. TRANSITION TO CHAOS.

Maps...

Let f be a diffeomorphism of the plane. A periodic point of f is a point P such that $f^n(P) = P$ for some $n > 0$. This periodic point is a saddle if the Jacobian matrix of f^n at P has one eigenvalue out of the unit circle, and one inside. A saddle has two invariant manifolds: the stable manifold of P is the set of points W_P^s whose orbits $\{f^m(P)\}$ converge to P as $m \to \infty$, the unstable manifold of P is the set of points W_P^u whose orbits converge to P as $m \to -\infty$. If f maps the 2-disk D^2 into itself, we call the restriction of f to D^2 a smooth embedding, and still denote by f the new map.

Chaotic Dynamics...

Let f be a smooth embedding of the 2-disk D^2, and P a saddle periodic point of f. A transverse homoclinic point Q to P, is a point where the stable and unstable manifolds of P intersect transversally, i.e. the tangent vectors to W_P^s and W_P^u at Q are not colinear.

These points have been first discovered by Poincaré who noticed also their connection with complicated dynamics: by invariance of the stable and unstable manifolds, transverse homoclinic points come in infinite bunches (see Figure 1).

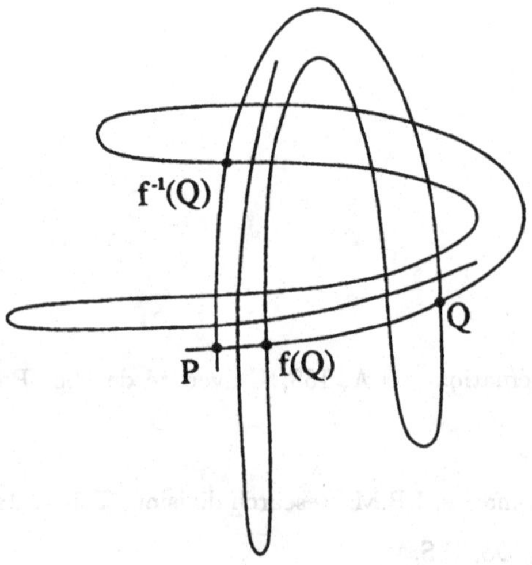

Figure 1.

This connection has been made more precise first by Birkhoff who proved that transverse homoclinic points imply the existence of infinitly many periodic orbits. Later, Smale described how transerve homoclinic points imply that the action of high iterates of the map, on a small rectangle based on the stable manifold, can be modeled by the Horseshoe maps (see Figure 2), and Katok proved that the existence of transverse homoclinic points is necessary, for a smooth enough map, to be chaotic.

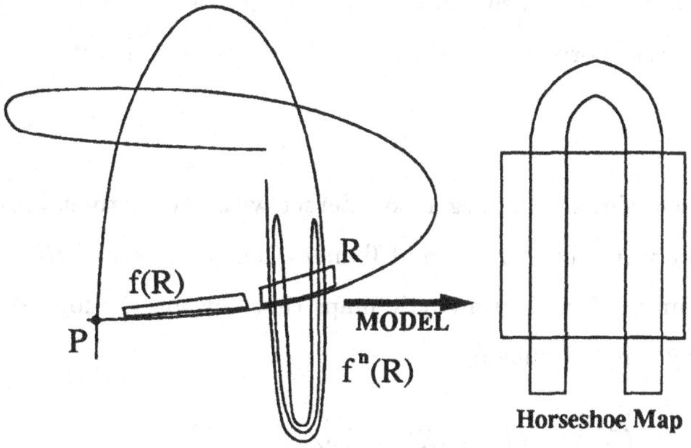

Figure 2.

More precisely, one has the following results:

Theorem 1 [Sma]:

A C^1 embedding of the 2-disk has some iterate acting as a horseshoe map on a subset of D^2, if and only if it has homoclinic points.

Theorem 2 [Kat]:

A $C^{1+\epsilon}$ embedding of the 2-disk has positive topological entropy if and only if it has homoclinic points.

Consequently, the question of how horseshoes are created arose naturally in bifurcation theory. The question can be stated as follows:

Q1. How does the dynamical behaviour evolve when varying the parameter of a continuous family of smooth embeddings $\{f_\mu\}_{\mu\in[0,1]}$ of the 2-disk D^2, such that f_0 is a contraction and f_1 possesses a horseshoe?

This question contains the following more precise ones:

Q2. In the space of $C^{1+\epsilon}$ embeddings of the disk, how can one describe the boundary between maps with and without transverse homoclinic points, and how are the maps lying on this boundary?

Remark that the condition on the smoothness is fundamental in this type of problem: in the C^0 topology, homeomorphisms of the disk generically have infinite topological entropy.

The Hénon mappings...

In the sixties, numerical simulations began to interact with the mathematics of dynamical systems. Following works by Pomeau and Ibanẽz, Hénon wrote in 1976 an equation describing a two-parameter family of analytic maps that develop the formation of a Horseshoe. Despite its very simple formula:

$$(X,Y) \rightarrow (1 - aX^2 + Y, bX),$$

this family is difficult to analyse from a dynamical point of view and far to be understood: actually, questions Q1 and even Q2 remain still mysterious. Nevertheless, an interesting point in the Hénon family is that, for $b = 0$, we are reduce to the iteration of the endomorphism of the interval:

$$X \rightarrow 1 - aX^2.$$

and this, we know much better!

There exists indeed the analogue of the questions Q1 and Q2 for spaces of endomorphisms of the interval, where no smoothness at all is needed to insure the equivalence of positive topological entropy with the existence of homoclinic points [Mis]. There, the answers are much more satisfying, specially in the case of Q2, even if they are still partial. In particular we know that a continuous one parameter family of smooth maps will go to positive topological entropy by a cascade of period doubling bifurcations [B.H.].

Boundary of chaos...

Many numerical experiments have been done on the Hénon models. Combining some small tests of ours with the huge collection of fine datas published by the Toulouse group (see e.g. [EH.M]) it seems that the boundary of positive topological entropy $B_{a,b}$, in the two parameter space (a,b) has the very particular shape, sketched on Figure 3 and that we are going to explain now.

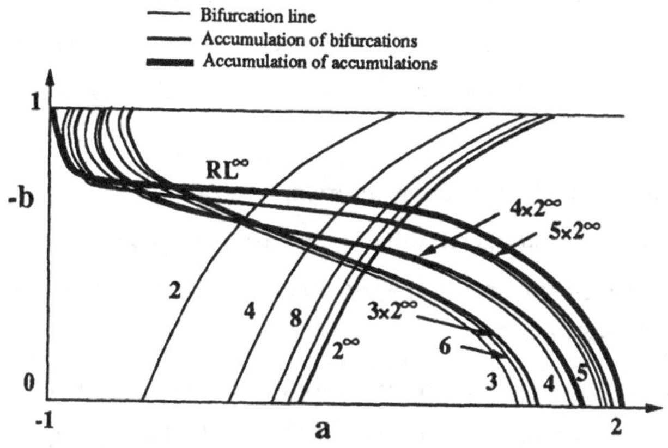

non-linear scales are used for better visualisation !

Figure 3.

For $b = 0$, we observe a sequence of increasing values a_{2^n} of the parameter a, corresponding to successive period doubling (or flip) bifurcations. There are no other bifurcations before the a_{2^n}'s accumulate to a value a_{2^∞}, but this is hard to prove. The next non-trivial fact is that this parameter value $(a_{2^\infty}, 0)$ is on the boundary $B_{a,b}$, because for a bigger than a_{2^∞} we know that the endomorphisms $X \to 1 - aX^2$ have positive topological entropy.

For $b < 0$, which is the case corresponding to orientation preserving maps and on which we are going to focalize, we can see lines l_{2^n} in the (a, b) space, emanating from the bifurcation points $(a_{2^n}, 0)$. On these lines a flip bifucation occurs where a sink with period 2^n tranforms into a saddle with period 2^n and where a sink with period 2^{n+1} appears. Until

now, nothing very surprising: this is just a consequence of the implicit function theorem! What is much more exciting is that the accumulation set l_{2^∞} of this set of lines l_{2^n} seems to be a line emanating from the point $(a_{2^\infty}, 0)$. For parameter values on l_{2^∞} with $-b$ small enough, the maps have a periodic orbit with period 2^n for all n, of saddle type and no other periodic orbit. This means that this line l_{2^∞} is not cut by any other bifurcation curves of periodic orbits, at least for small values of $-b$. This is the first phenomenon we want to focus on here. It is somehow surprising because of the following sequence of known facts.
(i) For $b = 0$, infinitely many values of the parameter $a > a_{2^\infty}$ corresponding to saddle node and flip bifurcations of the endomorphism $X \to 1 - aX^2$, accumulate on $(a_{2^\infty}, 0)$ from above.
(ii) Again thanks to the implicit function theorem, bifurcation lines in the two parameter space emanate from these parameter values. The astonishing fact is that none of these curves is going to cross l_{2^∞} for small $-b$, while it is known that they have mutual crossings abitrarily close to the axis $b = 0$.

The situation is even more surprising endeed since, as $-b$ increases, seemingly the **first** bifurcation line that crosses the line l_{2^∞}, is the line l_3 that corresponds to the birth, by saddle node bifurcation, of a periodic orbit with period 3. This line l_3 meets the axis $b = 0$ at a value $a = a_3$, where it is known (Sarkovskii [Sar]) that the endomorphism $X \to 1 - a_3X^2$ possesses already positive topological entropy, and periodic orbits of all the periods! To summarize, if the cycle 3 appears very late in the bifurcation sequence of the endomorphism $X \to 1 - aX^2$, it is the first one which appears on the set l_{2^∞} as $-b$ increases. After this first high point which will furnish the background of our mathematical discussion, let us describe for completness , the rest of what one can observe numerically, and that we hope to prove in later work.

l_3 is followed by a series of lines $l_{3.2^n}$ that correspond to a flip bifurcation from a cycle 3.2^n to a cycle 3.2^{n+1}, and that accumulate on a line $l_{3.2^\infty}$. The maps corresponding to parameter value on $l_{3.2^\infty}$ have a periodic orbit of saddle type of period 3.2^n for all n. Again $l_{3.2^\infty}$ is cut by a line l_4 corresponding to the creation of a cycle with period four. l_4 emanates from the axis $b = 0$ where it corresponds to the saddle node bifurcation of the cycle RL^3 (we use here the standard notation of symbolic dynamics of unimodal endomorphisms of the interval [C.E]).Quickly after its creation, this cycle 4 is going to make a cascade of period doubling that accumulate on a line $l_{4.2^\infty}$, which, in turn, is crossed by a line l_5 corresponding to the saddle-node of the 5-cycle RL^4, etc, etc...

The conjectural picture of the boundary $B_{a,b}$ is that it is piecewise smooth and constituted by segments of the successive lines $l_{n.2^\infty}$, starting from $(a_{2^\infty}, 0)$ with l_{2^∞} and ending, when n goes to infinity, at $(-1,-1)$.

Results and conjecture...

The description we gave of the boundary of positive topological entropy in the Hénon model suggests two main questions:

1-How much of it is true?

2- What parts of the description really depend on the model and is there something general or universal in this picture?

The state of our knowledge in two dimensional dynamics does not yet allow complete answers to these questions. However, in this paper we shall make an attempt to explain why part of this figure is true, and even universal for a large class of models including the Hénon one. More precisely, we are going to explain why the cycle 3 is the first one which cuts the line l_{2^∞} in the Hénon model and why this phenomenom is quite general. As a motivation, let us finish this long introduction with the following conjecture, directly induced by the previous discussion, and that describe what we would like to know about the boundary of chaos.

Conjecture:

In the space of C^k orientation preserving embeddings of the two disk, with $k \geq 1+\epsilon$, which are area contracting, generically, maps which belong to the boundary of positive topological entropy, have a set of periodic orbits which except for a finite subset, is made of an infinite number of periodic orbits with periods $k.2^n$ for a given k and all $n > 0$.

2. RENORMALIZATION

A short trip in the one dimensional theory...

Let M be the space of real-analytic functions $g : [-1,1] \to [-1,1]$ equipped with the supremum norm and such that for every g in M:

i) g has a unique critical point at 0 with g increasing on $[-1,0[$ and decreasing on

$]0,1]$,

ii) $g(0) = 1$ and $g''(0) \neq 0$,

iii) $g(1) < 0$.

In M we can define a functional operator R by

$$R(g(x)) = (g(1))^{-1}.gog(g(1).x)$$

The map $g \to Rg$ is called the renormalisation operator for period doubling, and we shall recall now what we need to know about this theory (a nice review on this subject can be found in [C.E]). The first main result of this theory goes as follows:

Theorem [Lan]:

The functional operator R has a fixed point $\phi \in M$ satisfying:

-ϕ is an analytic function defined in a neighborhood U of $[-1,1]$,

-ϕ is a symmetric function, $\phi(x) = f(x^2)$,

-The linearized operator DR_ϕ is a compact operator with one eigenvalue δ with modulus greater than one, the remainder of the spectrum being inside the open unit disk.

It follows from the theory of invariant manifolds, that there exists a **local** stable manifold $W^s_{loc}(\phi)$ in a neighborhood of ϕ in M , and that $W^s_{loc}(\phi)$ has codimension 1. Standard arguments from the theory of endomorphisms of the interval show that all maps in $W^s_{loc}(\phi)$ belong to the domain of all iterates of R. This implies that they must have the same topological dynamics .More precisely, they have a periodic orbit with period 2^n for each n, a single invariant Cantor set, on which the dynamics is well understood, and nothing else in the non wandering set. Since maps on this manifold converge to the fixed point ϕ under the iteration of R, it follows furthermore that the geometry of the invariant Cantor set is unique. This means that, whatever the map you choose on $W^s_{loc}(\phi)$, you will observe asymptotically the same ratios when comparing successive scales of the invariant Cantor set. $W^s_{loc}(\phi)$ is the accumulation set of a sequence of local manifolds $W^{2^n}_{loc}$ which correspond to the flip bifurcation where a 2^n cycle gets unstable and a 2^{n+1} is created.

Until now everything has been local; we just described the situation in the neighborhood of the fixed point ϕ. A deep contribution of Sullivan has been to obtain the following global picture [Sul] (we just give here a simple version of his result):

Theorem:

Any map in M which possesses the topological dynamics of the accumulation of period doubling, converges to the map ϕ under iteration of the renormalization operator. In particular,the orbit of the map $X \rightarrow 1 - a_{2\infty}X^2$ under the action of renormalization, converges to the fixed point ϕ.

Consequently, for all maps in M, the Cantor sets have the same asymptotical ratios. In other words, topology implies geometry.

And in the two dimensional theory...

For $\Delta > 0$, let us consider the disk:

$$C(\Delta) = \{(x,y) \in C^2 s.t. \exists x_0 \in [0,1] \text{satisfying} |x - x_0| + |y| < \Delta\}$$

and the disk

$$R(\Delta) = C(\Delta) \cap I\!R^2$$

Let $H(\Delta)$ be the set of of analytic functions from $C(\Delta)$ to C^2 mapping the reals on the reals and let M^* be the subset of $H(\Delta)$ of maps of the form:

$$(x,y) \rightarrow (g(x),0)$$

,where g is a map in M. We are going to consider now the set $E_{\epsilon,\Delta}$ of maps of $H(\Delta)$ that are ϵ-close, in the C^0 topology, to a map in M^* and whose restriction to $R(\Delta)$ is an embedding of $R(\Delta)$.

For ϵ sufficiently small, there exists in $E_{\epsilon,\Delta}$ a "big" ball B_ϕ surrounding the map $\phi*$: $(x,y) \rightarrow (\phi(x),))$ in which the real dynamics is as follows (see figure 4):

Any maps G in B_ϕ possesses a unique fixed point P_G which is hyperbolic and of saddle type. Let W_1^s be the connected component of the stable manifold in $R(\Delta)$. There exist, on the left of W_1^s, two other components of the stable manifold, W_3^s and W_2^s, such that:

-W_3^s is mapped in W_2^s,

-W_2^s is mapped in W_1^s,

and W_3^shas no preimage in $R(\Delta)$.

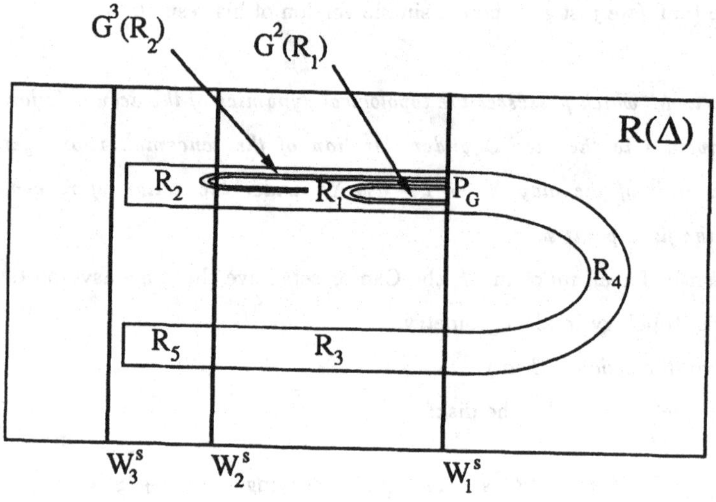

Figure 4.

Consider now the closed rectangle R_1, containing the fixed point on its boundary, whose vertical sides are included in W_2^s and W_1^s, and whose vertical sides are determined by the image of the boundary of $R(\Delta)$. The ball B_ϕ is chosen so that that $GoG(R_1) \subset R_1$, and consequently we can define a renormalization operator N:

$$R(\Delta) \rightarrow R(\Delta)$$

$$(x, y) \longmapsto N(G) = \Lambda^{-1} oGoGo\Lambda(x, y)$$

where Λ is a map sending the boundary of $R(\Delta)$ to the boundary of R_1. Of course, here the map Λ is not uniquely defined. We would have to add some supplementary condition to determine it completly. However, this two dimensional theory of renormalization has been rigorously developped by Collet , Eckmann and Koch [C.E.K] and its geometrical content has been described in [GST]. It follows from these two papers that for a somehow different renormalization operator N', there exists, in $H(\Delta)$ and in a neighborhood of ϕ^*, a local codimension 1 manifold $W^s_{loc}(\phi^*)$, containing ϕ^*, on which the maps have the following real dynamics:they have a periodic orbit with period 2^n for all n, no other periodic orbits and an invariant Cantor set. If we start with a map in $B_{\phi^*} \cap W^s_{loc}(\phi^*)$, its orbit under the iteration of the renormalization operator N, converges to a singular map, which maps the disk $R(\Delta)$ onto a line where the x dynamics is given by ϕ. Thus we can consider, in B^*_ϕ, the set $W^s(\phi)$ of maps which after iterations of the renormalization operator, belong eventually to $W^s_{loc}(\phi^*)$:

$$W^s(\phi^*) = B^*_\phi \cap \bigcup_{n=0 to \infty} N^{-n}(W^s_{loc}(\phi^*)).$$

3. CYCLE 3 IS NOT ALWAYS THE BAD GUY.

We are not able to describe the geometry of the set $W^s(\phi^*)$ in the large. For instance not all the maps in $W^s(\phi^*)$ have the property of having no other periodic orbit than the orbits with period 2^n for all n, as it is the case on $W^s_{loc}(\phi^*)$. We can nevertheless describe how this simple picture get spoiled when moving away from ϕ^* on $W^s(\phi)$, by the following result:

Theorem:

Let G be a map in $W^s(\phi^*)$ such that $N(G)$ has only a periodic orbit with period 2^n for all n and no other periodic orbit and assume that G has a periodic orbit which is not a periodic orbit of $N(G)$, then G has a periodic orbit with period 3.

Proof:

Let us make a partition of the disk $G(R(\Delta))$ in 5 regions R_1, R_2, R_3, R_4 and R_5, as shown in figure 3. If G has a periodic orbit O which is not a periodic orbit of $N(G)$, then this orbit cannot meet R_1. It cannot meet R_5 neither, because $GoG(R(\Delta)) \cap R_5 = \emptyset$. It follows that O has to remain in $R_2 \cup R_3 \cup R_4$. But $G(R_2) \subset R_3$ and $G(R_3) \subset R_4$, thus the period of O is a multiple of 3. Consider now the restriction h of the map G^3 to R_2. $G^3(R_2) \cap R_2 \neq \emptyset$ because it contains at least some part O_2 of the periodic orbit O, and obviously $G^3(R_2) \not\subset R_2$. We can complete the map h to get an orientation preserving embedding h^* of a disk D (see figure 5).

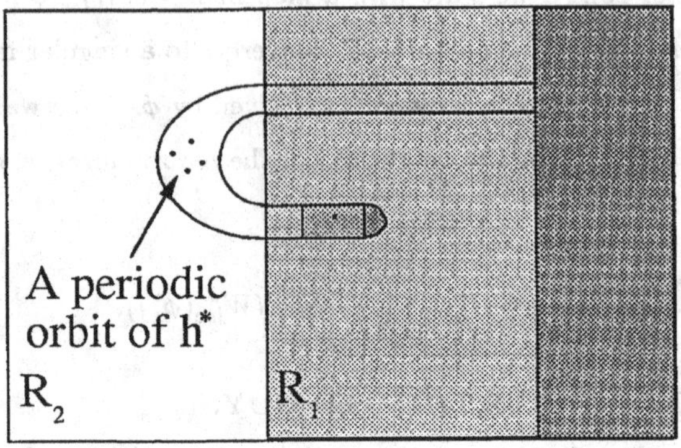

Figure 5.

We are going to assume that h^* has no fixed point in R_2 and get a contradiction. Let p be a fixed point of h^*, this fixed point (which necessarily exists thanks to Brouwer theorem) must be in the extra region we add to R_2 to get the map h^*.Consequently the boundary ∂R_2 of R_2 is a simple closed curve which is mapped isotopically to itself in the punctured disk $D - O_2 - \{p\}$. But we know [Gam], [Kol], that for an orientation preserving embedding of the disk with a periodic orbit O, there exists always a fixed point q such that any simple closed curve, bounding a disk that contains O and not q, cannot be sent isotopically to itself in the punctured disk $D - O_2 - \{q\}$. This yields a contradiction and thus h^* has a fixed point in R_2. Consequently G^3 has a fixed point in R_2 and G has a periodic orbit with period 3. Q.E.D.

REFERENCES

[B.H] L. Block and D. Hart, "The bifurcation of periodic orbits of one dimensional maps", *Erg. Th.. and Dynam. Syst.*, **2**, (1982) 125-129.

[C.E] P. Collet and J.P. Eckmannn, **Iterated maps of the interval as dynamical systems**, (Birkhauser, Basel) 1980.

[C.E.K.] P. Collet, J.P. Eckmannn and H. Koch: "Period doubling bifurcations for families of maps on $I\!R^n$", *J. Stat. Phys.* **25** (1980)1-15.

[EH.M] H. El Hamouly and C. Mira, "Lien entre les propriétés d'un endomorphisme de dimension un et celles d'un difféomorphisme de dimension deux", *C.R.Acad.Sc.Paris*,**293** (1981),525-528.

[Gam] J.M. Gambaudo " Linked fixed points of a C^1 orientation preserving embedding of D^2", to appear in *Proc. of the Cambridge Phil. Soc.*.

[Lan] O.E. Lanford III, " A computer assisted proof of the Feigenbaum conjectures", *B.A.M.S.*, **6**, (1982), 427-434.

[Kat] A. Katok, "Lyapunov exponents, entropy and periodic orbits for diffeomorphims", *Pub. Math. I.H.E.S.* **51** (1980), 137-174.

[Kol] B. Kolev " Point fixe lié à une orbite périodique d'un difféomorphisme de $I\!R^2$", to appear in the *Comptes rendus Acad. Sci. Paris.*

[Mis] M. Misiurewicz, " Horseshoes for mappings of the interval" *Bull. Acad. Pol. Ser. Sci. Math.* **27** (1979) 167-169.

[Sar] A. Sarkovskii, "Coexistence of cycles of a continuous map of the line into itself", (in russian), *Ukr. Mat. Z.*, **16**, (1964) 61-71.

[Sma] S. Smale, "Diffeomorphisms with many periodic points", in **Differential and combinatorial topology** (Princeton Univ. Press, Princeton 1965).

[Sul] D. Sullivan, " Quasi-conformal homeomorphisms in dynamics, topology, and geometry". **Proceedings of the I.C.M. Berkeley, California, U.S.A.** (1986).

Fine Structure of Universal Cantor Sets

Charles Tresser,
I.B.M., Thomas J. Watson Research Center,
P.O. Box 218, Yorktown Heights, N.Y. 10598.

I. INTRODUCTION.

Consider both the unique quadratic map at the boundary between zero and positive entropy, and any other smooth enough generic unimodal mapping with the same topological dynamics. If you look well enough at the invariant Cantor sets of these maps, not only do you see more and more similar geometries when looking at finer and finer scales, which is the main fact in this context, but it seems that you can also understand the way the small scales are organized, i.e. you can guess where to look for the various ratios describing the fine structure of the Cantor sets. It will take a fair amount of this paper to transform this single long statement to a long sequence of shorter and (hopefully) understandable ones. On the way, I shall give crude numbers and some remarks and theorems, some of which refer to deep questions but all of which are elementary. The very existence of some elementary mathematics in this context has been my main surprise these years; I "have known" (quotation marks often mean numerical evidence) the facts reported here (and some more details) since 1977, at the time Pierre Coullet and I discovered on our part universality and the role of the renormalization group in dynamics [CT] (i.e. before we learned of Mitchell Feigenbaum and of his slightly anterior but in any case, by then unpublished similar findings [Fe]). I shall be quite brief on generalizations to other renormalizable settings: I talked about these in Chile, but I shall only give very few hints here.

E. Tirapegui and W. Zeller (eds.), Instabilities and Nonequilibrium Structures III, 27–42.

II. CANTOR SETS.

Let us begin with some classical definitions, trying not to presuppose much more (nor much less) from the reader than Marston Morse and Gustav Hedlund did when writing their first joint paper, "Symbolic Dynamics", published in the American Journal of Mathematics in 1938.

A metric space M is:
- **compact** if each infinite sequence of points of M contains at least one sub-sequence which converges to a point of M,
- **totally disconnected** if its connected subsets reduce to points,
- **perfect** if no point is isolated,
- a **Cantor set** if it is compact, perfect and totally disconnected.

It is well known that all Cantor sets are homeomorphic. In particular, they all are homeomorphic to the **middle third Cantor set** $\mathcal{K}$ defined as follows:
- let $\mathcal{K}_0$ be another notation for the the unit interval I,
- $\mathcal{K}_n$ being defined as a union of 2^n disjoint subintervals $\mathcal{K}_{n,h}$ of I, define inductively $\mathcal{K}_{n+1}$ as the union of 2^{n+1} disjoint subintervals $\mathcal{K}_{n+1,k}$ of I, where $\mathcal{K}_{p+1,2q}$ and $\mathcal{K}_{p+1,2q+1}$ are the two sub-intervals which remain from $\mathcal{K}_{p,q}$ after taking out its middle third, and $\mathcal{K}_{p+1,2q+1}$ stands to the right of $\mathcal{K}_{p+1,2q}$ when the real line has its usual orientation (i.e. $\mathcal{K}_{p+1,2q+1} > \mathcal{K}_{p+1,2q}$),
- then let $\mathcal{K} = \bigcap_{n \geq 0} \mathcal{K}_n$.

Here we are in good shape to define a **doubling Cantor set** as the pair (i,ii) that we now specify:

i - a Cantor subset $\mathcal{D} = \bigcap_{n \geq 0} \mathcal{D}_n$ of the real line constructed as above except that:

- $\mathcal{D}_0$ is not necessarily the unit interval,
- there is still a single segment taken out from the inside of $\mathcal{D}_{p,q}$ to generate $\mathcal{D}_{p+1,2q}$ and $\mathcal{D}_{p+1,2q+1} > \mathcal{D}_{p+1,2q}$, but sizes depend on p and q,

ii - the **defining sequence** $\{\mathcal{D}_n\}_n$.

(*) Up to an orientation preserving affine transformation, a doubling Cantor set is determined by the doubly indexed sequence $\{\mathfrak{r}_{n,m}\}$ of ratios defined by $\mathfrak{r}_{p+1,a}=\dfrac{|\mathfrak{D}_{p+1,a}|}{|\mathfrak{D}_{p,q}|}$ where a is one of 2q and 2q+1 and |J| is the length of the interval J. The orientation preserving affine transformation corresponds to the choice of the initial interval $\mathfrak{D}_0$.

III. HOMEOMORPHISMS.

Given any doubling Cantor set $\mathfrak{D}$, we define two splitings of $\mathfrak{D}$ as the disjoint unions:

$$\mathfrak{D} = \cup_{n>0}\mathcal{A}_n \quad \text{and} \quad \mathfrak{D} = \cup_{n>0}\mathcal{H}_n$$

where $\mathcal{A}_0$ and $\mathcal{H}_0$ are just other notations for $\mathfrak{D}_0$, and:

- $\mathcal{A}_{n+1}$ is the right hand side half of $\mathcal{A}_0 \backslash \cup_{n\geq i>0}\mathcal{A}_i$ when n is even,
- $\mathcal{A}_{n+1}$ is the left hand side half of $\mathcal{A}_0 \backslash \cup_{n\geq i>0}\mathcal{A}_i$ when n is odd,
- $\mathcal{H}_{n+1}$ is the left hand side half of $\mathcal{H}_0 \backslash \cup_{n\geq i>0}\mathcal{H}_i$ for all n.

For any $m\geq 1$, let us define the map $F_m: \mathcal{A}_m \to \mathcal{H}_m$ as the unique monotone homeomorphism from $\mathcal{A}_m$ to $\mathcal{H}_m$ which:

- preserves orientation if m is even,
- reverses orientation if m is odd,
- respects the defining sequences structure.

There is a unique Cantor set homeomorphism $F: \mathcal{A} \to \mathcal{H}$ such that F_m is the restriction of F to $\mathcal{A}_m$, which we call the <u>period doubling minimal map</u> of $\mathfrak{D}$. One knows that this map is indeed <u>minimal</u> in the usual sense that every orbit is dense in $\mathfrak{D}$ (<u>hint:</u> for one proof of this fact, use § IV).

IV. PSEUDO-RENORMALIZATION.

Let $\mathfrak{D}$ stand for a doubling Cantor set, and F for its period doubling minimal map. We next remark that each of the sets $\mathbf{D}_{1,0}=\mathfrak{D}\cap\mathfrak{D}_{1,0}$ and $\mathbf{D}_{1,1}=\mathfrak{D}\cap\mathfrak{D}_{1,1}$ is also a doubling Cantor set. The restrictions of F^2 to both $\mathbf{D}_{1,0}$ and $\mathbf{D}_{1,1}$ are minimal self-homeomorphisms, but while the restriction to $\mathbf{D}_{1,1}$ is a period doubling minimal map, the restriction to $\mathbf{D}_{1,0}$ is not quite. However we get a new period doubling minimal map if beside restricting F^2 to $\mathbf{D}_{1,0}$, we also change the orientation of $\mathfrak{D}_{1,0}$. Altogether, we end up with two new period doubling minimal maps that we denote $\mathbf{R}_1(F)$ and $\mathbf{R}_0(F)$, and the operators $\mathbf{R}_1$ and $\mathbf{R}_0$ which transform F respectively to $\mathbf{R}_1(F)$ and $\mathbf{R}_0(F)$ we call **pseudo-renormalization operators**. We say "pseudo" since there is no "re-normalization" involved, i.e. the operators do not involve re-scaling; this is not needed since we follow the point of view expressed by (*) in §II.

From the preceding definitions and easy remarks, it immediately follows that one can perform successive pseudo-renormalizations as many times as one wants. This will help us to describe the geometrical properties of doubling Cantor sets.

V. RE-LABELINGS.

The main fact about the $\mathbf{R}_i$'s that we will use in this section is that, so to speak, $\mathbf{R}_1$ chooses the right of $\mathfrak{D}$ while $\mathbf{R}_0$ chooses the left hand half. Given any sequence $(s_1,..,s_n)$ of 0's and 1's, we will thus represent by $\mathbf{R}_{s_n}\text{o}...\text{o}\mathbf{R}_{s_1}(\mathfrak{D})$ the subinterval of $\mathfrak{D}$ (among the 2^n ones the which constitute $\mathfrak{D}_n$) which is the domain of the map $\mathbf{R}_{s_n}\text{o}...\text{o}\mathbf{R}_{s_1}(F)$. The ratios of the doubly indexed sequence $\{\mathfrak{r}_{p,q}\}$ will similarly be re-labeled, i.e. if $\mathbf{R}_{s_n}\text{o}...\text{o}\mathbf{R}_{s_1}(\mathfrak{D})$ or $\mathfrak{D}_{\{s_n,...,s_1\}}$ is the alternate (and more precise) notation for some $\mathfrak{D}_{n,m}$, then $\mathfrak{r}_{\{s_n,...,s_1\}}$ is the alternate name of the corresponding $\mathfrak{r}_{n,m}$.

VI. WHERE ARE THE INTERVALS?

To any sequence $(s_1,..)$ of 0's and 1's, we associate a dyadic rational (or a rational number with same bits if the sequence is finite) by the following algorithm where we "read" the given sequence $(s_1,..)$ and "write" a new one eventually to be considered a number:

- To start, in all cases write a zero, then a dot,
- keep reading the sequence $(s_1,..)$ while following the next instructions,
- assume the last written symbol is i, then:
    - if you read 0, write 1-i,
    - if you read 1, write i,
    - stop writing when there is nothing left to read.

If s stands for the original sequence, let v(s) stand for the number so generated, that we call the value of s. Values being numbers, the set of values carries the usual order (some readers may have recognized not only a version of a technique from kneading theory, but also the precise translation for R-L sequences for unimodal maps like $1- ax^2$: of course such background may help but is not assumed in the sequel).

The following result is quite easy:

PROPOSITION 1:

(i)- The intervals $\mathfrak{D}_{\{s_n,...,s_1\}}$ of $\mathfrak{D}_n$ are ordered along the real line in the order opposite from the one given by the values $v(s_1,..,s_n)$.

(ii)- Infinite sequences label all points of $\mathfrak{D}$.

(iii)- The points of $\mathfrak{D}$ are ordered as the inverses of the corresponding dyadic values.

VII. SIZES (AND RE-RE-LABELINGS !).

Let G_0 be the unique increasing homeomorphism from $\mathbf{D}_{1,0}$ to $\mathfrak{D}$, and G_1 be the unique decreasing homeomorphism from $\mathbf{D}_{1,1}$ to $\mathfrak{D}$. This defines the unique self map G of $\mathfrak{D}$ whose restriction to $\mathbf{D}_{1,i}$ is G_i. Let then H be a map from the interval $\mathfrak{D}_0$ to the real line whose restriction to $\mathfrak{D}$ is G. Then from [Su] we know that H can be made $C^{1+\alpha}$ for some α with $0<\alpha\leq 1$ if and only if the sequence of ratios $\mathbf{r}_{\{s_1\}}, \mathbf{r}_{\{s_2,s_1\}}, \mathbf{r}_{\{s_3,s_2,s_1\}}, \ldots, \mathbf{r}_{\{s_n,\ldots,s_3,s_2,s_1\}}$ converges exponentially fast when n goes to infinity. If the set of ratios has this property, we will say that $\mathfrak{D}$ is a **good** (doubling) Cantor set . The previous discussion generalizes to other inductive constructions of Cantor sets, whence the parentheses around the word doubling.

To any finite sequence $(s_1,..)$ of 0's and 1's, we will associate a second dyadic rational, the **size** $\mathbf{s}(s_1,..)$, which can also be understood as a rational number with same bits since now the sequence is finite (but we shoot for limits!). To get the size of $(s_1,..)$, we use the following algorithm, similar to the previous one described in §VI, except that now we will "read" the given sequence $(s_1,..)$ backward(i.e. we read s_n first, then s_{n-1} and so on) while "writing" $\mathbf{s}(s_1,..)$. To be precise, this goes as follows:

- to start, write in all case a zero, then a dot,
- read the sequence $(s_1,..)$ backward, while following the next instructions ,
- assume the last written symbol is i, then:
  - if you read 0, write 1-i,
  - if you read 1, write i,
  - stop writing when there is nothing left to read.

In these terms, we can say that $\mathfrak{D}$ is a good doubling Cantor set if and only if $\mathbf{r}_{\{s_n,\ldots,s_1\}}$ converges exponentially fast with $\mathbf{s}(s_1,\ldots,s_n)$.

Since sizes are in bijective correspondence with sequences, they can be used to write new labels: if $\mathbf{s}(s_1,...,s_n)=0.t_1...t_n$, then $\mathbf{r}_{\{s_n,...,s_1\}}$ can be designated as $\mathbf{r}_{\mathbf{s}^{-1}(0.t_1...t_n)}$.

VIII. UNIVERSALITY (BRIEF REMARKS).

The fact that period doubling homeomorphisms can be realized by smooth maps like some element of the quadratic family $q_a(x)=1-ax^2$ on some invariant Cantor set is not trivial but was known for a long time. That such Cantor sets are good was "discovered" at the same time as the much more surprising fact that the limit ratios do not depend on the function (hence the word "universality"), for a generic set of functions with the good topological dynamics. This simultaneity in the discoveries is linked to the facts that these properties were merely inferred from numerical observations, and that the renormalization group theory had been developed previously in statistical physics to explain similar universal features(1). The relevant phenomenology from [Fe, CT] has already been discussed in many books, and pieces of mathematical theory to explain some of these facts have been published by Massimo Campanino, Pierre Collet, Jean-Pierre Eckmann, Henri Epstein, Oscar Lanford, David Ruelle and others since on their tracks. All pieces of a more complete theory have been described to many audiences by Dennis Sullivan (to appear), although this is more a new beginning than the death of the subject (see e.g. [GST] for an application of universality to a classical mathematical question, and [OT] for more conjectures). While the rest of the paper will not presuppose any of this knowledge, even not at the phenomenological level, this might help.

(1)Renormalization group ideas came from quantum field theory (Stueckelberg & Petermann (1953), Gell-Man & Low (1954), Bogoliubov & Shirkov (1959) etc.: it's still alive!), and were adapted to a universality theory by Kadanoff (1966), di Castro & Jona-Lasinio (1969), Fisher (1972), etc. , and mainly Wilson (1971). Wilson's point of view is the one used in dynamical systems; for a tutorial approach in the frame of statistical physics, see e.g. [Ma], although as many, this book forgets the Italian team.

IX. NUMBERS.

The following numbers have been obtained from the approximation of a minimal Cantor set, computed from the function whose coefficients are given in [La] . This function is a good approximation of a real analytic function f such that $\mathfrak{D}$ and $\mathbf{D}_{1,0}$ are the same doubling Cantor set up to an orientation reversing affine transformation, with $F=\mathbf{R}_0(F)$ for the restriction F of f to $\mathfrak{D}$ (of course we are just trying to stick to what has been defined here: most readers could avoid the last long sentence). The table should be read as follows:

- The integers n from 1 to 5 refer to successive scales of the Cantor set,
- the first column on the right of n gives ratios $\mathfrak{r}_{\{s_n,\dots,s_1\}}$,
- the second column is made of the symbolic sequences $s_n,..,s_1$ without the comas, corresponding to the numbers on their left,
- the third column gives sizes as defined in § VII ordered in decreasing order,
- and the last column gives the ratios reordered according to the sizes immediately to their left.

1				
	.17225469	1	0.1	.39953530
	.39953530	0	0.0	.17225469
2				
	.16118324	11	0.11	.42116045
	.42116045	01	0.10	.39953536
	.39953536	00	0.01	.17225472
	.17225472	10	0.00	.16118324
3				
	.15986530	111	0.111	.42360342
	.42360342	011	0.110	.42116075
	.38897305	001	0.101	.39953566
	.17894857	101	0.100	.38897305
	.17225485	100	0.011	.17894857
	.39953566	000	0.010	.17225485
	.42116075	010	0.001	.16118336
	.16118336	110	0.000	.15986530

4

.15966663	1111	0.1111	.42396869
.42396869	0111	0.1110	.42360487
.38792838	0011	0.1101	.42116220
.17960347	1011	0.1100	.41899821
.17001441	1001	0.1011	.40340131
.40340131	0001	0.1010	.39953707
.41899821	0101	0.1001	.38897443
.16239762	1101	0.1000	.38792838
.16118392	1100	0.0111	.17960347
.42116220	0100	0.0110	.17894919
.39953707	0000	0.0101	.17225545
.17225545	1000	0.0100	.17001441
.17894919	1010	0.0011	.16239762
.38897443	0010	0.0010	.16118392
.42360487	0110	0.0001	.15986585
.15986585	1110	0.0000	.15966663

5

.15963769	11111	0.11111	.42403287
.42403287	01111	0.11110	.42397547
.38778389	00111	0.11101	.42361165
.17970026	10111	0.11100	.42325239
.16978105	10011	0.11011	.42180365
.40381246	00011	0.11010	.42116895
.41880695	01011	0.11001	.41900495
.16251109	11011	0.11000	.41880695
.16083594	11001	0.10111	.40381246
.42180365	01001	0.10110	.40340792
.39794690	00001	0.10101	.39954366
.17321672	10001	0.10100	.39794690
.17841349	10101	0.10011	.38985639
.38985639	00101	0.10010	.38898090
.42325239	01101	0.10001	.38793484
.16006641	11101	0.10000	.38778389
.15986843	11100	0.01111	.17970026
.42361165	01100	0.01110	.17960639
.38898090	00100	0.01101	.17895210
.17895210	10100	0.01100	.17841349
.17225824	10000	0.01011	.17321672
.39954366	00000	0.01010	.17225824
.42116895	01000	0.01001	.17001716
.16118651	11000	0.01000	.16978105
.16240023	11010	0.00111	.16251109
.41900495	01010	0.00110	.16240023
.40340792	00010	0.00101	.16118651
.17001716	10010	0.00100	.16083594
.17960639	10110	0.00011	.16006641
.38793484	00110	0.00010	.15986843
.42397547	01110	0.00001	.15966920
.15966920	11110	0.00000	.15963769

X. A "NEW" CONJECTURE.

The classical main conjecture of the universality theory can be stated as follows:
" The numbers which should be written on the right most column for n large, have an exponentially small dependence on the smooth function used to generate them, as long as this function has the right topological dynamics, is $C^{1+\alpha}$ and is unimodal with as single critical point which has to be generic (i.e. the graph there is a generic fold, i.e. the singularity is quadratic)."

The new results by Dennis Sullivan we alluded to before, contain a stronger version of this statement (other universality properties than the one associated to period doubling) as a main theorem. Let me just remark that the statement in quotation marks is not the best known form of the Main Conjecture, which uses generally a functional equation. This is nevertheless one of the ways Pierre Coullet and I understood it (up to a detail: see § XII), and essentially the way I described it to Dennis Sullivan, on the first occasion I had (the pleasure) to meet him. I was happily surprised when he later told me that he recalls this version of the "facts".

Now comes a formulation of the "new conjecture" (see § XII: this is not new, but it is new in print. I have described it for years to lots of people.) :

ORDERING CONJECTURE:

The numbers which should be written on the right most column for n large, are ordered according to the corresponding symbolic sizes , as long as the generating function has the right topological dynamics, is $C^{1+\alpha}$ and is unimodal with as single critical point.

In symbols this reads:

$$0.t_1 \ldots t_n > 0.u_1 \ldots u_n \Rightarrow \mathbf{r}_{\mathbf{S}^{-1}(t_1 \ldots t_n)} > \mathbf{r}_{\mathbf{S}^{-1}(u_1 \ldots u_n)} .$$

There is a wild conjectural picture in function space which I deliver sometimes in the same package as the ordering conjecture. I described it at the conference in Valparaiso, but (fortunately) there is no space here to develop this in print (this was conceived in discussions with Jean-Marc Gambaudo in 1987, and not yet quite ripe for writing).

For other renormalizations, one has a similar statement: what has to be changed is the rule to assign sizes to symbolic sequences. I do not know the extent one should give to the conjecture, but it should cover at least all n-tupling cascades with blocks like RL^m (period doubling corresponds to m=0). I described quickly period tripling (m=1) in Valparaiso, but I have already exceeded the number of pages a priori available in these proceedings.

I do not know how serious the genericity condition should be: should one put instead a non-flatness condition?

XI. SIMPLE CANTOR SETS AND SIMPLE MAPS.

Let us now define a (doubling) simple Cantor set as a doubling Cantor set with a finite number 2^r of ratios (we say r-simple when useful), and such that furthermore:

- $\mathfrak{D}$ and $\mathbf{D}_{1,0}$ are the same up to an order reversing affine map,
- the ratios read from $\mathfrak{D}_n \cap \mathfrak{D}_{1,1}$ and from $\mathfrak{D}_n \cap \mathfrak{D}_{1,0}$ for n>r are the same but placed in reverse order (so that $\mathbf{D}_{1,1}$ is as close as possible to be also an affine copy of $\mathfrak{D}$ under an order preserving map).

Figure 1 shows a 0-simple and a 1-simple Cantor sets together with doubling maps on them extended to small neighborhoods. We call simple maps the period doubling maps of simple Cantor sets, and also use "r-simple" when we want to be more specific. When trying to draw the graph of a simple map, you rather allow yourself to extend it to a small neighborhood, as we do in Figure 1.

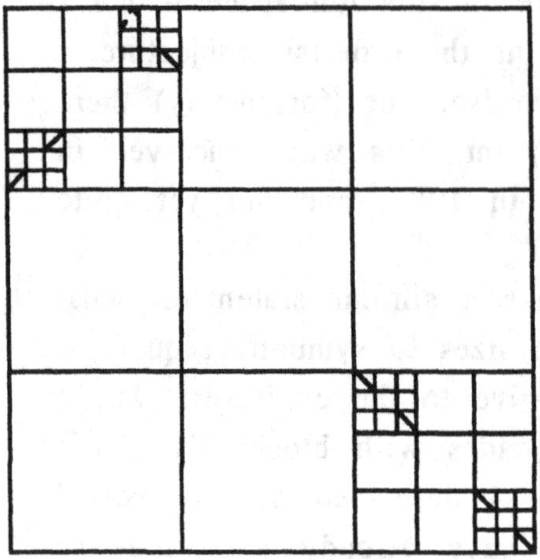

Figure 1-a: A 0-simple map.

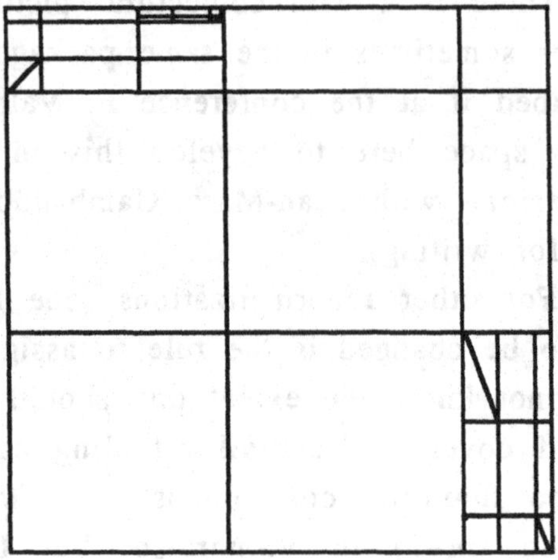

Figure 1-b: A 1-simple map.

XII. SOME SIMPLE RESULTS.

The first occurrence that I know of simple maps was in the hands of Alain Chenciner. He immediately got the following (unpublished) result, and then turned back to higher dimensional dynamics.

THEOREM 1 (CHENCINER 1977).

The period doubling homeomorphisms of the 0-simple Cantor sets (and in particular of the middle third Cantor set) cannot be extended to C^1 maps.

This is due to the fact that slopes are different on the left and right of the "critical point". Then what about piecewise C^1 maps?

THEOREM 2.

No 0-simple map (and in particular the period doubling homeomorphism of the middle third Cantor set) can be extended to a piecewise C^1 map.

In contrast, we have:

THEOREM 3 (a).

A 1-simple but not 0-simple map can be extended to a piecewise C^1 map. It can also be extended to a C^{1+1} map if $\mathbf{r}_{\{1\}}<\mathbf{r}_{\{0\}}$.

Simple Cantor sets were instrumental in suggesting the possibility of periodic behavior for the renormalization group as reported in [ACT] (although the paper is on the smooth functions' side: to get in a physics journal at a time when it was particularly difficult for us to write about period doubling, we avoided all combinatorial aspects). For a while, "knowing" empirically the above had Coullet and I thinking that the smoothness increases with r, so that universality was a C^∞ conjecture for us.

Also playing with the ratios can yield arbitrarily smoother function, even when r=1. More precisely:

THEOREM 3 (b) .

A 1-simple map can be extended to a C^{k+1} map if $\mathbf{r}_{\{1\}}<(\mathbf{r}_{\{0\}})^k$.

But this might be a WRONG point of view, although it has caught many. A GOOD (should I say better? see Q2 below) question (following in particular the work of Sullivan) goes a follows:

Q1- Assuming that the singularity at the critical point is x^y, and that the function f which extends the doubling map F is rewritten as $f(x)=g(x^y)$, what smoothness can one get for g when starting from r-simple maps?

THEOREM 4.

g is even not C^1.

This generalizes Theorem 2.

Q2-*What is the "good" question in higher dimension?*

The last results are still quite partial:

THEOREM 5.

If $\mathfrak{g}_1$ is a 1-simple concave map then:

$$\mathfrak{r}_{\mathbf{s}^{-1}(0.1)} > \mathfrak{r}_{\mathbf{s}^{-1}(0.0)}.$$

If $\mathfrak{g}_2$ is a 2-simple concave map then:

$$\mathfrak{r}_{\mathbf{s}^{-1}(0.11)} > \mathfrak{r}_{\mathbf{s}^{-1}(0.10)} > \mathfrak{r}_{\mathbf{s}^{-1}(0.01)} > \mathfrak{r}_{\mathbf{s}^{-1}(0.00)}.$$

If $\mathfrak{g}_3$ is a 3-simple concave map then:

$$\mathfrak{r}_{\mathbf{s}^{-1}(0.111)} > \mathfrak{r}_{\mathbf{s}^{-1}(0.110)} > \mathfrak{r}_{\mathbf{s}^{-1}(0.101)} > \mathfrak{r}_{\mathbf{s}^{-1}(0.100)} > \cdots$$
$$\cdots > \mathfrak{r}_{\mathbf{s}^{-1}(0.011)} > \mathfrak{r}_{\mathbf{s}^{-1}(0.010)} > \mathfrak{r}_{\mathbf{s}^{-1}(0.001)} > \mathfrak{r}_{\mathbf{s}^{-1}(0.000)}.$$

If $\mathfrak{g}_4$ is a 4-simple concave map then:

$$\mathfrak{r}_{\mathbf{s}^{-1}(0.1111)} > \mathfrak{r}_{\mathbf{s}^{-1}(0.1110)} > \mathfrak{r}_{\mathbf{s}^{-1}(0.1101)} > \mathfrak{r}_{\mathbf{s}^{-1}(0.1100)},$$
$$\mathfrak{r}_{\mathbf{s}^{-1}(0.1101)} > \mathfrak{r}_{\mathbf{s}^{-1}(0.1010)},$$
$$\mathfrak{r}_{\mathbf{s}^{-1}(0.1011)} > \mathfrak{r}_{\mathbf{s}^{-1}(0.1010)} > \mathfrak{r}_{\mathbf{s}^{-1}(0.1001)} > \mathfrak{r}_{\mathbf{s}^{-1}(0.1000)} > \cdots$$
$$\cdots > \mathfrak{r}_{\mathbf{s}^{-1}(0.0111)} > \mathfrak{r}_{\mathbf{s}^{-1}(0.0110)} > \mathfrak{r}_{\mathbf{s}^{-1}(0.0101)} > \mathfrak{r}_{\mathbf{s}^{-1}(0.0100)},$$
$$\mathfrak{r}_{\mathbf{s}^{-1}(0.0101)} > \mathfrak{r}_{\mathbf{s}^{-1}(0.0010)},$$
$$\mathfrak{r}_{\mathbf{s}^{-1}(0.0011)} > \mathfrak{r}_{\mathbf{s}^{-1}(0.0010)} > \mathfrak{r}_{\mathbf{s}^{-1}(0.0001)} > \mathfrak{r}_{\mathbf{s}^{-1}(0.0000)}.$$

THEOREM 6.

If $\{\mathfrak{f}_n\}_n$ is a sequence of n-simple concave maps such that the ratios of $\mathfrak{f}_{m+1}$ are obtained as small enough perturbations of the corresponding ratios from $\mathfrak{f}_m$,then the limit function $\mathfrak{f}_\infty$ exists and its ratios satisfy the Ordering Conjecture.

The proofs of all these results are quite elementary. The main ingredients are to use the picture of graphs of doubling maps, like in Figure 1, and the elementary consequence of concavity illustrated in Figure 2.

Part of the beauty of this subject is that a clever use of considerations barely more difficult than in Figure 2 can sometimes give truly deep mathematics, as e.g. in [MS].

$$\frac{Y_2 - Y_1}{X_2 - X_1} > \frac{Y_4 - Y_1}{X_4 - X_1} > \frac{Y_4 - Y_3}{X_4 - X_3}.$$

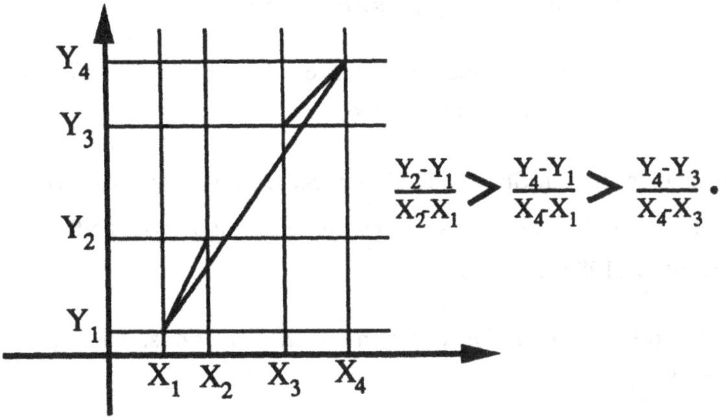

Fig. 2: Concavity in action.

FINAL REMARK.

The techniques of § VII are also useful when dealing with the scaling function of mappings topologically conjugate to $x \to 1-2|x|$, as described in [Ji], if they satisfy some concavity properties, close to our Theorem 6. It follows from [Su] that both questions are related.

ACKNOWLEDGEMENTS.

For this paper I have benefited over the years, from various interactions with many colleagues, some of whom are among my best friends. But my talk in Valparaiso would have been about a different subject without Thomas Sullivan, my kids Yuval, Ygaël & Liza, and my dear Léa. I want also to express my admiration for the efforts made by our Chilean friends to keep their students in touch with foreign scientists. Special thanks to Karen Brucks and Michael Shub for correcting a first version of this text.

A TINY BIBLIOGRAPHY.

[ACT] A. Arnéodo, P. Coullet and C. Tresser, A renormalization group with periodic behavior. Phys. Lett. 70A (1979) 74-76 .

[CT] P. Coullet and C. Tresser, Itérations d'endomorphismes et groupe de renormalisation. J. Phys. C5(1978) 25-28 , C. Tresser and P. Coullet, Itérations d'endomorphismes et groupe de renormalisation. C. R. Acad. Sc. Paris 287A (1978) 577-580, P. Coullet and C. Tresser, Critical transition to stochasticity for some dynamical systems. J. Physique Lettres 41 , (1980) L-255-L-258 .

[Fe] M. Feigenbaum, The universal metric properties of non-linear transformations, J. Stat Phys. 19 (1978) 25-52 . 21 (1979) 7669-706.

[GST] J.M. Gambaudo, S. van Strien and C. Tresser, Hénon-like maps with strange attractors: there exists a C^{∞} Kupka-Smale diffeomorphism on S^2 with neither sinks nor sources. Nonlinearity 2 (1989) 287-304.

[Ji] Y. Jiang, Generalized Ulam-Von Neumann Transformations. Thesis, C.U.N.Y. Graduate Center (1990).

[La] O.Lanford III, A computer assisted proof of the Feigenbaum conjectures. Bull. of Amer. Math. Soc. 6 (1982) 427-434.

[Ma] S.K. Ma, *Modern Theory of Critical Phenomena*, (Benjamin; Reading, MA 1976)

[MS] W. de Melo and S. van Strien, Schwarzian derivative and beyond. Bull. of Amer. Math. Soc. 18 (1988) 159-162.

[OT] M.V. Otero-Espinar and C. Tresser, Global complexity and essential simplicity: a conjectural picture of the boundary of chaos for smooth endomorphisms of the interval. Physica 39D (1989) 163-168.

[Su] D. Sullivan, Differentiable structure on fractal like sets, determined by intrinsic scaling functions on dual Cantor sets. Proceedings of Symposia in Pure Mathematics Volume 48 (1988).

HAMILTONIAN CHAOS IN A LASER DAMAGE MODEL

W. Becker, M. Fuka, J.K. McIver
Department of Physics
University of New Mexico

M. Orszag and R. Ramírez
Facultad de Física
Universidad Católica de Chile
Casilla 6177, Santiago 22, Chile

ABSTRACT. We review a simple model of laser damage in the context of Hamiltonian Chaos. A novel method of finding the last KAM is presented, based on a convergence theorem of continued fractions. The critical value of the field corresponding to the destruction of the last KAM is obtained. This value compares very well with the one obtained from the phase space plots.

Introduction

In a very simple model for laser damage, an electron is trapped in a metallic inclusion (modelled by a square well) in a transparent dielectric, in the presence of an external sinusoidal electric field (Fuka *et al.* and Becker and McIver (1987)).

It turns out that a classical mechanics treatment of this non-integrable problem is suitable to find a threshold value of the electromagnetic field for which the KAM surfaces break. A procedure which is applicable for the one dimensional version of this model is shown later in this work.

We present an introductory description of the KAM theory for a simple problem with two degrees of freedom (a pair of coupled harmonic oscillators). We show, that for a weak coupling, in a sense(to be specified more accurately below), the natural frequencies of the system, for most initial conditions, are only slightly altered by the presence of such a weak non-integrable perturbation. The argument presented here is based on perturbation theory. If we assume a Hamiltonian of the form:

$$H = H_0 + V, \qquad (1)$$

where H_0 represents an integrable system and V a weak non-linear and non-integrable perturbation, then historically two types of solutions have been found (Walker and Ford (1969):

(1) Poincare (1957) and others found, via perturbation expansions, that a weak perturbation changes only slightly the unperturbed motion, in the sense that the frequencies of the system suffer only minor changes and they also observed the appearence of harmonics.

E. Tirapegui and W. Zeller (eds.), Instabilities and Nonequilibrium Structures III, 43–57.

(2) Fermi (1923), on the other hand, found that even small non-integrable perturbations can produce dramatic effects on the unperturbed motion. He even observed ergodic behavior in a totally deterministic problem.

These divergent views can be unified by the so called KAM theorem. But before doing so, let's define integrability. A Hamiltonian system is called integrable if one can find a canonical transformation such that if the original Hamiltonian is expressed in terms of action-angle variables, the new Hamiltonian depends only on the new actions. If one has N degrees of freedom, then we find N actions which are constants of motion, and the corresponding angles that are linear functions of time. If we take a simple example of an integrable system with two degrees of freedom, the phase space is four dimensional and the trajectories move on a two-torus, where the two radii are J_1 and J_2.

Closed orbits occur only when the ratio of the two frequencies is a rational number, that is

$$\frac{\omega_1}{\omega_2} = \frac{m}{n}.$$

In this case, let's assume that θ_1 performs a complete cycle and $\delta\theta_2$ is the corresponding shift in θ_2, i.e.

$$\delta\theta_2 = 2\pi\frac{\omega_2}{\omega_1}.$$

On the other hand, if the ratio of the two frequencies is an irrational, the orbit never repeats itself and approaches every point on the torus.

The KAM theorem was originally stated by Kolmogorov (1954) and was proven later, independently, by Arnold (1963) and Moser (1962). There are several good review articles in this subject (Arnold (1978), Berry (1978) and Helleman (1980))

We give a simple account of this theorem following the approach Walker and Ford (1969). Let's assume a physical system with two degrees of freedom, with a Hamiltonian given by:

$$H = H_0(J_1, j_2) + V(J_1, J_2, \phi_1, \phi_2), \tag{2}$$

expressed in terms of angle-action variables.

If $V = 0$, J_1 and J_2 are constants of motion, then the system has two characteristic frequencies given by $\omega_i = \partial H_0/\partial J_i$ and$\phi_i = \omega_i t + \phi_{io}$ (i =1,2). As mentioned earlier, the trajectories move on a two-torus with radii J_1 and J_2 and rotation angles ϕ_1 and ϕ_2. The KAM theorem addresses the question of what happens to the invariant tori of the integrable Hamiltonian H_0, when V is small, but different from zero. The theorem says that most of the unperturbed tori with incommensurate frequencies ($\omega_1/\omega_2 =$ irrational number) continue to exist, being only slightly distorted by the perturbation. On the other hand, the tori bearing periodic motion or very nearly periodic motion, with commensurate frequencies ($\omega_1/\omega_2 = m/n$)

are greatly deformed by the perturbation, no matter how small is it and the trajectories no longer remain close to the unperturbed tori. The KAM theorem states that for the vast majority of the initial conditions (in the sense of measure theory), for V small, the trajectories lie on the preserved tori. However, there is a small set of initial conditions, for which the trajectories are not in the preserved tori, but rather pathologically interspersed between the preserved tori. We say that the tori corresponding to conmensurate frequencies are "destroyed" and ergodic motion is generated in small regions of phase space. Now, we will study the conditions under which the tori are destroyed, based on small denominator arguments. Now we expand V in Fourier series, keeping only one component:

$$H = H_0(J_1, J_2) + f_{mn}(J_1, J_2)\cos(m\phi_1 + n\phi_2) + ... \tag{3}$$

We try now, to eliminate the angle dependence of the particular Fourier term using a generating function:

$$F = j_1\phi_1 + j_2\phi_2 + B_{mn}(j_1, j_2)\sin(m\phi_1 + n\phi_2), \tag{4}$$

which only differs slightly from the identity transformation ($F = j_1\phi_1 + j_2\phi_2$) by the term proportional to B_{mn}.

The canonical transformation is:

$$(J_1, J_2, \phi_1, \phi_2) \longrightarrow (j_1, j_2, \theta_1, \theta_2) \tag{5}$$

With the generating function given in eq(4), we get:

$$J_1 = \frac{\partial F}{\partial \phi_1} \longrightarrow J_1 = j_1 + mB_{mn}\cos(m\phi_1 + n\phi_2),$$

$$J_2 = \frac{\partial F}{\partial \phi_2} \longrightarrow J_2 = j_2 + nB_{mn}\cos(m\phi_1 + n\phi_2),$$

$$\theta_1 = \frac{\partial F}{\partial j_1} \longrightarrow \theta_1 = \phi_1 + \frac{\partial B_{mn}}{\partial j_1}\sin(m\phi_1 + n\phi_2),$$

$$\theta_2 = \frac{\partial F}{\partial j_2} \longrightarrow \theta_2 = \phi_2 + \frac{\partial B_{mn}}{\partial j_2}\sin(m\phi_1 + n\phi_2). \tag{6}$$

Now we can also expand $H_0(J_1, J_2)$ in terms of the new action variables:

$$H_0(J_1, J_2) = H_0(j_1, j_2) + \frac{\partial H_0}{\partial j_1}\delta j_1 + \frac{\partial H_0}{\partial j_2}\delta j_2 + ..., \tag{7}$$

or:

$$H_0(J_1, J_2) = H_0(j_1, j_2) + [-(m\omega_1 + n\omega_2)B_{mn}]\cos(m\theta_1 + n\theta_2) \tag{8}$$

where $\omega_1 = \partial H_0/\partial j_1$, $\omega_2 = \partial H_0/\partial j_2$ and $\delta j_1, \delta j_2$ were obtained from eq.(6).

The total Hamiltonian now becomes:

$$H = H_0(j_1, j_2) + [-(m\omega_1 + n\omega_2)B_{mn} + f_{mn}]\cos(m\phi_1 + n\phi_2). \tag{9}$$

The idea is to eliminate the $m-n$ Fourier component (resonance) by setting the coefficient of the cosine term in eq.(9) to zero, and H becomes once more, phase independent. Therefore:

$$B_{mn} = \frac{f_{mn}}{m\omega_1 + n\omega_2}. \tag{10}$$

Now we recognize the two opposite cases mentioned in the introduction.

If $| m\omega_1(J_1, J_2) + n\omega_2(J_1, J_2) | \gg f_{mn}$, B_{mn} is small and the generating function required to eliminate the $m-n$ Fourier component of the perturbation, differs only slightly from the identity transformation. This is the case when the ratio of the two frequencies is an irrational number or a rational with very large m, n.

The opposite case, that is when:

$$| m\omega_1(J_1, J_2) + n\omega_2(J_1, J_2) | \ll f_{mn}, \tag{11}$$

is called a resonance ($m-n$ resonance).

When f_{mn} (the perturbation) is small, there is a small range of frequencies ω_1 and ω_2 (or a small range of initial conditions) for which the inequality (11) is satisfied. This defines a narrow band or layer in phase space. We will call this band, the *stochastic layer*. *As it turns out, the presence of stochastic layers seems to be a universal property of non-integrable Hamiltonians.*

In order to illustrate this point with an example, we show in Figure 1(a), the phase space trajectories of the well known pendulum. The separatrix, separates oscillations from rotations (Chernikov *et al.* 1988). It also contains points of unstable equilibrium, which are the intersections of the separatrix with the x-axis (these points are often called hyperbolic points). Now, in Figure 1(b), we perturb the pendulum with a non-linear, non-integrable perturbation. We immediately notice the appearence of a stochastic layer replacing the separatrix. If the pendulum has an initial condition within the stochastic layer, *it will be trapped in the layer, in its subsequent motion.*

In Figure 1(b), we also observe curves above and below the stochastic layer, that look very similar to the unperturbed ones. These correspond to irrational ratios of the frequencies and their shapes are only slightly disturbed. We will refer to them as KAM curves or KAM surfaces.

There are, in every non-integrable Hamiltonian problem, KAM surfaces and stochastic layers between them. The KAM surfaces act as "barriers" that prevent the particles to escape from a given stochastic layer.

A popular view of the Hamiltonian chaos that goes beyond the validity of the KAM theory and it is mainly supported by a large body of numerical work, goes as follows: when the perturbation is small, the phase space contains many thin stochastic layers, corresponding to various (m,n) resonances and KAM surfaces which isolate these resonances from each other. A given particle, trapped by its initial condition, in one of these layers, will stay in that layer and cannot cross a KAM surface to a different layer. This trapping is in the vertical direction of the phase space (p) and these particles can gain or loose a very limited amount of momentum.

To summarize, for systems with two degrees of freedom, KAM surfaces block the flow of trajectories over wide ranges of the action (momentum) variables.

However, as the size of the perturbation increases, the inequality (11) corresponds to a broader zone and, according to Chirikov (1979), the destruction of the KAM surfaces and the ulterior onset of global chaos is due to the growth and eventual overlap of these resonances. Later on, Escande and Dorveil (1981), pointed out that the resonance zones are self-similar, and developped a renormalization group scheme, in order to predict the critical values of the perturbation, for which the KAM surfaces get destroyed.

What do we mean by the destruction of a KAM surface?

It means, that a particle, originally trapped in a stochastic layer, can now *cross* the KAM surface to a different resonance. The KAM's, as seen in a Poincare surface section, appear to have a distribution of holes with a Cantor set structure.

There is a very interesting conjecture put forward by Greene (1979). He addressed the problem of the standard map of a plane onto itself, representing, for example, the Hamiltonian problem of two coupled oscillators. The main hypothesis, here, is that the dissapearence of a KAM surface is associated to the change in stability of nearby(in phase space) periodic orbits.

He explored extensively, by numerical methods, the relation between the KAM surfaces and the periodic orbits. If we call $\alpha = \omega_1/\omega_2$ the rotation number and since the KAM's correspond to irrational rotation numbers, while the periodic orbits have rational α, the problem here reduces to how to approach an irrational number by a sequence of rationals.

A good way of representing irrationals is by continued fractions. If α is an irrational, it has a unique continued fraction representation which in shorthand notation can be written as

$$\alpha = [a_1, a_2, a_3 \ldots], \tag{12}$$

where the a_n's are positive integers.

The succesive iterates of a continued fraction r_n/s_n, where a_n is the last term taken, approximates α well in the sense that no other r/s is closer to α, when $s \leq s_n$.

Now, if we want to explore the destruction of the last KAM, we would expect it to correspond to an irrational furthest away from a rational, or the one that converges the slowest. That irrational has been known for a long time with the name of ***golden-mean:***

$$\alpha_I = [1,1,1,...] = \frac{\sqrt{5}-1}{2} = 0.6180\ldots \tag{13}$$

For the standard mapping, where the resonances have equal amplitudes, frequencies and phases, there is plenty of numerical evidence that the last KAM to be destroyed corresponds to the golden-mean. Unfortunately, in other maps, where the resonances have different amplitudes and frequencies, the last KAM does not correspond to the golden-mean. This will be the case of the example discussed in the next section of this article (Lichtenberg and Lieberman(1983)).

One-Dimensional Model

The problem we address in this work is the following: We consider a particle of mass m trapped in a one-dimensional infinitely deep square potential well of diameter $2a$ and subject to an externally applied sinusoidally varying electric field with amplitude $(-\varepsilon/e)$. As was stated before this system may have some physical relevance as a model of a metallic inclusion in a transparent dielectric subject to a laser field. It allows to study the conditions under which an electron initially trapped in the inclusion gains sufficient energy from the field to reach the empty conduction band of the dielectric. Such an electron, after further acceleration on the conduction band may then excite an additional electron from the valence band onto the conduction band, thus starting an electron avalanche which in turn may lead to visible damage of the dielectric material.

The Hamiltonian of the system can be written as $(a = 1, m = 1/2)$

$$H = p^2 + b[\theta(x-1) + \theta(-x-1)] + \varepsilon x \cos\omega t \tag{14}$$

where $b \to \infty$ and $\theta(x)$ denotes the Heaviside step function.

The trajectory of the particle in between the walls can be written down as

$$x_s(t) = a_s + b_s t + \kappa \cos\omega t \qquad (t_s \le t \le t_{s+1})$$
$$p_s(t) = \frac{1}{2}(b_s - \kappa\omega \sin\omega t) \tag{15}$$

where $\kappa = 2\varepsilon/\omega$ and the t_s $(s = 0,1,2,\ldots)$, denotes the times when the particles hits one of the walls. The a_s and b_s are determined by the initial conditions. In all of the following calculations the particle will start at time $t_o = 0$ from the right wall. Moreover, we will restrict ourselves to situations where the particle makes alternate walls contacts. This is guaranteed when $|b_s| > |\kappa\omega|$ for all s. The trajectories are then recursively determined in terms of the momentum p_o at time t_o by solving the equations:

$$x_s(t_s) = (-1)^s, \qquad p_o(t_o) = p_o < 0, \tag{16a}$$
$$x_s(t_{s+1}) = (-1)^{s+1}, \tag{16b}$$
$$x_{s+1}(t_{s+1}) = x_s(t_{s+1}), \qquad p_{s+1}(t_{s+1}) = -p_s(t_{s+1}). \tag{16c}$$

The only step that cannot be carried through analytically is the determination of contact times t_s which requires solving the trigonometric equation (16b,c).

In the absence of the field the particle bounces back and forth between the walls with constant energy $E_o = p_o^2$. If the field is turned on, then depending on the initial momentum some orbits will be more strongly affected than others. A particular important type of orbits are the periodic orbits: We call any orbit such that

$$x_{s+2n}(t + NT) = x_s(t) \tag{17}$$

a (n, N) cycle. Here $T = 2\pi/\omega$ is the period of the field and n and N are integers with no common divisors. The ratio

$$\nu = n/N \tag{18}$$

defines the winding number of the periodic orbit. In the limit of a vanishing external field the initial momentum that yields a (n, N) cycle is

$$p_o = \frac{2n}{NT} = \frac{\omega}{\pi}\nu \qquad (\varepsilon = 0) \tag{19}$$

For a nonzero field, p_o, viz the momentum of the particle when it starts at $t = 0$ at the right wall, will become a function of the field, viz $p_o = p_o^{(\nu)}(\varepsilon)$. The (n, N) cycle are particularly important since they form a dense set in the phase space; also the curves $p_o^{(\nu)}(\varepsilon)$ with rational ν are dense in the (ε, p_o) space.

The curves $p_o^{(\nu)}(\varepsilon)$, can be calculated numerically and, for some simple cycles, even analytically. Whether or not, however, the particle actually performs such a cycle depends upon whether or not this cycle, for a given ε, is stable against small perturbations. As an example, let us discuss the $(1, N)$ cycle. By following pictorially the motion of the particle whose initial momentum is slightly perturbed from the value given by $p_o^{(\frac{1}{N})}(\varepsilon)$ one can convince oneself that the 1-cycles with odd N are all unstable for any value of $\varepsilon > 0$, while they are stable in some region $-|\varepsilon_c^{(\frac{1}{N})}| < \varepsilon \leq 0$. In the action-angle representation of the Hamiltonian (Escande and Dorveil (1981)), the $(1, N)$ cycles with odd N are readily identified as the primary resonances (Lin and Reichl (1986)). Our model differs from the standard map which has been very extensively investigated (Green (1979)) by the fact that it exhibits an infinite sequence of primary resonances. On the other hand, one can convince oneself in the same way that the $(1,N)$ cycles with even N are stable within a symmetrical interval $-|\varepsilon_c^{(\frac{1}{N})}| < \varepsilon < |\varepsilon_c^{(\frac{1}{N})}|$

An economic and illuminating way of displaying the phase-space trajectories in a system with periodic external forces is provided by a stroboscopic plot, that is, the position of the particle in phase space is recorded at integer multiples of the period T of the field, starting with $t = 0$ and continuing for a very large number of periods. Stroboscopic phase-space plots are displayed in Figures 2 and 3. These plots have been generated by starting the particle at $t_o = 0$ at the right wall with different values of p_o and calculating its trajectory from Eqns. (15) and (16). The field strength ε is fixed in each plot. We recognize several distinctly different types of trajectories: (i) the (n,N) cycles discussed above show in the stroboscopic phase-space plot as a set of N discrete points* then as n and N become large, this set of discrete points may seem to approach a smooth continuous curve; (ii) suppose we consider, for $\varepsilon = 0$, a value of p_o which is via Eqn. (19), related to an irrational winding number ν. Then by increasing ε from zero to a finite value, we choose the initial momentum as $p_o = p_o^{(\nu)}(\varepsilon)$. The corresponding particle orbit shows in the stroboscopic phase-plot as a continuous line. It is the manifestation of a KAM torus. We will cavalierly refer to the continuous lines in the stroboscopic phase-plot as KAM surfaces or KAMs. We discriminate betwen two types of KAMs, global KAMs which extend all the way from left to the right wall, and local KAMs which do not, but rather form localized islands. Global KAMs are of particular importance since they prevent difusion of a particle in phase space from lower to higher momenta (or vice versa). We will discuss the question of how in a computer- generated plot the distinction between periodic orbits and KAMs, i.e. between rational and irrational number, is actually realized. (iii) Finally, the strobe plots exhibit areas with an apparent random distribution of points. These areas are either in between KAMs in which case they correspond to stochastic layers or, for small p, no periodic orbits nor KAMs are left at all. This area is referred as the stochastic sea.

The most unstable orbits are the $(1,N)$ cycles with odd N, i.e., the primary resonances. The area in phase space between two primary resonances is called a resonance zone. If one moves away from the primary resonances, the periodic orbits and the KAMs become, in general, more stable, i.e. they remain stable up to higher values of $|\varepsilon|$. (This is notwithstanding the fact that there are always cycles interspersed which are unstable for all $\varepsilon > 0$, along with the associated stochastic layers). The most stable objects in each resonance zone is the respective $(1,N)$ cycle with even N. Somewhere, in each resonance zone, is the "last KAM", i.e.

*In order to see N points we would have to display the phase space for both positive and negative momentum and to identify points on the left or right wall with opposite values of the momentum. Since the phase-space plots are symmetric upon $p \to -p$ they are only shown for $p < 0$.

the global KAM that is stable up to the highest value of $|\varepsilon|$. For the standard map, the last KAM is distinguished by a winding number equal to the golden-mean (Lichtenberg and Lieberman (1983)). We will see, that this is not so in our model. Locating the last KAM, and most importantly, finding the critical value of ε beyond which the last KAM becomes unstable is the central problem. The principal and practical importance of this critical value of ε becomes obvious if we return to the physical problem that partially motivated the study of this model. If the last KAM is destroyed, the particle can diffuse in phase space into the next resonance zone, thus gaining a significant amount of energy. The critical value of ε may therefore, in principle, be related to threshold values of laser damage (Fuka *et al.*)

In what follows we will concentrate on the problem of locating the last KAMs and finding the critical values of ε when they cease to be stable. We start from a nonlinear recurrence relation between the contact times t_s, t_{s+1} and t_{s+2} which can be derived from Eqns. (16),

$$2\omega\kappa \sin \omega t_{s+1} = \frac{2(-1)^s - \kappa(\cos \omega t_{s+2} - \cos \omega t_{s+1})}{t_{s+2} - t_{s+1}} - \frac{2(-1)^s + \kappa(\cos \omega t_{s+1} - \cos \omega t_s)}{t_{s+1} - t_s} \tag{20}$$

Given ε, and therefore κ, and $t_o = 0$, we calculate t_1 from Eqn. (16b), for given p_o. The recurrence relation (20) can then be used to compute the contact times t_s with $s \geq 2$. If, for given ε, we choose p_o to correspond to some rational or irrational winding number ν, viz $p_o = p_o^{(\nu)}(\varepsilon)$, then we are computing the contact times for the corresponding (n, N) cycle or KAM, repectively. Let us denote the corresponding solution of Eqn (20) by $t_s^{(\nu)}$.

In order to learn whether or not the orbit with the winding number ν is stable we performed a linear stability analysis. Let us consider a slightly perturbed orbit whose contact times are given by

$$\omega t_k = \omega t_k^{(\nu)} + \epsilon_k \tag{21}$$

Introducing the perturbed contact times (21) into the nonlinear recurrence relation (7), and linearizing with respect to the perturbations ϵ_k we obtain a linear recurrence relation for the ϵ_k

$$\epsilon_{k+1} = B_k \epsilon_k + A_k \epsilon_{k-1} \qquad (k = 1, 2, 3, \ldots) \tag{22}$$

where $B_k = \alpha_k / \beta_{k+1}$, $A_k = -\gamma_{k-1}/\beta_{k+1}$, and

$$\begin{aligned}\alpha_k \equiv& 2\kappa \cos\omega t_k - \frac{2(-1)^{k-1} - \kappa(\cos\omega t_{k+1} - \cos\omega t_k)}{[\omega(t_{k+1} - t_k)]^2} + \frac{\kappa \sin\omega t_k}{\omega(t_{k+1} - t_k)}\\ &- \frac{\kappa \sin\omega t_k}{\omega(t_k - t_{k-1})} - \frac{2(-1)^{k-1} + \kappa(\cos\omega t_k - \cos\omega t_{k-1})}{[\omega(t_k - t_{k-1})]^2}\\ \beta_k \equiv& \frac{\kappa \sin\omega t_k}{\omega(t_k - t_{k-1})} - \frac{2(-1)^k - \kappa(\cos\omega t_k - \cos\omega t_{k-1})}{[\omega(t_k - t_{k-1})]^2}\\ \gamma_k \equiv& -\frac{\kappa \sin\omega t_k}{\omega(t_{k+1} - t_k)} - \frac{2(-1)^k + \kappa(\cos\omega t_{k+1} - \cos\omega t_k)}{[\omega(t_{k+1} - t_k)]^2}\end{aligned} \tag{23}$$

All of the t_k on the right-hand side of Eqn. (23) are actually the $t_k^{(\nu)}$, for convenience, the superscript (ν) has been dropped.

A linear three-term recurrence relation (3TRR) has two linearly independent solutions, $\epsilon_k^{(1)}$ and $\epsilon_k^{(2)}$, say, such that any solution ϵ_k can be written as a linear combination of the two, viz $\epsilon_k = c_1\epsilon_k^{(1)} + c_2\epsilon_k^{(2)}$. The orbit whose perturbations the ϵ_k represent will be stable iff all solutions of the 3TRR (22) are bounded, i.e. $|\epsilon_k| < M$ for all k and all solutions ϵ_k.

Let us first discuss the stability of the (n, N) cycles. In that case, $A_{k+2n} = A_k$ and $B_{k+2n} = B_k$ and we have a periodic 3TRR with period $2n$. It can be shown that, in analogy with Floquet theory, for second-order differential equations, two linearly independent solutions can be found and that

$$\epsilon_{k+2nm}^{(1,2)} = \mu_{1,2}^m \epsilon_k^{(1,2)} \tag{24}$$

for all k and integer m. The μ_1 and μ_2 are called the characteristic coefficients. For the case of interest to us (Eqns. (22) and (23)) it can be shown that (Fuka *et al.*).

$$\mu_1\mu_2 = 1 \tag{25}$$

This leaves two possibilities: (i) the μ's are complex and $\mu_1 = \mu_2^*$, i.e. $|\mu_1| = |\mu_2| = 1$. In this cases, all solutions are bounded, and the orbit is stable. The second possibility is that (ii) the μ's are real and $\mu_2 = \mu_1^{-1}$. Then either $|\mu_1|$ or $|\mu_2|$ is larger than unity, say $|\mu_2|$, and any solution ϵ_k which is not orthogonal to $\epsilon_k^{(2)}$ will grow without limit. Hence the orbit is unstable. Criteria can be formulated in terms of the coefficients A_k and B_k to decide which is realized (Fuka *et al.*). However, in the following we will consider a different method which is not restricted to periodic orbits, but it can also be used for KAMs, i.e. for irrational winding numbers ν.

To this end, define the minimal solution of a linear 3TRR: $\epsilon_k^{(0)}$ is called a minimal solution iff

$$\lim_{k \to \infty} \frac{\epsilon_k^{(0)}}{\epsilon_k} = 0 \tag{26}$$

for any solution ϵ_k which is linearly independent of $\epsilon_k^{(0)}$. A given 3TRR may or not have a minimal solution. In our case, obviously a minimal solution exists ($\epsilon_k^{(1)}$) in case (ii) when the characteristic coefficients are real. In the opposite case (i) where $|\mu_1| = |\mu_2| = 1$ there is no minimal solution. The question of whether or not a given 3TRR of the form (22) has a minimal solution is answered by Pincherle's theorem (Jones and Thron (1980)).

A minimal solution exists iff the infinite continued fraction

$$\cfrac{A_1}{B_1 + \cfrac{A_2}{B_2 + \cfrac{A_3}{B_3 + \cdots}}} \tag{27}$$

converges. Convergence means that the nth approximant of the continued fraction (27), viz the finite continued fraction

$$f_n = \cfrac{A_1}{B_1 + \cfrac{A_2}{B_2 + \cfrac{A_3}{B_3 + \cdots \cfrac{A_n}{B_n}}}} \tag{28}$$

converges to a (finite or infinite) limit, viz $\lim_{n \to \infty} f_n = f$. Otherwise, the continued fraction (27) is divergent. Hence, the orbit under consideration is stable, iff the continued fraction (27) constructed from the coefficients of the 3TRR (9) diverges.

For a (n, N) cycle, the continued fraction is periodic with period $2n$. For periodic continued fractions, necessary and sufficient convergence criteria are available, which, however, are cumbersome (and equivalent to other methods of assessing the stability of the periodic orbits which we alluded to above). The advantage of the continued-fraction criterion is that we can apply it to KAMs too. That is, we conjecture that for irrational winding numbers, too, divergence of the continued fraction (14) implies stability, and vice versa.

Notice, that this is a conjecture which we did not prove. The proven equivalence between divergence and stability only holds for periodic orbits, since the proof hinges on Eqn. (25) which has a meaning only for periodic orbits. Equation (25) is intimately connected with the Hamiltonian character of our problem. It rules out, for example, the possibility that both $|\mu_1| < 1$ and $|\mu_2| < 1$. If this were the case, the corresponding orbit would be an attractor which does not exist in a Hamiltonian problem.

In order to approach the problem of finding the last KAM and its breakdown field, we will proceed as follows. Inspection of the stroboscopic phase-space plots shows that the last KAM is somewhere between the (2,3) cycle and the (3,4) cycle, fairly close to the (5,7) cycle (which is unstable for $\varepsilon > 0$). Hence, its winding number ν_l should satisfy $2/3 < \nu_l < 3/4$. The phase-space plots also suggest that the last KAM is destroyed between $\varepsilon = 0.14$ and 0.15. Fig. 4 gives a more detailed view of the situation. It indicates for each point of a grid in (ε, p_o) space whether the cycle or KAM, which corresponds to this point is stable or unstable. (The winding number ν of this cycle or KAM is determined by the equation $p_o = p_o^{(\nu)}(\varepsilon)$). The continued fraction method described above has been used to obtain the answer for each point. The figure exhibits a highly complicated boundary between the stable and the unstable region which has a fractal structure. The (2,3) cycle and the (3,4) cycle are immediately identified as the pronounced areas of stability which extend farthest to the right. On the other hand, the (5,7) cycle which is unstable for $\varepsilon > 0$ shows as the area of instability which goes farthest to the left. (To what extent a "narrow" unstable cycle becomes visible depends, of course, on the resolution). Notice, that the last KAM has nothing to do with the aformentioned pronounced areas of stability. These are related to local KAMs associated with the (2,3) cycles and the (3,4) cycle, while the "last KAM" is by definition global, unrelated to cycles of low n and N. Rather, the last KAM ought to pass through one of the two valleys on either side of the deep 'valley" of instability associated with (5,7) cycle. The average "bottom" of these valleys is roughly at $\varepsilon = 0.148$. This provides an estimate of the breakdown field of the last KAM, in agreement with the estimate gained from inspection of the phase-space plots.

A tentative value of the winding number of the last KAM can be arrived at as follows. First, notice that the last KAM is definitely not characterized by the golden-mean winding number.

As discussed above, this is conspicuously evident from the phase-space plots. One might, however, expect that the last KAM should still be related to a higher-order golden-mean. The conjecture

$$\nu = \cfrac{1}{1 + \cfrac{1}{2 + \cfrac{1}{1 + \cdots}}} = \frac{1}{2}\left(1 + \frac{1}{\sqrt{5}}\right) = 0.7236\ldots, \tag{29}$$

has as its first rational approximants

$$\nu = 1, \frac{2}{3}, \frac{3}{4}, \frac{5}{7}, \frac{8}{11}, \frac{13}{18}, \ldots, \tag{30}$$

which are the winding numbers of some of the most noticeable cycles in Fig. 4. This conjecture locates the last KAM closely below the (5,7) cycle. Further investigations are required to corroborate or refute this conjecture.

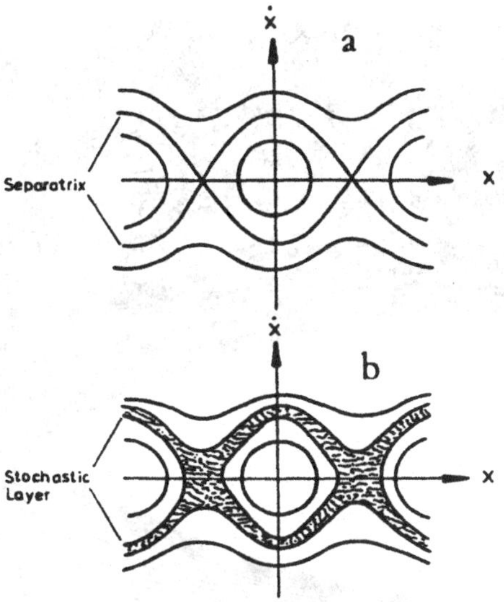

Figure 1. Phase portraits of (a) unperturbed and (b) perturbed pendulums. The separatrices divide the oscillatory from the rotational trajectories. As it is shown in (b), the perturbation destroys the separatrix and a stochastic layer is formed in its vicinity. The region occupied by the stochastic layers has holes that a stochastic trajectory cannot enter. There is an infinite number of stability islands, within which, an infinite number of still thinner stochastic layers reside.

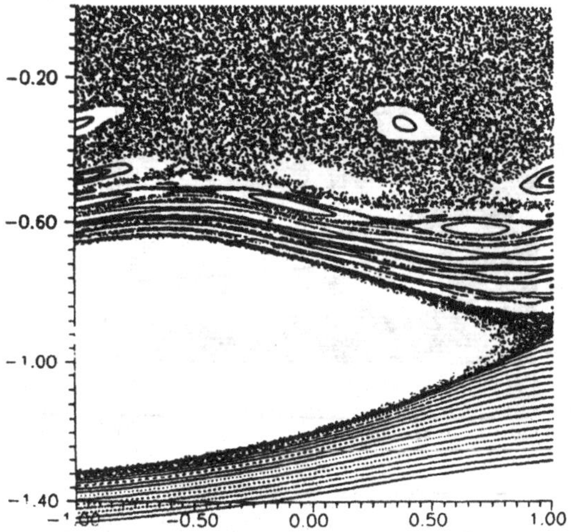

Figure 2. Phase space plot (p vs. x) for $\varepsilon = 0.14$ for $-1 \leq x \leq 1$. The particle starts at $t = 0$ from the right wall with different values of p_o. We can still see some KAM curves above resonance.

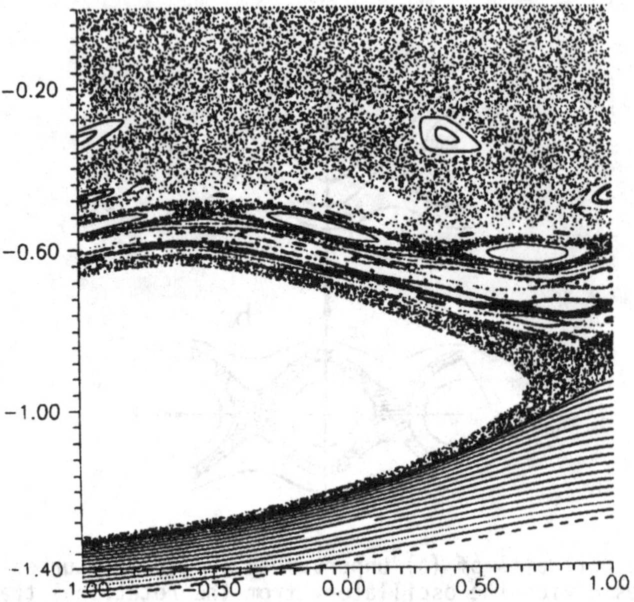

Figure 3. Phase space plot (p vs. x) for $\varepsilon = 0.15$ for $-1 \leq x \leq 1$. The particle starts at $t = 0$ from the right wall with different values of p_o. We can no longer find KAM curves above resonance.

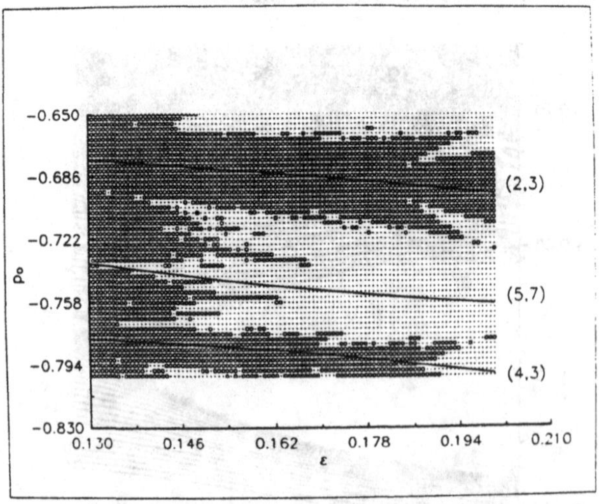

Figure 4. Stable (circles) and unstable (dots) regions are shown in the (ε, p_o) space. The figure exhibits a highly complicated boundary between stable and unstable regions, which has a fractal structure. We show the (2,3) and the (3,4) cycles in a region os stability and the (5,7) unstable cycle.

References

Fuka, M., Orszag, M., Becker, W., Ramírez, R. and McIver, J.K. (in preparation)

Becker, W. and McIver, J. K. (1987) Phys. Lett. **121**, 286

Walker, W. H. and Ford, J. (1969) Phys. Rev. **188**, 416

Poincare, H. (1957) Methodes Nouvelles de la Mechanique Celeste, Dover. Publ. Inc. NY

Fermi, E. (1923) Z. Physik, **24**, 261

Kolmogorov, A. N. (1954) Dokl. Akad. Nauk, SSSR,**98**, 527

Arnold, V. I. (1963) Russian. Math. Surveys, **18**, 9

Moser, J. (1962) Nachr. Akad. Wiss, 2, Math. Physik **K1**,1

Arnold, V. I. (1978) Mathematical Methods of Classical Mechanics, Springer, Heildelberg.

Berry, M. V. (1978) in S.Jorna (ed) Topics on Nonlinear Dynamics, Am. Inst. Phys. Proc. **46**

Helleman, R.H.G. (1980) in E.G.D. Cohen(ed) Fundamental Problems in Statistical Mechanics, **5** North Holland, Amsterdam

Chernikov, A. A., Sagdeev, R.Z., Zaslavsky, G.M. (1988) Physics Today, **41**, 27

Chirikov, B.V. (1979) Phys. Rep., **52**, 263

Escande, F. and Dorveil, F. (1981) J. Stat. Phys, **26**, 257

Greene, J.M. (1979) J. Math. Phys, **20**, 1183. A previous related work by the same author is: Greene, J.M. (1968) J. Math. Phys, **9**, 760

Lichtenberg, A.J., and Lieberman, M.A. (1983) Regular and Stochastic Motion, Springer, NY.

Lin, W.A., and Reichl, L.E. (1986) Physica D, **19**,145

Jones, W.B. and Thron, W.J. (1980) Continued Fraction. Analytical Theory and Applications, Addison-Wesley, Mass.

PERIODIC MODULATION OF A LOGISTIC OSCILLATOR

MIGUEL KIWI
Facultad de Física, Universidad Católica de Chile
Casilla 6177, Santiago 22, Chile

EDMUNDO LAZO
Departamento de Física, Universidad de Tarapacá
Casilla 7–D, Arica, Chile

JAIME RÖSSLER
Departamento de Física, Facultad de Ciencias,
Universidad de Chile, Casilla 653, Santiago

ABSTRACT. We study the effect of a periodic modulation on the bifurcation parameter of the logistic oscillator. Remarkable results are obtained, including early chaos, resonance phenomena and the disappearance of period–doubling bifurcations. Also, long term unpredictable dynamics appears for *negative* values of the Lyapunov exponent; the latter occurs when a resonant modulation is applied within the period–3 window.

I Introduction

The effect of perturbations in non–linear oscillator dynamics has been extensively treated in the literature. One approach to this problem is given by circle maps, [2] where the oscillator is locked to the external driver when the system lies in the so called "Arnold tongues" of parameter space. The coupled non–linear oscillator is another example where resonance phenomena may occur; in particular, the case of two coupled logistic maps was studied in Ref. 2.

In another context, a logistic map with a time dependent bifurcation parameter A_t was studied in Ref. 3, where the particular case of A_t assuming only two values was investigated. Chaotic behaviour was reported for some cases where A_t is well below the critical range threshold $A = 3.5699\ldots$. This remarkable result was denoted as "early chaos". Finally, we mention experimental work, like the modulation of chemical [4] or mechanical [5] non–linear oscillators.

E. Tirapegui and W. Zeller (eds.), Instabilities and Nonequilibrium Structures III, 59–64.

II The Model

We consider the traditional logistic map [6], $f(X_t) = A\,X_t(1 - X_t) = X_{t+1}$ and modulate it by super–imposing a periodic time dependence on the parameter A

$$X_{t+1} = f_t(X_t) = A(\Phi_t)\,X_t\,(1 - X_t)\ , \tag{1}$$

where

$$A(\Phi_t) = A(\Phi_t + 1) \quad \text{and} \quad \Phi_{t+1} = \Phi_t + \omega\ . \tag{2}$$

In this paper we choose $A(\Phi) = A_0 + D(2|\Phi| - \frac{1}{2})$; $-\frac{1}{2} \le \Phi \le \frac{1}{2}$. Here A_0 is the mean value of A_t, while D, Φ_t and ω are the amplitude, phase and frequency of the modulation, respectively. The latter will be the key parameter in the present study. However, other modulation shapes (as a sinusoidal one) lead to the same qualitative results than the triangular modulation introduced above.

The application defined by Eqs. (1) and (2) may be interpreted either as a time dependent one–dimensional map for X_t, or as a time independent bidimensional map; say $(X_t, \Phi_t) \rightarrow (X_{t+1}, \Phi_{t+1})$. However, both interpretations lead to the same value of the Lyapunov exponent. [2] The bidimensional attractor associated to $\{X_t\ ,\ \Phi_t\}$, together with the Lyapunov exponent λ, are useful tools in the understanding of the system dynamics.

In spite of its simplicity our model contains the essential elements to comprehend the effect of periodic perturbations in more complex non–linear systems. In particular, it poses many important questions, among them: *(i)* Is the "early chaos" phenomenon [3] also present in this system? *(ii)* What is the effect of a moderately low frequency perturbation over the bifurcation cascade? [7] *(iii)* If the bifurcation parameter $A(\Phi_t)$ is constrained inside a period–P cycle, what is the effect of a resonant modulation of frequency $\omega \sim 1\,/\,P$?

III Results

In order to obtain a general insight on the dynamics of the perturbed logistic oscillator, we study the Lyapunov exponent $\lambda(\omega)$ for several values of A_0 and D. The initial value of the phase, Φ_0, is irrelevant as long as the frequency ω is not a simple rational number. In fact, an irrational value of ω assures the ergodicity condition for Φ_t. We also notice that since $A(\phi) = A(-\phi)$, it holds that $\lambda(\omega) = \lambda(-\omega)$. Thus, we limit the range of ω to the interval $[0,\, 0.5]$.

a) Resonant Modulation

Now we consider a resonant perturbation of a stable cycle; therefore we impose that A_t must remain completely inside the stability interval of a periodic cycle of the (time independent) logistic map.

In Fig. 1*a* the Lyapunov exponent $\lambda(\omega)$ versus the modulation frequency, is shown for $A_0 = 3.2$, $D = 0.36$, so that A_t stays in the period two cycle. Thus, in this case, the resonant modulation frequency is $\omega_0 = \frac{1}{2}$. The graph indicates non–chaotic behavior (inherited from the non–modulated case), *i.e.* $\lambda(\omega) \leq 0$. A sharp peak at $\omega \sim 0.4591\cdots$ is the most prominent detail; there, $\lambda(\omega)$ vanishes at the top of the peak. This is an enigmatic feature, since the meaning of a "bifurcation" is not obvious in the context of our model.

On choosing lower values for D the peak converges to $\omega = \frac{1}{2}$, *i.e.* the resonant frequency value, thus suggesting a relation between the peak and a resonance phenomenon. The convergence is fulfiled for a critical modulation D_c; the peak disappears for $D < D_c$. In the case of Fig. 1, ($A_0 = 3.2$), $D_c \approx 0.1$. As A_0 increases within a given cycle, D_c also increases; on the other hand, an arbitrarily small modulation can lead to resonant behaviour at the beginning of a given cycle.

We have considered other periodic cycles, obtaining similar features of the Lyapunov exponent graph, like deep cavities surrounded by two large peaks around each resonant value of the frequency, $\omega = Q/P$. Here P is the period of the non–modulated case, and Q is an odd integer number for the first bifurcation cascade. The case of the period–3 window is shown in Fig. 1*b*, where $A_0 = 3.8345$ and $D = 0.012$. The peaks around the resonant frequency $\omega = \frac{1}{3}$ are quite apparent and in contrast to Fig. 1*a*, the Lyapunov exponent now does not vanish.

In order to obtain a deeper insight into the meaning of the Lyapunov peaks around the resonant modulation frequencies, we have analyzed the attractor of the bidimensional map $(X_t, \Phi_t) \rightarrow (X_{t+1}, \Phi_{t+1})$. A general characteristic is that when A_t lies within the domain of a P–cycle of the static logistic map, and the modulation D is not too large, then the attractor $\{X_t, \Phi_t\}$ usually contains P curves, say $\{\xi_\nu(\Phi), \nu = 1, \cdots, P\}$. Each of them is related to one of the P fixed points of the static case. A point X_t visits a given curve $\xi_\nu(\Phi)$ with a time period P. Thus, using a period P stroboscope, and a suitable temporal phase t_0, (say, $t = t_0 + lP$, $l = integer$) the point X_t can be assigned to that curve. A breakdown of these general features constitutes a clue which suggests that a very peculiar phenomenon is present.

We have analyzed the attractor $\{X_t, \Phi_t\}$ for the parameters of Fig. 1*a*, associated to a bi–cycle. When $\omega < 0.4591\cdots$ the attractor shows two curves, thus confirming the previous statements. However, when ω approaches the prominent peak of Fig. 1*a* the two curves $\xi_\nu(\Phi)$ approach each other, as can be seen in Fig. 2*a*. When ω just crosses the maximum of $\lambda(\omega)$ the attractor undergoes a qualitative modification, as the two curves merge. This is shown in Fig. 2*b*, where $\omega = 0.4592\cdots$ lies slightly to the right of the peak.

In order to understand this surprising attractor behaviour for $\omega \sim .5$, we have analyzed the exact resonant limit $\omega = \frac{1}{2}$ by means of the composite application

$$F(\Phi_0, X) = f_{2t+1}[f_{2t}(X)] \; , \tag{3}$$

where Φ_0 is the initial phase, which in this limit is a relevant parameter. $F(\Phi_0, X)$ reproduces the effect of a period two stroboscope, as $X_{2t+2} = F(\Phi_0, X_{2t})$. The fixed points of $F(\Phi_0, X)$ give rise to two curves, say $\{\eta_\nu(\Phi_0), \nu = 1, 2\}$. But, each curve shows a break in some interval of Φ_0, where the associated fixed point disappears [3].

Now we return to the dynamical problem and consider a nearly resonant frequency $\omega = \frac{1}{2} + \varepsilon$, with $\varepsilon \approx 0$. As time flows, the phase of the stroboscopic image X_{2t} slowly increases, $\Phi = 2t\varepsilon$, and X_{2t} closely follows a branch of fixed points of application (3), say $\eta_1(\Phi)$. But, when point X_{2t} reaches the break in the curve $\eta_1(\Phi)$, the associated attractor disappears, and X_{2t} becomes unstable for a short time interval until it reaches the other curve $\eta_2(\Phi)$. Thereafter, X_{2t} walks through $\eta_2(\Phi)$ until it attains the break which exists on that curve. Thereafter X_{2t} returns to the original band, by crossing through the repulsive zone of the logistic map, thus closing the cycle. However, point X_{2t} has described a single curve, as each value of Φ_{2t} is univocally associated to a specific branch, say, η_1 or η_2. The point X_{2t+1} describes the same curve, although with a phase delay equal to $\frac{1}{2}$.

If we increase the detuning $|\varepsilon|$, the velocity of X_{2t} over a given branch also increases; when $|\varepsilon|$ surpasses some critical value, the point X_{2t} reaches a velocity which is large enough to jump across the gap, without transferring to the other branch. Thus, the two bands dissociate from each other and we return to the non-resonant regime.

b) Resonance in Periodic Windows of the Chaotic Regime.

We have also analyzed attractors in the resonant regime of the 3-cycle window of Fig. 1*b*, say, for $\omega \approx \frac{1}{3}$. A period-three stroboscope has been used, in spite of which three branches appear. In resemblance to application (3) each branch has a small gap, and the point X_{3t} can jump from one branch to another, through these gaps, by crossing the repulsive zones of the logistic map.

But, in this 3-cycle window case, ***it is not possible to predict*** the new band where X_{3t} does stabilize its orbit; in fact, the uncertainty in the knowledge of the initial conditions (X_0, Φ_0), no matter how small, is strongly amplified during its unstable trajectory. This is a remarkable feature (not found in the first bifurcation cascade), as it corresponds to long term unpredictable dynamics, in spite of the fact that ***the Lyapunov exponent has a negative value.***

c) Low Frequency Modulation

In order to understand the effect of a low frequency modulation of the bifurcation cascade, we consider the particular case $A_0 = 3.26$, $D = 0.6$, where A_t ranges between the periods 1 and 8 of the first bifurcation cascade. In Fig. 3 we display the associated Lyapunov exponent $\lambda(\omega)$. For very low frequencies, say $\omega < .001$, the attractor $\{X_t, \Phi_t\}$ follows quite closely the fixed points of the static logistic map with bifurcation parameter $A(\Phi_t)$. However, there is an appreciable delay in the appearance of bifurcations [7]. For a slight increase in frequency the 8-branch features, associated to the underlying 8-cycle, become blurred in the attractor. Upon a further increase of ω the details of the 4-branches also disappear, being replaced by fast oscillations. The latter is shown in Fig. 4, where only two branches can be seen over the whole phase

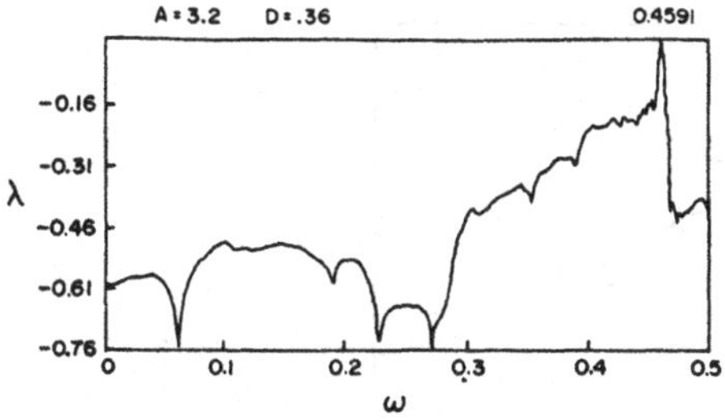

Figure 1a.

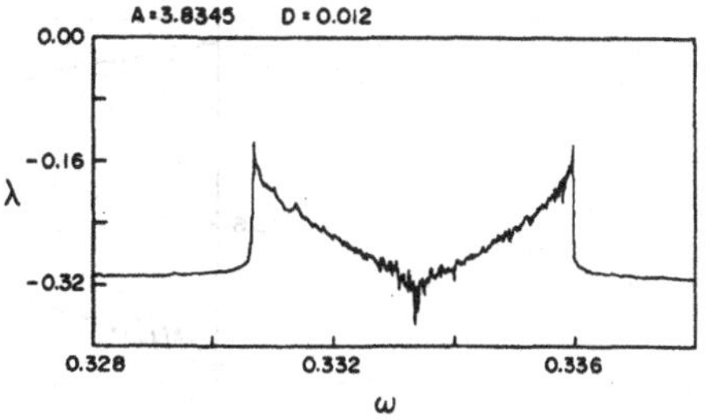

Figure 1b.

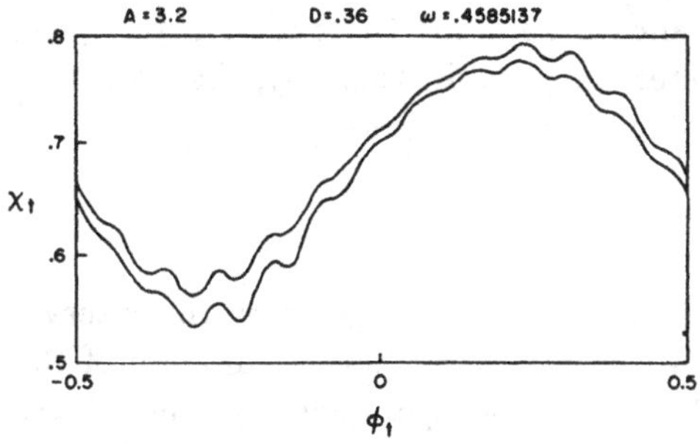

Figure 2a.

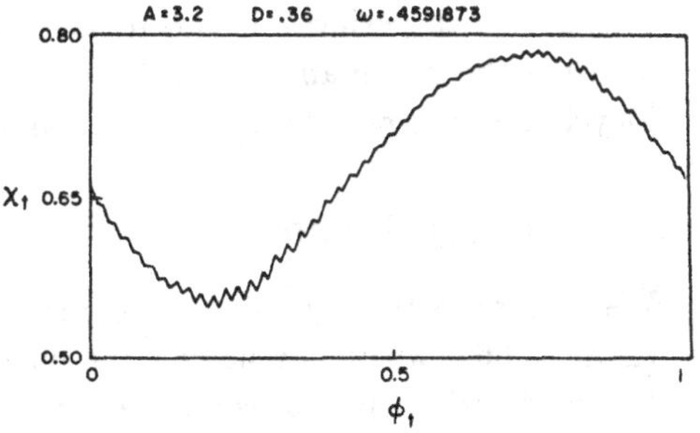

Figure 2b.

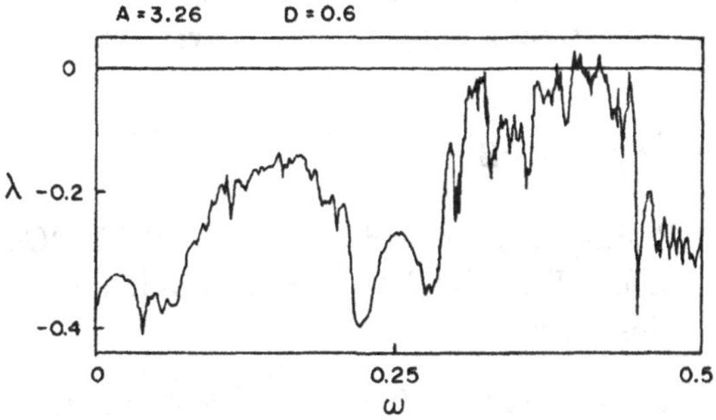

Figure 3.

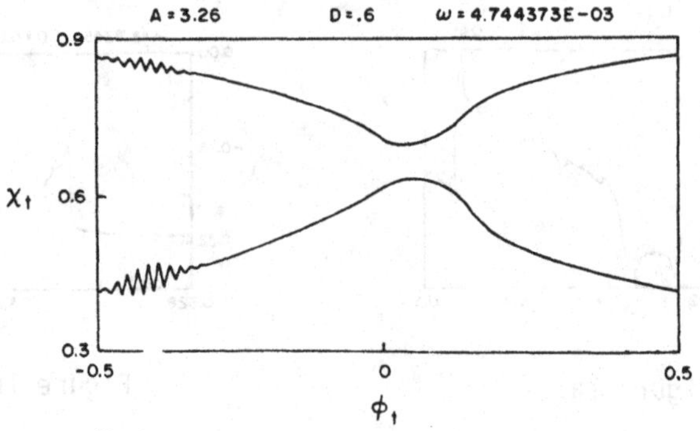

Fig. 4

range, including that associated to the monocycle (say, for $A(\Phi) < 3$). The point X_t oscillates with period 2 between these two branches, recalling the underlying bicycle, which is the dominant period for the parameters in play. In this way, the dominant bicycle has covered the realms of monocycle, 4–cycle and 8–cycle.

d) Early Chaos

We return to the Lyapunov exponent of Fig. 3, where A_t ranges over the interval $[2.96\,,\,3.56]$, without reaching the chaotic threshold $A = 3.5699\cdots$. In spite of the latter, chaotic behavior is apparent for $\omega \sim 0.4$, where the Lyapunov exponent shows a large peak which reaches positive values. Consequently, the "early chaos" phenomenon [3] is also a feature of this model. In the chaotic regime the attractor $\{X_t, \Phi_t\}$ has the appearance of two blurred laced ribbons, associated to the underlying dominant period two. However, the point X_t performs out–of–phase oscillations within this bi–cycle; in fact, on applying a period two stroboscope, we have verified that the attractor shape remains invariant.

References

[1] P. Bak, T. Bohr and M. Jensen (1988) *"Circle maps, Mode–locking and Chaos"*, in DIRECTIONS IN CHAOS, vol.2, edited by Hao Bai–Lin, World Scientific, Singapore.

[2] T. Hogg and B. Huberman (1984) *Phys. Rev.* **A29**, 275.

[3] J. Rössler, M. Kiwi, B. Hess and M. Markus (1989) *Phys. Rev.* **A39**, 5954.

[4] M. Eiswirth, P. Möller and G. Ertl (1989) *Surface Sci.* **208**, 13.

[5] O. Maldonado, M. Markus and B. Hess (1990) *Phys. Letts.* **A.** (To be published).

[6] Hao Bai-Lin (1984) *Chaos*, World Scientific, Singapore.

[7] R. Kapral and P. Mandel (1985) *Phys. Rev.* **A32**, 1076.

QUANTUM SPIN CHAINS WITH RESIDUAL ENTROPY

B. Nachtergaele [1]

Depto de Física, Universidad de Chile, Casilla 487-3, Santiago de Chile

Abstract

We extend a recently discovered class of Quantum Antiferromagnets with exactly computable ground states. To study these so-called VBS-models we use a new technique for constructing states of Quantum Spin Systems with a kind of markovianity property. In particular in this paper we present a class of Quantum Antiferromagnets which violate the Third Law of Thermodynamics, i.e. they exhibit the phenomenon of non-vanishing entropy density at zero temperature (residual entropy). A sketch of the explicit construction of the ground states and the determination of the non-vanishing entropy density is given.

1 Onderzoeker I.I.K.W. Belgium, on leave from Universiteit Leuven, Belgium

E. Tirapegui and W. Zeller (eds.), Instabilities and Nonequilibrium Structures III, 65–73.

1. Introduction

Consider a Quantum Spin Chain of spin s. This means that at any site i of the one-dimensional lattice $\mathbf{Z}$, we have as the algebra of single-site observables, a copy of M_d, the complex $d \times d$ matrices, where $d = 2s+1$. We are interested in models which have translation and rotation (SU(2)) invariant nearest neighbour interactions. It is easy to write down the most general Hamiltonian with these properties: let D_s be the $2s+1$-dimensional irreducible representation of SU(2), acting on the one-site Hilbert space $\mathbf{C}^d$ (we put $d = 2s+1$ in all what follows). One then has the well-known Clebsch-Gordan series

$$D_s \otimes D_s \cong D_0 \oplus D_1 \oplus \cdots \oplus D_{2s} \tag{1}$$

decomposing the representation $D_s \otimes D_s$ into $2s+1$ irreducible ones. The representation space is decomposed accordingly and we denote the corresponding orthogonal projections by $P^{(k)}, k = 0, 1, \ldots, 2s$. A translation invariant nearest neighbour interaction for a spin s model is of the form: on the interval $[-L, L] \subset \mathbf{Z}$ the Hamiltonian is given by

$$H_L = \sum_{i=-L}^{L-1} h_{i,i+1}$$

where the $h_{i,i+1}$ are copies at the nearest neighbour pairs $\{i, i+1\}$ of a given self-adjoint element $h \in M_d \otimes M_d$. To require rotation invariance, i.e.

$$(D_s(g)^* \otimes D_s(g)^*)h(D_s(g) \otimes D_s(g)) = h \quad \text{for all } g \in \text{SU}(2)$$

one concludes that h has a decomposition

$$h = \sum_{k=0}^{2s} J_k P^{(k)} \tag{2}$$

where the $P^{(k)}$ are the spin-projections mentioned above and the J_k are real coupling constants. Without loss of generality one can assume that all $J_k \geq 0$. It also follows from the representation theory of SU(2) that the $P^{(k)}$ can be written as a linear combination of powers of the Heisenberg interaction $\sum_\alpha S^\alpha \otimes S^\alpha$, where α runs through $\{x, y, z\}$ and the S^α are the generators of D_s, normalised such that $\sum_\alpha (S^\alpha)^2 = s(s+1)\mathbf{1}$. This enables one to rewrite the interaction (2) in terms of more physical quantities:

$$h = \sum_{l=0}^{2s} C_l \left(\sum_\alpha S^\alpha \otimes S^\alpha\right)^l \tag{3}$$

where the C_l are again real constants which are in one-to-one correspondence with the J_k.

Now, the so-called VBS-models have interactions of the form (2) with

$$0 = J_0 = J_1 = \cdots = J_s, \text{ and } J_{s+1} \geq 0, \cdots, J_{2s} \geq 0 \tag{4}$$

As far as the ground states of these models are concerned, we observe the following particularities:

i) The ground states can be constructed explicitely and fall into the class of the Finitely Correlated States. The latter are a class of states with a very nice and handsome mathematical structure, showing a lot of similarities with the Markov processes of Classical Probability [FNW2]. This general structure is the following: denote by $\langle\cdot\rangle$ such a state. Then, the n-point correlation, with observables $X_1, \ldots, X_n$ at sites $1, \ldots, n$, can be expressed as follows

$$\langle X_1 \otimes \ldots \otimes X_n \rangle = \mathrm{Tr}_{\mathbb{C}^k}\, \rho E_{X_1} \circ \cdots \circ E_{X_n}(\mathbf{1}) \tag{5}$$

where k is a positive integer, ρ is a $k \times k$ density matrix and for all $X \in M_d$

$$E_X : M_k \to M_k$$

is a linear transformation, depending also linearly on X.

ii) Typical two-point correlation functions behave as exact exponentials in the distance between the points, and the correlation length can be calculated exactly; e.g. there are constants $C_{\alpha,\beta}, \xi > 0$ and a periodic function of $\mathbf{Z}$, $\phi(r), r \in \mathbf{Z}$, such that for all $\alpha, \beta \in \{x, y, z\}$;

$$\langle S_0^\alpha S_r^\beta \rangle = C_{\alpha,\beta} \phi(r) e^{-\frac{|r|}{\xi}} \tag{6}$$

For a class of antiferromagnetic models one simply has $\phi(r) = (-1)^r$ (see e.g [FNW1]).

iii) For the VBS-models it has been demonstrated rigorously that there is a gap in the spectrum immediately above the ground state [FNW2].

2. Models with residual entropy

In this paper we are interested in the phenomenon of zero-temperature entropy. The entropy density of a translation invariant state ω of a Quantum Spin Chain, is defined as follows:

$$s(\omega) = -\lim_{L\to\infty} \frac{1}{L+1} \mathrm{Tr}\, \rho_L \ln \rho_L \tag{7}$$

where ρ_L is the density matrix of the restriction of the state ω to the finite volume $[0, L] \subset \mathbf{Z}$. For a model with translation invariant finite range interaction we know that for all $\beta \geq 0$ there is a unique infinite volume equilibrium state at inverse temperature β, ω_β, of the system. The limit $\lim_{\beta\to\infty} \omega_\beta = \omega_\infty$ exists and is a ground state of the model. The "Third Law of Thermodynamics" states that

$$\lim_{\beta\to\infty} s(\omega_\beta) = s(\omega_\infty) = 0$$

Although this 'law' might be generically true, it is violated for some specific interactions. In [AL] Aizenman and Lieb have shown that this happens if and only if the number of ground states (for suitable boundary conditions) grows exponentially with the volume.

The main result we want to announce here is that there is a whole family of models of the type (4) with residual entropy, i.e. with a ground state ω such that $s(\omega) > 0$. In these models both rotation and translation invariance are spontaneously broken in the ground state. Models of this type (often called VBS-models) were introduced in [K,ALKT] and they are of interest exactly because their ground states are in some sense highly disordered, they are 'liquids' at zero temperature. This feature is a essential ingredient in a possible explanation of high T_c superconductivity [A].

To be concrete, consider a spin s quantum chain with an interaction of the form:

$$H_L = \sum_{i=-L}^{L-1} \sum_{k=s_0}^{2s} J_k P^{(k)}_{i,i+1} \tag{8}$$

where $P^{(k)}_{i,i+1}$ is the orthogonal projection on the spin k subspace on the nearest neighbour pair $\{i, i+1\}$ and the J_k are non-negative coupling constants. s_0 indicates that the couplings in the general interaction (2) are chosen to be zero for all $k < s_0$. If s is integer, $s_0 = s+1$ and all $J_k > 0$, then the model has a unique infinite volume ground state, which correlations can be expressed exactly by a formula of the type (5) [AAH,FNW1]. For halfinteger $s = 3/2, 5/2, \ldots$ and with $s_0 \geq s + 3/2$, the models have a ground state, which can be constructed by our techniques and with strictly positive entropy density. The simplest example of the latter class is the one with $s = 3/2$ and the (up to trivial constants unique) Hamiltonian:

$$H_L = \sum_{i=-L}^{L-1} \{495 + 972\vec{S}_i \cdot \vec{S}_{i+1} + 464(\vec{S}_i \cdot \vec{S}_{i+1})^2 + 64(\vec{S}_i \cdot \vec{S}_{i+1})^3\}$$

We now indicate how the maps E_X in (5) have to be choosen to produce the desired ground state for the models described in (7) with $s \geq 3/2$ and half-integer and $s_0 \geq s + 3/2$. In some specific sense this ground state contains all other ground states of the same Hamiltonian [FNW3]. We then sketch briefly how the entropy density of this state can be computed or at least estimated.

For any $s \geq 3/2$ and half-integer, we define

$$j = 1/2(s - 1/2)$$

Then one takes a linear map

$$V : \mathbf{C}^{2j+1} \to \mathbf{C}^{2s+1} \otimes \mathbf{C}^2 \otimes \mathbf{C}^{2j+1}$$

which satisfies

$$V^* V = \mathbf{1} \tag{9}$$

and for all $g \in SU(2)$

$$V D_j(g) = D_s(g) \otimes D_{1/2}(g) \otimes D_j(g) V \tag{10}$$

Such a map is uniquely determined by the equations (9) and (10), up to a phase. One then defines for all $X \in M_{2s+1}$ and all $B \in M_{2j+1}$:

$$E_X(B) = V^*(X \otimes \mathbf{1}_2 \otimes B) V \tag{11}$$

The density matrix $\rho \in M_{2j+1}$ in (5) has to be taken to be $1/(2j+1)\mathbf{1}$. It follows then that (5) defines a state of the spin chain with local density matrices

$$\begin{aligned} \rho_L = & \frac{1}{2j+1} \mathrm{Tr}_{\mathbf{C}^{2j+1}} \mathrm{Tr}_{(\otimes \mathbf{C}^2)^{L+1}} \\ & \times ((\otimes \mathbf{1})^L \otimes V)((\otimes \mathbf{1})^{L-1} \otimes V) \cdots (\mathbf{1} \otimes V) V \mathbf{1}_{2j+1} V^* (\mathbf{1} \otimes V^*) \cdots ((\otimes \mathbf{1})^L \otimes V^*) \end{aligned} \tag{12}$$

In [FNW3] it is shown that thus one obtains indeed a ground state of the models under consideration.

To estimate the entropy density, observe first that by convexity, for any translation invariant state ω with local density matrices ρ_L, one has:

$$s(\omega) \geq -\frac{1}{2} \lim_{L \to \infty} \ln \mathrm{Tr}\, \rho_L^2 \tag{13}$$

The special structure of the defining formulae (12) for ρ_L makes it possible to make an estimate of the following kind:

$$\operatorname{Tr} \rho_L^2 \leq C e^{-\gamma L} \tag{14}$$

with $C > 0$ and $\gamma > 0$ positive constants; combining (13) and (14) one gets

$$s(\omega) \geq \frac{1}{2}\gamma > 0 \tag{15}$$

For the determination of the value of γ see [FNW3].

For the residual entropy of a model, using the basic result of [AL], one can go one step further and improve (15) to obtain

$$s(\omega) \geq \gamma > 0$$

We conjecture that for the soluble models presented in this paper it is possible, at least in principle, to calculate the optimal γ for which the bound (14) holds, and that with this optimal value

$$s(\omega) = \gamma.$$

Acknowledgement

The author acknowledges support from the Fondo Nacional de Desarrollo Científico y Tecnológico (Chile, Fondecyt project Nr 90-1156).

References

[A] P. Anderson, Science **235**, 1196 (1987)

[AL] M. Aizenman and E.H. Lieb, J. Stat. Phys. **24**, 279 (1981)

[AKLT] I. Affleck, E.H. Lieb, T. Kennedy, H. Tasaki, Commun. Math. Phys. **115**, 477 (1988)

[AAH] D.P. Arovas, A. Auerbach, F.D.M. Haldane, Phys.Rev.Lett. **60**, 531 (1988)

[FNW1] M. Fannes, B. Nachtergaele, R. Werner, Europhys. Lett. **10**, 633 (1989)

[FNW2] M. Fannes, B. Nachtergaele, R. Werner, *Finitely Correlated States of Quantum Spin Chains*, in preparation

[FNW3] M. Fannes, B. Nachtergaele, R. Werner, *Generalized Majumdar-Ghosh models, Haldane's conjecture, and ground states with residual entropy*, in preparation

[K] D.J. Klein, J.Phys.A:Math.Gen. **15**, 661 (1982)

Quantum Fluctuations and Linear Response Theory

A. VERBEURE
Instituut voor Theoretische Fysica
Katholieke Universiteit Leuven
B-3030 Leuven, Belgium

In this lecture we focus on the mathematical characterization of fluctuations for the local observables in a quantum system. We apply the results to a model for which we prove that the linear response theory of Kubo [1,2] becomes exact.

We start with a microscopic dynamical system [3] consisting of the triplet $(\mathcal{A}, \omega, \alpha_t)$ where $\mathcal{A} = \cup_{L \in \mathbb{Z}^\nu} \mathcal{A}_\Lambda$ is the quasi-local algebra of observables, $\mathcal{A}_\Lambda = \otimes_{i \in \Lambda} (M_n)_i$; M_n the $n \times n$ matrices; $\mathbb{Z}^\nu$ stands for the ν-dimensional square lattice; clearly $[\mathcal{A}_{\Lambda_1}, \mathcal{A}_{\Lambda_2}] = 0$ if $\Lambda_1 \cap \Lambda_2 = \phi$, $\mathcal{A}_{\Lambda_1} \subset \mathcal{A}_{\Lambda_2}$ if $\Lambda_1 \subset \Lambda_2$.

Denote by $\tau_x, x \in \mathbb{Z}^\nu$ the space translation automorphism of the translation over the distance x i.e. $\tau_x \mathcal{A}_{\{y\}} = \mathcal{A}_{\{x+y\}}$.

Let ω be a state of the algebra of observables $\mathcal{A}$. We assume that it is homogeneous:

$$\omega . \tau_x = \omega \text{ for all } \ x \in \mathbb{Z}^\nu$$

and space clustering:

$$\lim_{|x| \to \infty} \omega(A \tau_x B) = \omega(A) \omega(B)$$

for all $A, B \in \mathcal{A}$.

Finally we suppose that the dynamics α_t is described by the set of local Hamiltonians $\{H_\Lambda\}_{\Lambda \subset \mathbb{Z}^\nu}$ describing finite range interactions between the particles on the lattice $\mathbb{Z}^\nu$. As usual we consider α_t as the norm limit of the local dynamics:

$$\alpha_t = \lim_{\Lambda \to \mathbb{Z}^\nu} e^{itH_\Lambda} . e^{-itH_\Lambda}$$

Remark that in general $\alpha_t \mathcal{A} \subsetneq \mathcal{A}$. The state ω is supposed to be time translation invariant.

$$\omega . \alpha_t = \omega \text{ for all } t \in \mathbb{R}$$

In the following we report on work in collaboration with D. Goderis and P. Vets [4,5,6].

1 Weak law of large numbers

We start from the microscopic dynamical system $(\mathcal{A}, \omega, \alpha_t)$ and try to specify the macroscopic dynamical system obtained as a result of the weak law of large numbers.

For any $A \in \mathcal{A}$, consider the local mean

$$m_\Lambda(A) = \frac{1}{|\Lambda|} \sum_{x \in \Lambda} \tau_x A$$

E. Tirapegui and W. Zeller (eds.), Instabilities and Nonequilibrium Structures III, 75–80.

It is trivially checked that

$$\text{weak} \lim_{\Lambda} m_\Lambda(A) = m(A)$$

exists in the weak operator topology induced by the state ω; $m(A)$ is called the mean of A.

The following properties are immediate: let $m(\mathcal{A}) = \{m(A)|A \in \mathcal{A}\}$ then;

(i) $[m(\mathcal{A}), \mathcal{A}] = 0$ i.e. $m(\mathcal{A})$ is a set of observables at infinity (see [7]).

(ii) $m(\mathcal{A})$ is an abelian algebra, hence the states on the algebra $m(\mathcal{A})$ are probability measures.

(iii) $m(A) = \omega(A), A \in \mathcal{A}$; i.e. $m(\mathcal{A})$ is trivial.

(iv) the map $m : \mathcal{A} \to m(\mathcal{A})$ is not injective; this is the mathematical expression of coarse graining under the law of large numbers.

(v) the macro dynamics induced by the micro-dynamics α_t on $m(\mathcal{A})$ is trivial:

$$m(\alpha_t A) = \omega(\alpha_t A) = \omega(A) = m(A), A \in \mathcal{A}.$$

These properties yield the conclusion that the macrodynamical system obtained by applying the weak law of large numbers is given by the triplet $(m(\mathcal{A}), \omega =$ prob. measure on $m(\mathcal{A}), \tilde{\alpha}_t =$ id.) It is a trivial dynamical system.

2 Central limit theorem

Again we start from the micro system $(\mathcal{A}, \omega, \alpha_t)$ and we try now to specify the macrosystem obtained as a result of the central limit theorem.

For any local $A \in \mathcal{A}_{sa}$, the set of selfadjoint elements of $\mathcal{A}$, consider the local fluctuation

$$b_\Lambda(A) = \frac{1}{|\Lambda|^{1/2}} \sum_{x \in \Lambda} \tau_x A - \omega(A)$$

One proves the following central limit theorem:

If

$$\sum_{x \in \mathbb{Z}^\nu} \alpha_\omega(|x|) < \infty$$

where

$$\alpha_\omega(d) = \sup_{\Lambda, \tilde{\Lambda}} \sup_{\substack{A \in \mathcal{A}_\Lambda \\ B \in \mathcal{A}_{\tilde{\Lambda}}}} \{\frac{\omega(AB) - \omega(A)\omega(B)}{\| A \| \| B \|}; d \le d(\Lambda, \tilde{\Lambda})\}$$

then

$$\lim_{\Lambda} \omega(e^{ib_\Lambda(A)}) = e^{-\frac{1}{2}s_\omega(A,A)}, A \in \mathcal{A} \tag{1}$$

where

$$s_\omega(A,B) = Re \sum_x \omega\left((\tau_x A - \omega(A))(B - \omega(B))\right)$$

This result establishes the meaning of the limit $\lim_{\Lambda} b_\Lambda(A) \equiv b(A)$, which we call the fluctuatioin of A. If the system satisfies (1), then we call the system having normal fluctuations.

The proof of (1) is performed using a non commutative generalization of the Bernstein argument [8].

Consider now the vectorspace $\mathcal{A}_{sa}$, denote by σ_ω the bilinear, anitsymmetric, degenerate form :

$$\sigma_\omega(X,Y) = -i\omega\left(\sum_y [X, \tau_y Y]\right)$$

Remark that σ_ω is a symplectic form on $\mathcal{A}_{sa}$. Consider the Weyl algebra $W(\mathcal{A}_{sa}, \sigma_\omega)$, generated by the Weyl operators $W(X) = \exp ib(X)$ where $b(X)$ is the Boson field operator, acting on the representation space of the quasi-free state $\tilde{\omega}$ defined by

$$\tilde{\omega}(W(X)) = \exp -\frac{1}{2}s_\omega(X,X)$$

This definition is possible because we have the inequality

$$\frac{1}{4}|\sigma_\omega(X,Y)|^2 \leq s_\omega(X,X)s_\omega(Y,Y)$$

as a consequence of Schwartz inequality.

Identification of $\tilde{\omega}(W(X))$ with formula (1) yields a probabilistic basis of the canonical commutation relations. Also we obtained a characterization of the transition of the micro system $(\mathcal{A}_{sa}, \omega)$ to the macro system $(W(\mathcal{A}_{sa}), \tilde{\omega})$, via the central limit theorem.

Remark also that the map b is not injective. This is a mathematical way of expressing the phenomenon of coarse graining due to the limit theorem.

Next we discuss the dynamics of the fluctuation algebra $W(\mathcal{A}_{sa}, \sigma_\omega)$. One proves that if $\sum_x \alpha_\omega(|x|) < \infty$ and if the potential has finite range interactions then also the following central limits exist :

$$\lim_{\Lambda} b_\Lambda(\alpha_t X) \quad \text{for all} \quad X \in \mathcal{A}_{sa}\,. \tag{2}$$

This result allows us to define a dynamics $\tilde{\alpha}_t$ of the fluctuation algebra by the formula

$$\tilde{\alpha}_t b(X) = b(\alpha_t X)\,.$$

For each value of the time parameter t, $\tilde{\alpha}_t$ is a quasi-free map i.e.

$$\tilde{\alpha}_t(b(X_1)\ldots b(X_n)) = b(\alpha_t X_1)\ldots b(\alpha_t X_n)\,.$$

The α_t-time invariance of ω implies the $\tilde{\alpha}_t$-time invariance of the macro state $\tilde{\omega}$. Also if ω is $(\alpha_t, \beta) - KMS$ then $\tilde{\omega}$ is $(\tilde{\alpha}_t, \beta) - KMS$, which means that the KMS-property is stable for the central limit.

3 Linear Response theory for systems with normal fluctuations

One considers the perturbed dynamics

$$\alpha_t^{P_\Lambda} = e^{it(H+P_\Lambda)} \cdot e^{-it(H+P_\Lambda)}$$

where H is the Hamiltonian of α_t and P_Λ the local fluctuation

$$P_\Lambda = \frac{1}{|\Lambda|^{1/2}} \sum_{x\in\Lambda} \tau_x P - \omega(P) \ ; \ P \in \mathcal{A}_{sa} .$$

The Dyson sequence yields : for any $A \in \mathcal{A}_{sa}$;

$$\alpha_t^{P_\Lambda}(A_\Lambda) = \alpha_t(A_\Lambda) + \sum_{n=1}^{\infty} i^n \int_0^t ds_1 \ldots \int_0^{s_{n-1}} ds_n$$

$$[\alpha_{s_n}(P_\Lambda), \ldots [\alpha_{s_1}(P_\Lambda), \alpha_t(A_\Lambda)] \ldots] .$$

In the Kubo theory [1,2] one is interested in the following expectation value :

$$\lim_\Lambda \omega\left(\alpha_t^{P_\Lambda}(A_\Lambda)\right) .$$

One proves

$$\begin{aligned} \lim_\Lambda \omega\left(\alpha_t^{P_\Lambda}(A_\Lambda)\right) &= \tilde{\omega}\left(\tilde{\alpha}_t^P b(A)\right) \\ &= \omega(A) - \int_0^t ds\, \sigma_\omega(P, \alpha_s A) \\ &= \omega(A) - i \int_0^t ds\, \tilde{\omega}([b(P), \tilde{\alpha}_s b(A)]) \end{aligned} \tag{3}$$

i.e. we prove that the response function is linear. Furthermore the first equality yields the existence of a perturbed macro dynamics

$$\tilde{\alpha}_t^P = e^{it(\tilde{H}+b(P))} \cdot e^{-it(\tilde{H}+b(P))}$$

where $\tilde{\alpha}_t = e^{it\tilde{H}} \cdot e^{-it\tilde{H}}$ is the dynamics described above. The formulae of above tell that the response and linear response theory do not introduce irreversibility as such. The micro reversible dynamics is transported to a macro reversible dynamics. The linearity of the response function is also not directly linked to the system being in equilibrium or not.

On the other hand if one wants to recover the Duhamel two-point function of the Kubo formula, one has to assume that the micro state is in a $(\alpha_t, \beta) - KMS$ state. In that case one obtains the following differential form of the Kubo formula

$$\frac{d}{dt}\tilde{\omega}\left(\tilde{\alpha}_t^P b(A)\right) = \frac{d}{dt}\int_0^\beta du\, \tilde{\omega}\left(b(\alpha_t P)\tilde{\alpha}_{iu} b(A)\right) .$$

4 Small oscillations around equilibrium

In section 3 we constructed the perturbed macro dynamics $\tilde{\alpha}_t^P$. Now it is a natural question to ask for the perturbed equilibrium states i.e. we are looking for the $(\tilde{\alpha}_t^P, \beta)$-equilibrium state $\tilde{\omega}^P$. The corresponding KMS-equations yield the following solution :

$$\tilde{\omega}^P\left(e^{ib(A)}\right) = \tilde{\omega}\left(e^{ib(A)}\right) e^{-i\beta(b(P),b(A))_\sim} \tag{4}$$

where

$$(b(P), b(A))_\sim = \frac{1}{\beta}\int_0^\beta du\, \omega(b(P)\alpha_{iu}b(A))$$

is the Duhamel two-point function. As $\tilde{\alpha}_t^P$ is a perturbation of $\tilde{\alpha}_t$ by a fluctuation, equation (4) yields the stability of the equilibrium state under fluctuations. We call this property the fluctuation stability of equilibrium.

Using formula (4) one gets that

$$\frac{d}{dt}\tilde{\omega}\left(\tilde{\alpha}_t^P b(A)\right) = \frac{d}{dt}\tilde{\omega}^{-P}\left(\tilde{\alpha}_t b(A)\right)$$

expressing the duality between the perturbation of the dynamics and of the corresponding equilibrium state.

Denote by $\Gamma = \{\tilde{\omega}^P | P \in \mathcal{A}_{sa}\}$ the set of perturbed states around the equilibrium state $\tilde{\omega}$. The relative entropy of $\tilde{\omega}^P$ with respect to $\tilde{\omega}$ can be computed and is given by :

$$S(\tilde{\omega}^P|\tilde{\omega}) = -\frac{\beta^2}{2}(b(P), b(P))_\sim . \tag{5}$$

This relative entropy function is a distance function on Γ, it is quadratic in the perturbation observable P. This shows that our model is reproducing exactly the harmonic approximation of the phenomenological Onsager theory. Equation (5) is an explicit expression of the harmonic potential in thermodynamics.

Futhermore the relative entropy is conserved under the dynamics $\tilde{\alpha}_t$:

$$S(\tilde{\omega}^P \cdot \tilde{\alpha}_t|\tilde{\omega} \cdot \tilde{\alpha}_t) = S(\tilde{\omega}^P|\tilde{\omega}) .$$

Hence the macro dynamics $\tilde{\alpha}_t$ of the fluctuation algebra is describing small oscillations around equilibrium. The small oscillations conserve the thermodynamic harmonic potential.

The relative entropy functional (5) yields also a characterisation of the macro equilibrium itself, because it can be shown that

$$S(\tilde{\omega}^P|\tilde{\omega}) = 0 \quad \text{implies} \quad P = 0 .$$

Hence

$$S(\tilde{\omega}^P|\tilde{\omega}) \leq 0 \quad \text{for all} \quad P \in \mathcal{A}_{sa}$$

is a variational principle for $\tilde{\omega}$.

References

1. R. Kubo; J. Phys. Soc. Japan **12**, 570 (1957).

2. R. Kubo, M. Toda, N. Hashitsume; Statistical Physics II, Nonequilibrium Statistical Mechanics; Springer-Verlag 1985.

3. O. Bratteli, D.W. Robinson; Operator Algebras and Quantum Statistical Mechanics I,II, Springer-Verlag, N.Y. 1979/1981.

4. D. Goderis, A. Verbeure, P. Vets; Prob. Th. Rel. Fields **82**, 527 (1989).

5. D. Goderis, A. Verbeure, P. Vets; Dynamics of Fluctuations for Quantum Lattice Systems; Preprint KUL-TF 89/6; to appear in Comm. Math. Phys.

6. D. Goderis, A. Verbeure, P. Vets; About the exactness of the Linear Response Theory, Preprint KUL-TF 89/34.

7. O.E. Lanford, D. Ruelle; Comm. Math. Phys. **13**, 194 (1969).

8. S.N. Bernstein; Math. Ann. **97**, 1 (1926).

NON-COMMUTATIVE LARGE DEVIATIONS AND APPLICATIONS

G.A. Raggio
CONICET and Fac. of Mathematics, Astronomy and Physics,
National University of Córdoba;
5000 Córdoba, Argentina.

ABSTRACT. Results on equilibrium statistical mechanics of inhomogeneous quantum mean-field systems are described. These can be seen as non-commutative extensions of classical large deviation theory.

INTRODUCTION

The applications of large deviation theory, more specifically Varadhan's Large Deviation Principle and its consequences for the asymptotics of integrals ([1], [2]), to statistical mechanics ([3], [4], [5]) are familiar. Here I propose to briefly describe some recent results that can be regarded as extensions to the non-commutative domain, and their applications in quantum statistical mechanics.

The setting is a C^*-algebra $\mathcal{A}$ with unit $\mathbf{1}$. The probability measures are replaced by a state (i.e. a positive, normalized linear functional) on $\mathcal{A}$. The exponential perturbation $\exp\{h(x)\}d\mu(x)$ of the prob. meas. μ is replaced by Araki's [6] perturbation of the state ρ by the self-adjoint element $h\in\mathcal{A}$, written ρ^h. When $\mathcal{A}$ is the bounded linear operators $\mathcal{B}(\mathcal{H})$ over a Hilbert space $\mathcal{H}$, and ρ is given by $\rho(\cdot)=\mathrm{tr}(D\cdot)$, where D is a non-singular density operator on $\mathcal{H}$, then the density operator corresponding to ρ^h is $\exp\{\log(D)+h\}$. Notice that ρ^h is not normalized. In fact, the number

$$F(\rho,h) = \log(\rho^h(\mathbf{1})) ,$$

is the negative increment in free-energy of the renormalized perturbed state $\rho^h/\rho^h(\mathbf{1})$ w.r.t. the state ρ. An important and useful result due to Petz [7], states that $F(\rho,\cdot)$ is the Legendre-Fenchel transform of the relative entropy $S(\rho,\cdot)$ introduced in [8] as a generalization to arbitrary C^*-algebras of the definition

E. Tirapegui and W. Zeller (eds.), Instabilities and Nonequilibrium Structures III, 81–85.

$$S(\rho,\varphi) = -\log\{tr(D_\varphi[\log(D_\varphi)-\log(D_\rho)])\} \quad ,$$

for the case of $\mathcal{B}(\mathcal{H})$. Indeed,

(i) $F(\rho,h) \geq \varphi(h) - S(\rho,\varphi)$, for all states φ on $\mathcal{A}$, with equality iff $\varphi=\rho^h/\rho^h(1)$;

(ii) $S(\rho,\varphi) = \sup\{\varphi(h)-F(\rho,h):\ h^*=h \in \mathcal{A}\}$.

This duality theorem is the backbone of all results to be described. The thermodynamic limit of the free-energy is established via the formula

$$F(\rho,h) = \sup\{\varphi(h) - S(\rho,\varphi):\ \varphi \text{ state on } \mathcal{A}\} \ ,$$

by controlling the contributions of the internal energy, and relative entropies.

The first result I want to mention is proved in [10], and can be regarded as a non-commutative version of Cramér's Theorem ([9]). I first remind you of the classical result (see [2]). Let $x_1,x_2,\cdots$ be a sequence of independent identically distributed random variables with distribution μ. Let μ_n be the distributions of their averages $x^{(n)}=(x_1+x_2+\cdots x_n)/n$. Cramér established that the convergence of $\{\mu_n\}$ to μ (weak law of large numbers) is exponential with a local rate given by the function I which is the Legendre-Fenchel transform of the logarithm of the moment-generating function $M(t)=\int\exp(tu)d\mu(u)$ of μ; i.e.

$$I(u) = \sup\{tu - \log(M(t)):\ t\in\mathbb{R}\} \ .$$

Suggestively, $d\mu_n(u) \approx \exp(-nI(u))d\mu(u)$; more precisely:

$$\limsup_{n\to\infty} n^{-1}\log(\mu_n[F]) \leq -\inf\{I(u):\ u\in F\} \text{ for } F \text{ closed;}$$

$$\liminf_{n\to\infty} n^{-1}\log(\mu_n[G]) \geq -\inf\{I(u):\ u\in G\} \text{ for } G \text{ open.}$$

In general, for a sequence of Borel probability measures $\{\mu_n\}$ on a complete separable metric space $\mathcal{X}$, the existence of I (the rate-function) satisfying these two inequalities, and having compact level-sets, insures ([1]) that for a continuous function f,

$$\lim_{n\to\infty} n^{-1}\log\left(\int_{\mathcal{X}}\exp(nf(u))d\mu_n(u)\right) = \sup\{f(u) - I(u):\ u\in\mathcal{X}\} \ . \qquad (*)$$

Hence the applications to classical statistical mechanics of models of mean-field type of large deviations. In going to the non-commutative domain, and as announced in the previous paragraph, the random variables become self-adjoint elements of an algebra together with a state of this algebra. Let $x^*=x$ be in $\mathcal{A}$, and let x_i be a copy of x at the i-th position of the n-fold tensor-product $\otimes^n\mathcal{A}$. Put $x^{(n)}=(x_1+x_2+\cdots+x_n)/n$. Let ρ

be a separating state of $\mathcal{A}$. Consider the perturbation of $\otimes^n\rho$ by $nf(x^{(n)})$, defined by the spectral calculus. It is proved in [10] that

$$\lim_{n\to\infty} n^{-1}F(\otimes^n\rho, nf(x^{(n)})) = \sup\{f(\varphi(x)) - S(\rho,\varphi):\ \varphi\in\mathcal{S}(\mathcal{A})\}$$

$$= \sup\{f(u) - I_\rho^x(u):\ u\in\mathbb{R}\} \quad .$$

The first supremum is over the state space $\mathcal{S}(\mathcal{A})$ of $\mathcal{A}$. The second one is completely analogous to (*); the rate-function I_ρ^x is given by the Legendre-Fenchel transform of the function $t \to \log(\rho^{tx}(1))$, $t\in\mathbb{R}$, again in complete analogy with the classical case. The proof however is quite different. The applications to quantum mean-field systems are again quite straightforward (see [10]).

(INHOMOGENEOUS) MEAN-FIELD MODELS

The result of [10] has been extended in [11]; the connection between permutational symmetry and "mean-field nature" is exploited fully leading to a considerably larger class of hamiltonians. A homogeneous mean-field system is specified, first of all, by the underlying algebra $\mathcal{A}$ (observables of the single system) and a separating state ρ on it. The noninteracting system consisting of n single subsystems is characterized by the tensor-product state $\otimes^n\rho$ on the n-fold tensor product $\otimes^n\mathcal{A}$. The interaction is introduced by a sequence of hamiltonian densities $\{h_n\}_{n\geq n_o}$ which are selfadjoint elements of $\otimes^n\mathcal{A}$. The state $\otimes^n\rho$ is exponentially perturbed by nh_n, which is the hamiltonian for n systems. The mean-field character of the model enters through the requirement that $\{h_n\}$ be approximately symmetric:

for every $\varepsilon>0$, there exists an integer m such that for every $n>m$, $\|h_n - \mathrm{sym}_n(h_m\otimes 1_{n-m})\| \leq \varepsilon$.

Here sym_n denotes the operation of resymmetrization w.r.t. to the permutations $\mathcal{S}_n$ (e.g.: $\mathrm{sym}_2(a\otimes b)=(1/2!)(a\otimes b+b\otimes a)$). Up to a sign, and incorporating the temperature in the hamiltonian, the change in free-energy per "particle" due to the interaction is then $n^{-1}F(\otimes^n\rho, nh_n)$. The result is the following Gibbs Variational Principle:

$$\lim_{n\to\infty} n^{-1}F(\otimes^n\rho, nh_n) = \sup\{\lim_{n\to\infty}\phi(h_n) - S_M(\otimes^\infty\rho,\phi):\ \phi\in\mathcal{S}_s(\otimes^\infty\mathcal{A})\}$$

$$= \sup\{U(\varphi) - S(\rho,\varphi):\ \varphi\in\mathcal{S}(\mathcal{A})\} \ .$$

Here, the first supremum is over the permutation-symmetric states, $\mathcal{S}_s(\otimes^\infty\mathcal{A})$, of the infinite tensor-product of $\mathcal{A}$ for which $\lim_{n\to\infty}\phi(h_n)$ exists;

and S_M denotes the mean-relative entropy. The second supremum is over the states of the single system, $\mathcal{S}(\mathcal{A})$, and $U(\varphi)=\lim_{n\to\infty}(\otimes^n\varphi)(h_n)$. The variational problem reduces further when the hamiltonian densities h_n are given by a fixed function ℓ of k symmetric sequences as follows. Let $x^{(1)},\cdot\cdot,x^{(k)}$ be k self-adjoint elements of $\mathcal{A}$. Let $X_n^{(j)} = \mathrm{sym}_n(x^{(j)}\otimes 1_{n-1})$, j=1,2,··,k. Then

$$\lim_{n\to\infty} n^{-1}F(\otimes^n\rho, n\ell(X_n^{(1)},\cdot\cdot,X_n^{(k)})) = \sup\{\ell(\underline{u})-I_\rho^x(\underline{u}) : \underline{u}\in\mathbb{R}^k\}.$$

This is a non-commutative multidimensional version of Cramér's result; I_ρ^x is the Legendre-Fenchel transform of $\underline{t} \to \log(\rho^{\underline{t}\cdot\underline{x}}(1))$, $\underline{t}\in\mathbb{R}^k$, with $\underline{x}=(x^{(1)},\cdot\cdot,x^{(k)})$.

The next step was then the extension of the mean-field results to what were called <u>inhomogeneous</u> mean-field systems ([12]). Here, the hamiltonian has mean-field scaling, but the coupling constants are "site-dependent". A prototype is the full BCS-model, the thermodynamic equilibrium properties of which were established rigorously only recently in [13] by other methods. These results now follow by a straightforward application of the general theorems of [12]. An inhomogeneous mean-field system is specified by the algebra $\mathcal{A}$ and its separating state ρ which describes the non-interacting composite of n single systems. The interaction enters by perturbing $\otimes^n\rho$ with $nh_n(\xi_{n,1},\xi_{n,2},\cdot\cdot,\xi_{n,n})$ where the n parameters $\xi_{n,1},\xi_{n,2},\cdot\cdot,\xi_{n,n}$ take values in a compact space $\mathcal{X}$, and h_n is a continuous function from $\times^n\mathcal{X}$ to $\otimes^n\mathcal{A}$, i.e. an element of the C^*-algebra $\otimes^n\mathcal{C}(\mathcal{X},\mathcal{A}) \cong \mathcal{C}(\times^n\mathcal{X},\otimes^n\mathcal{A})$. The sequence $\{h_n\}$ is assumed to be approximately symmetric. The sequence of n-tupels $\xi_n \equiv (\xi_{n,1},\xi_{n,2},\cdot\cdot,\xi_{n,n})$ must admit a limiting density; i.e., there exists a probability measure μ on $\mathcal{X}$ such that

$$n^{-1}\sum_{i=1}^{n}\delta(\xi_{n,i}) \to \mu .$$

The following Gibbs Variational Principle holds for inhomogeneous mean-field systems:

$$\lim_{n\to\infty} n^{-1}F(\otimes^n\rho, nh_n(\xi_n))$$

$$= \sup\{\lim_{n\to\infty}\phi(h_n) - S_M(\otimes^\infty(\mu\otimes\rho),\phi) : \phi\in\mathcal{S}_s^\mu(\otimes^\infty\mathcal{C}(\mathcal{X},\mathcal{A}))\}$$

$$= \sup\{U(\varphi) - S(\mu\otimes\rho,\varphi) : \varphi\in\mathcal{S}(\mathcal{C}(\mathcal{X},\mathcal{A})),\ \varphi|\mathcal{C}(\mathcal{X})=\mu\} .$$

The first supremum is over the permutation-symmetric states,

$\mathcal{S}^{\mu}_{s}(\otimes^{\infty}\mathcal{C}(\mathcal{X},\mathcal{A}))$, of $\otimes^{\infty}\mathcal{C}(\mathcal{X},\mathcal{A})$ which have the property that their restrictions to $\otimes^{\infty}\mathcal{C}(\mathcal{X})$ are equal to $\otimes^{\infty}\mu$. The second supremum is over the states of $\mathcal{C}(\mathcal{X},\mathcal{A})$ which are equal to μ when restricted to $\mathcal{C}(\mathcal{X})$. This variational problem can be handled using the fact that every state φ of $\mathcal{C}(\mathcal{X},\mathcal{A})$ can be written uniquely (provided $\mathcal{A}$ is separable) as

$$\varphi = \int_{\mathcal{X}}^{\oplus} \mu_{\varphi}(dx)\ \varphi_x \quad ,$$

with a regular Borel probability measure μ_{φ} on $\mathcal{X}$, and function $x \to \varphi_x$ on $\mathcal{X}$ taking values in the state-space of $\mathcal{A}$. Then, $S(\mu\otimes\rho,\varphi) = \int_{\mathcal{X}}\mu(dx)S(\rho,\varphi_x)$ if $\mu_{\varphi}=\mu$. The integral decomposition can also be used to rewrite the $U(\cdot)$-term.

Finally, I remark that the general result can be applied to "random-site" mean-field models if one assumes ergodicity. I refer to [12] for more details, and specific applications.

REFERENCES

[1] Varadhan, S.R.S.: *Asymptotic probabilities and differential equations.* Commun. Pure Appl. Math. **19**, 261-286 (1966).
[2] Varadhan, S.R.S.: *Large Deviations and Applications.* Philadelphia: Society for Industrial and Applied Mathematics 1984.
[3] Ellis, R.S.: *Entropy, Large Deviations and Statistical Mechanics.* New York, Berlin, Heidelberg, Tokio: Springer 1985.
[4] Berg, M. van den, Lewis, J.T., and Pulé, J.V.: *The Large Deviation Principle and some models of an interacting Bose gas.* Commun. Math. Phys. **118**, 61-85 (1988).
[5] Dorlas, T.C., Lewis, J.T., and Pulé, J.V.: *The Yang-Yang thermodynamic formalism and large deviations.* Commun. Math. Phys. (1990) ?.
[6] Araki, H.: *Relative hamiltonian for faithful normal states of a von Neumann algebra.* Publ. Res. Inst. Math. Sci. **9**, 165-209 (1973).
[7] Petz, D.: *A variational expression for the relative entropy.* Commun. Math. Phys. **114**, 345-349 (1988).
[8] Araki, H.:*Relative entropy of states of von Neumann algebras I & II.* Publ. Res. Inst. Math. Sci. **11**, 809-833 (1976) & **13**, 173-192 (1977).
[9] Cramér, H.: *On a new limit theorem in the theory of probability.* In: Colloq. on the Theory of Probability. Paris: Hermann 1937.
[10] Petz, D., Raggio, G.A., and Verbeure, A.:*Asymtpotics of Varadhan-Type and the Gibbs Variational Principle.* Commun. Math. Phys. **121**, 271-282 (1989).
[11] Raggio, G.A., and Werner, R.F: *Quantum statistical mechanics of general mean field systems.* Helvet. Phys. Acta. **62**, 980-1003 (1989).
[12] Raggio, G.A., and Werner, R.F: *The Gibbs Variational Principle for inhomogeneous mean-field systems.* Preprint DIAS, March 1990.
[13] Duffield, N.G, and Pulé, J.V.: *Thermodynamics and phase transitions in the Overhauser model.* J. Stat. Phys. **54**, 449-475 (1989).

SQUEEZING IN THE SU(1,1) GROUP

J. ALIAGA, G. CRESPO AND A.N. PROTO
Laboratorio de Física
Comisión Nacional de Investigaciones Espaciales
Av. Libertador 1513, (1638) Vicente López, Argentina

ABSTRACT. We present a density matrix method, based on information-theory, which makes it possible to find non-zero temperature coherent and squeezed states for the harmonic oscillator (HO) with time-dependent frequency, the Kanai-Caldirola (KC) Hamiltonian, and isomorphic ones. We establish a connection between the appearance of squeezed states and the relevant operators included in the density matrix and compare our results with previous ones that were obtained using wave functions.

Introduction

Recently,[1] we have analyzed the generation and properties of squeezed states in terms of the Maximum Entropy Principle (MEP) approach to the density matrix.[2-5] The so-called squeezed states satisfy

$$(\Delta \hat{O}_i)^2 < \frac{1}{2} \left| <[\hat{O}_i ; \hat{O}_j]> \right| \tag{1}$$

or

$$(\Delta \hat{O}_j)^2 < \frac{1}{2} \left| <[\hat{O}_i ; \hat{O}_j]> \right|, \tag{2}$$

if we consider dimensionless operators, $\hat{O}_i$, $\hat{O}_j$.

The density matrix $\hat{\rho}$ is expressed, within the MEP framework, by

$$\hat{\rho} = \exp(-\lambda_0 \hat{I} - \sum_{j=1}^{M} \lambda_j \hat{O}_j) , \tag{3}$$

in terms of the M+1 Lagrange multipliers, λ_k, k=0,1,...,M. Since the operator $\hat{\rho}(t)$ obeys[3-5] the Liouville equation and due to the fact that the entropy is a constant of the motion, it is found that the (relevant) operators entering Eq. (3) are those that close a partial Lie algebra under commutation

E. Tirapegui and W. Zeller (eds.), Instabilities and Nonequilibrium Structures III, 87–94.

with the Hamiltonian $\hat{H}$,

$$[\hat{H}, \hat{O}_k] = i\hbar \sum_{j=0}^{q} \hat{O}_j G_{jk} \tag{4}$$

where the G_{jk} are the elements (c-numbers) of a qxq matrix $\underline{G}$ (which may depend upon the time if $\hat{H}$ is time dependent).

We shall study the presence of squeezed states in $\hat{q}$ and $\hat{p}$ for the HO with either constant or time-dependent frequency and for the K-C Hamiltonian. For all these cases, the associate Lie algebra has two finite subalgebras of relevant operators, obtained taken into account Eq. (4) (note that those are the only finite subalgebras for the HO): The Heisenberg Algebra H_4, i.e.$\{ \hat{1}, \hat{q}, \hat{p} \}$, and SU(1,1), i.e. $\{ \hat{q}^2, \hat{p}^2, \hat{l} \}$, where $\hat{l} = \frac{1}{2} \{\hat{q}\hat{p} + \hat{p}\hat{q}\}$. As we want to obtain states with a very special dispersion relation between $\hat{q}$ and $\hat{p}$, we need to include information regarding not only these operators, but also their squares. Thus, we consider a general finite algebra composed by both subalgebras.

As it is well known, the Hamiltonian for the HO with variable frequency reads

$$\hat{H}_{HO} = \frac{\hat{p}^2}{2m_0} + \frac{1}{2} m_0\, \omega(t)^2 \hat{q}^2 = \hbar\, \omega(t) \left(\hat{a}^\dagger \hat{a} + \frac{1}{2}\right), \tag{5}$$

were $\omega(t)$ is the time-dependent frequency. The KC Hamiltonian is given by[6-8]

$$\hat{H}_{KC} = \hat{H}(\hat{q}, \hat{p}, t) = \frac{\hat{p}^2}{2m_0} e^{-2\alpha t} + \frac{1}{2} m_0\, e^{2\alpha t}\, \omega_0^2\, \hat{q}^2 . \tag{6}$$

It can be shown that the $\hat{H}_{KC}$ is canonically related to

$$\hat{H}_1 = \hat{H}_1(\hat{Q}, \hat{P}, t) = \frac{\hat{P}^2}{2m_0} + \frac{1}{2} m_0\, \omega_0^2\, \hat{Q}^2 + \alpha\, \hat{L} \tag{7}$$

where

$$\hat{Q} = \exp(\alpha t)\ \hat{q} \tag{8a}$$

$$\hat{P} = \exp(-\alpha t)\ \hat{p}, \tag{8b}$$

and with

$$\hat{H}_1(\hat{A}, \hat{A}^\dagger) = \hbar\Omega \left(\hat{A}^\dagger \hat{A} + \frac{1}{2}\right) \tag{9}$$

where

$$\hat{Q} = \left(\frac{\hbar}{2 m_0 \omega_0}\right)^{1/2} \left(\hat{A}^\dagger + \hat{A}\right) \equiv \gamma_1 \left(\hat{A}^\dagger + \hat{A}\right), \tag{10a}$$

$$\hat{P} = -\left(\frac{\hbar m_0 \omega_0}{2}\right)^{1/2} \left(u^* \hat{A}^\dagger + u \hat{A}\right) \equiv -\gamma_2 \left(u^* \hat{A}^\dagger + u \hat{A}\right), \tag{10b}$$

$$u = \frac{\alpha}{\Omega} + i, \tag{10c}$$

$$\Omega^2 = \omega_0^2 - \alpha^2. \tag{10d}$$

Following the literature, we consider, for the KC Hamiltonian, $\hbar = 1$, $m_0 = 1$.

Results

From Eqs. (5) and (9) it can be seen that both Hamiltonians can be interpreted as harmonic oscillators given in terms of different operators and with different frequency. The Lie algebra's for the transformed KC Hamiltonian are $\{\hat{1}, \hat{Q}, \hat{P}\}$, and $\{\hat{Q}^2, \hat{P}^2, \hat{L}\}$.

The density matrix is given by [2-5]

$$\hat{\rho} = \exp\left(-\lambda_0 \hat{1} - \lambda_q \hat{q} - \lambda_p \hat{p} - \lambda_{q2} \hat{q}^2 - \lambda_{p2} \hat{p}^2 - \lambda_1 \hat{1} - \beta \hat{H}_{HO}\right) \tag{11a}$$

for the HO Hamiltonian, or

$$\hat{\rho} = \exp\left(-\lambda_0 \hat{1} - \lambda_Q \hat{Q} - \lambda_P \hat{P} - \lambda_{Q2} \hat{Q}^2 - \lambda_{P2} \hat{P}^2 - \lambda_L \hat{L} - \beta \hat{H}_1\right) \tag{11b}$$

for the KC Hamiltonian.

As we consider a set of non-commuting relevant operators, we need to diagonalize the logarithm of the density matrix, $\hat{\rho}$, before evaluating mean values. This diagonalization is made[1] generalizing a method of "quasiphoton" number operator developed in [9]. We define a new creation operator

$$\hat{b}^\dagger = |\cosh r| e^{i\varphi} \hat{a}^\dagger + |\sinh r| e^{-i\theta} \hat{a} + |\gamma| e^{-i\psi} \tag{12a}$$

for the HO Hamiltonian, or

$$\hat{b}^\dagger = |\cosh r| e^{i\varphi} \hat{A}^\dagger + |\sinh r| e^{-i\theta} \hat{A} + |\gamma| e^{-i\psi} \tag{12b}$$

for the KC Hamiltonian. Then, the density matrix can be written in terms of the new operators $\hat{b}^\dagger$

and $\hat{b}$ as

$$\hat{\rho} = 2\sinh(\beta\hbar\omega(t)/2)\exp(-\beta\hbar\omega(t)/2)\exp\left(-\beta\hbar\omega(t)\,\hat{b}^{\dagger}\hat{b}\right), \tag{13a}$$

or

$$\hat{\rho} = 2\sinh(\beta\hbar\Omega/2)\exp(-\beta\hbar\Omega/2)\exp\left(-\beta\hbar\Omega\,\hat{b}^{\dagger}\hat{b}\right), \tag{13b}$$

if the Lagrange multipliers verify

$$\lambda_q\gamma_1 = \beta\hbar\omega(t)|\gamma|\left\{|\sinh r|\cos(\psi-\theta) + |\cosh r|\cos(\psi+\varphi)\right\}, \tag{14a}$$

$$\lambda_p\gamma_2 = \beta\hbar\omega(t)|\gamma|\left\{-|\sinh r|\sin(\psi-\theta) + |\cosh r|\sin(\psi+\varphi)\right\}, \tag{14b}$$

$$\lambda_1\gamma_1\gamma_2 = \beta\hbar\omega(t)|\sinh r||\cosh r|\sin(\theta+\varphi), \tag{14c}$$

$$\lambda_{q^2}\gamma_1^2 + \lambda_{p^2}\gamma_2^2 = \beta\hbar\omega(t)(\sinh r)^2, \tag{14d}$$

$$\lambda_{q^2}\gamma_1^2 - \lambda_{p^2}\gamma_2^2 = \beta\hbar\omega(t)|\sinh r||\cosh r|\cos(\theta+\varphi), \tag{14e}$$

or

$$\frac{\lambda_Q}{(2\Omega)^{1/2}} + \lambda_P u(\Omega/2)^{1/2} = \beta\Omega|\gamma|\left\{|\sinh r|e^{i(\psi-\theta)} + |\cosh r|e^{-i(\psi+\varphi)}\right\}, \tag{15a}$$

$$\frac{\lambda_Q}{(2\Omega)^{1/2}} + \lambda_P u^*(\Omega/2)^{1/2} = \beta\Omega|\gamma|\left\{|\sinh r|e^{-i(\psi-\theta)} + |\cosh r|e^{i(\psi+\varphi)}\right\}, \tag{15b}$$

$$\frac{1}{2}\left(\frac{\lambda_{Q^2}}{\Omega} + \lambda_{P^2}\Omega|u|^2 - \frac{\lambda_L(u+u^*)}{2}\right) = \beta\Omega(\sinh r)^2, \tag{15c}$$

$$\frac{1}{2}\mathrm{Re}\left(\frac{\lambda_{Q^2}}{\Omega} + \lambda_{P^2}\Omega u^2 - \lambda_L u\right) = \beta\Omega|\sinh r||\cosh r|\cos(\theta+\varphi), \tag{15d}$$

$$\frac{1}{2}\mathrm{Im}\left(\frac{\lambda_{Q^2}}{\Omega} + \lambda_{P^2}\Omega u^2 - \lambda_L u\right) = -\beta\Omega|\sinh r||\cosh r|\sin(\theta+\varphi), \tag{15e}$$

respectively. The restrictions imposed by these equations on the Lagrange multipliers dual space are,

$$\lambda'^2_1 = \hbar\,\omega(t)\left(\frac{\lambda'_{q^2}}{\gamma_1^2} + \frac{\lambda'_{p^2}}{\gamma_2^2}\right) + 4\,\lambda'_{q^2}\,\lambda'_{p^2} \geq 0, \tag{16a}$$

$$\lambda'_{q^2}\gamma_1^2 + \lambda'_{p^2}\gamma_2^2 \geq 0, \tag{16b}$$

$$(\lambda'_L + \alpha)^2 = \alpha^2 + 2\,\lambda'_{Q^2} + 2\,\omega_0^2\,\lambda'_{P^2} + 4\,\lambda'_{Q^2}\,\lambda'_{P^2} \geq 0, \tag{16c}$$

$$k \equiv 2\,\Omega^2\,(\sinh r)^2 = \lambda'_{Q^2} + \lambda'_{P^2}\,\omega_0^2 - \lambda'_L\,\alpha \geq 0. \tag{16d}$$

being the consequence of the quantum statistical character of the system under study (see [5]). We have defined (see [1-2]) $\lambda'_k = \lambda_k/\beta$. The inclusion of the Hamiltonian as a relevant operator does not modify the temporal evolution of the system, but transforms it in a thermodynamical one,[3] where β is the inverse of the temperature.

We can find that the operators' dispersions are given by

$$(\Delta\hat{q})^2 = \gamma_1^2\left(1 + \frac{4\,\gamma_2^2}{\hbar\,\omega(t)}\,\lambda'_{p^2}\right)\coth\!\big(\beta\,\hbar\,\omega(t)/2\big), \tag{17a}$$

$$(\Delta\hat{p})^2 = \gamma_2^2\left(1 + \frac{4\,\gamma_1^2}{\hbar\,\omega(t)}\,\lambda'_{q^2}\right)\coth\!\big(\beta\,\hbar\,\omega(t)/2\big), \tag{17b}$$

$$\Delta\hat{q}\;\Delta\hat{p} = \frac{\hbar}{2}\left(1 + \frac{\lambda'^2_1}{\omega(t)^2}\right)^{1/2}\coth\!\big(\beta\,\hbar\,\omega(t)/2\big), \tag{17c}$$

$$\Delta^2\hat{Q} = \frac{1}{2\Omega}\big(1 + 2\,\lambda'_{P^2}\big)\coth\!\big(\beta\,\Omega/2\big), \tag{18a}$$

$$\Delta^2\hat{P} = \frac{\Omega}{2}\left(1 + \frac{\alpha^2}{\Omega^2} + \frac{2\,\lambda'_{Q^2}}{\Omega^2}\right)\coth\!\big(\beta\,\Omega/2\big), \tag{18b}$$

$$\Delta\hat{Q}\;\Delta\hat{P} = \frac{1}{2}\left(1 + \frac{(\lambda'_L + \alpha)^2}{\Omega^2}\right)^{1/2}\coth\!\big(\beta\,\Omega/2\big). \tag{18c}$$

Notice that these results were obtained using the transformation given in Eq. (12), having been *unnecessary to solve explicitly* the evolution equations of the mean values and the Lagrange multipliers.

Using Eqs. (17-18), we can analyze the appearance of coherent or squeezed states, as it has been done, from a completely different point of view, in [10-13] and [14-18] for the HO and the KC Hamiltonian, respectively. The dynamical evolution of the system in the dual space of the Lagrange multipliers that allows to examine the appearance of coherent or squeezed states, is shown in Figs. 1-2. Their captions are self-explanatory.

Conclusions

It is important to point out that (a) the appearance (or not) of squeezing is a direct property of the dynamical Lie group structure (i.e. the Hamiltonian and the relevant operators taken into account); and (b) the thermodynamical approach to quantal systems that can be achieved within the MEP formalism allows for the straightforward analysis of squeezing for quantal systems with non-zero temperature. It can be said that the MEP approach makes evident the intrinsic connection among the appearance of coherent and squeezed states, the different "knowledge" introduced in a general density matrix (necessary in order to evaluate mean values properly) and the Hamiltonian dynamics.

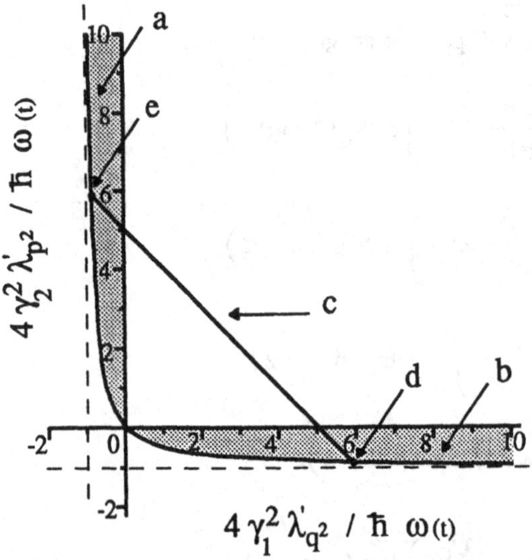

FIG. 1. HO Hamiltonian. Space of independent Lagrange multipliers: λ'_{q^2} and λ'_{p^2}. The gray zone, named "a" ("b"), shows the region where squeezing in $\hat{q}$ ($\hat{p}$) is obtained. The evolution of the system is shown by line "c". It is bounded by the hyperbola given by Eq. (3.13.a) (i.e. the uncertainty principle). Point "d" ("e") corresponds to the minimum dispersion in $\hat{p}$ ($\hat{q}$) for a given value of the invariant of motion or, equivalently, r.

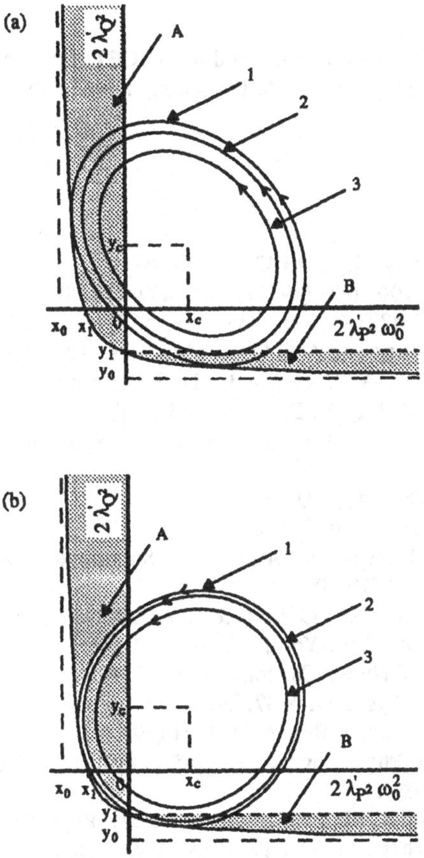

FIG. 2. KC Hamiltonian. Space of independent Lagrange multipliers: λ'_{Q^2} and λ'_{P^2}. The gray zone, named "A" ("B"), shows the region where squeezing in $\hat{Q}$ ($\hat{P}$) is obtained. The evolution of the system runs on an ellipse in counter-clockwise fashion. It is bounded by the hyperbola given by Eq. (16c) (i.e. the uncertainty principle). The asymptotes of the hyperbola are $x_0 = y_0 = -\omega_0^2$, and the intersections of the hyperbola with the x and y axes correspond to $x_1 = y_1 = -\alpha^2$. In Fig. 2.a (or 2.b) we consider the case $\omega_0^2 > 2\alpha^2$ (or $\omega_0^2 < 2\alpha^2$), for which the major axis of the ellipse has slope -1 (or 1). We show three different ellipses in each figure: two of them with squeezing in both coordinates (ellipse 1 and 2) and another one with squeezing only in $\hat{Q}$ (ellipse 3). The ellipse labeled as 1 intersects the hyperbola twice, and therefore those points are of minimum uncertainty product. All the ellipses depicted have been chosen with the same center, determined by $x_c = y_c = (k\,\omega_0^2/\Omega)$.

Acknowledgments

J. A. thanks the Argentine National Research Council (CONICET) for its support. A. N. P. and G. C. acknowledge support from the Comisión de Investigaciones Científicas de la Provincia de Buenos Aires (CIC).

References

1. J. Aliaga and A.N. Proto, Phys. Lett. **A142**, 63 (1989).
2. E.T. Jaynes, Phys. Rev. **106**, 620 (1957); **108**, 171 (1957), Y. Alhassid and R.D. Levine, J. Chem. Phys. **67**, 4321 (1977); Phys. Rev. **A18**, 89 (1978).
3. J. Aliaga, D. Otero, A. Plastino and A.N. Proto, Phys. Rev. **A37**, 918 (1988).
4. J. Aliaga, M. Negri, D. Otero, A. Plastino and A.N. Proto; Phys. Rev. **A36**, 3427 (1987).
5. A.N. Proto, J. Aliaga, D. R. Napoli, D. Otero, and A. Plastino, Phys. Rev. **A39**, 4223 (1989); G. Crespo, D. Otero, A. Plastino, and A. N. Proto, Phys. Rev. **A 39**, 2133 (1989).
6. R. W. Hasse, Rep. Prog. Phys. **41**, 1027 (1978).
7. E. Kanai, Prog. Theor. Phys. **3**, 440 (1948).
8. G. Crespo, D. Otero, A. Plastino and A.N. Proto, Submitted to Phys. Rev. **A**; E. H. Kerner, Can. J. Phys. **36**, 371 (1958).
9. H. P. Yuen, Phys. Rev. **A 13**, 2226 (1976); W.H. Louisell, Quantum Statistical Properties of Radiation, J. Wiley & Sons, New York, 1973, pp.264-266.
10. A. Jannussis and V. Bartiz, Il Nuovo Cimento 102 B, 33 (1988).
11. H.-Y. Fan and H.R. Zaidi, Phys. Rev. **A 37**, 2985 (1988).
12. Xin Ma and William Rhodes, Phys. Rev. **A 39**, 1941 (1989).
13. J. Janszky and Y. Yushin, Phys. Rev. **A 39**, 5445 (1989); H.R. Zaidi and H.-Y. Fan, Phys. Rev. **A 39**, 5447 (1989).
14. Christopher C. Gerry, Philip K. Ma and Edward R. Vrscay, Phys. Rev. **A 39**, 668 (1989); K.H. Yeon, C.I. Um and Thomas F. George, Phys. Rev. **A 36**, 5287 (1987).
15. G. Dattoli, C. Centioli and A. Torre, Il Nuovo Cimento **101 B**, 557 (1988).
16. L.C. Papaloucas, Il Nuovo Cimento **102 B**, 315 (1988); A. Jannussis and V. Bartzis, Il Nuovo Cimento **102 B**, 33 (1988).
17. M. Sebawe Abdalla, Phys. Rev. **A 33**, 2870 (1986).
18. E.N.M. Borges, O.N. Borges and L.A. Amarante Ribeiro, Can. J. Phys. **63**, 600 (1985).

FULL SCALING AND REAL-SPACE RENORMALIZATION*

Patricio Cordero
Departamento de Física
Facultad de Ciencias Físicas y Matemáticas
Universidad de Chile, Casilla 487–3, Santiago, Chile.

The freedom in choosing the weight function in a Niemeijer van Leeuwen real space renormalization transformation (RSRT) is analized. The analysis refers to Ising-like systems invariant with respect to the spatial group of the lattice. It is shown that the matrix of parameters that define the RSRT satisfies an eigenvalue type of equation with a specific eigenvalue related to the linear scale and with an eigenvector built with the correlation functions. From it follows a delicate relation between the RSRT and the critical exponents.

1. Introduction

The idea of real space renormalization transformations (RSRT) is usually tight to *locality* [1-10]. The transformations are designed by a map from the states of a small (thus local) neighborhood of spins onto the space of states of (usually) one "cell spin" of the new system. The characteristic handicap behind RSRT is that in principle they introduce infinitely many couplings which force, in any practical scheme, to truncate, that is, to neglect all but a few selected couplings which results in loosing control over the possible accuracy of the results.

Swendsen [11] designed a numerical experiment for the 2D Ising model using the Monte Carlo Renormalization Group to conjecture that it is possible to find a RSRT which adds no new couplings. The price he paid was loosing locality.

2. The symmetry group.

In the following we shall consider an Ising system defined on a finite (N sites) and regular lattice with periodic boundary conditions. The set of Ising variables is $S \equiv \{s_1, \ldots, s_N\}$. The system has 2^N states. The most general function of the N Ising variables is a polynomial on them. These functions naturally belong to a linear space L_N expanded by the monomials such as: $m = s_3 s_{12} s_{47}$.

Defining the scalar product

$$(f,g) = 2^{-N} \sum_S f(S) g(S). \tag{2.1}$$

where the sum is over all the possible configurations S of the system. It is noticed that the monomials form an orthonormal basis.

*This work was partially supported by grants: *FONDECYT* 1240-90 and *DTI* E-2853-8922

E. Tirapegui and W. Zeller (eds.), Instabilities and Nonequilibrium Structures III, 95–99.

The space group G_N of the lattice (a semi-direct product of the point group of the lattice times the discrete translation group), is of order γN, where γ is the order of the point group. The space L_N is a huge linear representation space of G_N.

The monomials can be divided in equivalent classes according to

$$m_1 \equiv m_2 \Longleftrightarrow \exists O \in G_N \text{ such that } m_1 = O(m_2). \tag{2.2}$$

Examples of equivalent classes are: the class of pairs $[s_a s_b]$ where a and b are nearest neighbors, or the class of minimal plaquetts $[s_a s_b s_c s_d]$ on a square lattice etc.

A class μ has n_μ elements and a typical monomial $m_{\mu a}$ belonging to it is left invariant by a subgroup of order γ_μ of G_N. It is quite trivial to check that

$$n_\mu = \frac{\gamma N}{\gamma_\mu}. \tag{2.3}$$

In this way one can see, for example, that the order $\gamma_{n.n.}$ of invariance group of a nearest neighbour class is $\gamma_{n.n.} = \frac{2\gamma}{q}$ where q is the coordination number of the lattice. The γ_μ's are usually finite numbers (i.e., independent of the size of the lattice).

3. The symmetric hamiltonian.

To deal with systems invariant with respect to G_N the trivial representations of the group on L_N play a special role. In other words, the invariants in L_N are of special interest. The subspace V_N of all the invariant functions is expanded by the (complete set of) functions

$$W_\mu(S) = \frac{1}{\sqrt{n_\mu}} \sum_a m_{\mu a}. \tag{3.1}$$

The summation index a runs through the equivalence classes μ. By definition of our notation, $W_0 = 1$ while all the rest are denoted W_A.

These invariant functions form an orthonormal base of V_N,

$$(W_\mu, W_\nu) = \delta_{\mu\nu}. \tag{3.2}$$

The most general G_N −invariant hamiltonian is of the form

$$H(K,S) = \sum_A \sqrt{n_A} K_A W_A(S) \tag{3.3}$$

where K denotes the set of coupling constants multiplied by $-1/kT$.

Since H is an invariant function, also $\exp[H]$ is invariant, and therefore can be expanded in term of the basis W,

$$\exp[H(K,S)] = 2^{-N} \sum_\mu Z_\mu(K) W_\mu(S). \tag{3.4}$$

In particular Z_0 is the partition function of the system while the Z_A can be written as,

$$Z_A = \frac{1}{\sqrt{n_A}} \frac{\partial Z_0}{\partial K_A} \tag{3.5}$$

which makes easier to see that the average of W_A is

$$< W_A >= \frac{Z_A}{Z_0}. \tag{3.6}$$

Defining the correlation functions

$$g_A(K) = \frac{1}{\gamma N} \frac{Z_A}{Z_0} \tag{3.7}$$

it is seen that

$$g_A(K) = \frac{\gamma_A}{\gamma} \frac{\partial f(K)}{\partial K_A} \tag{3.8}$$

where f is the specific free energy:

$$f(K) = \ln(Z_0)/N. \tag{3.9}$$

4. Real Space Transformations.

The Niemeijer van Leeuwen RSRT [1-3] are defined through a weight function $P(S', S)$ as,

$$\exp[G(K) + H'(K', S')] = \sum_S P(S', S) \exp[H(K, S)] \tag{4.1a}$$

where, as usual,

$$\sum_{S'} P(S', S) = 1. \tag{4.1b}$$

Equations (4.1) in principle give $G(K)$, the renormalized hamiltonian H', the renormalization equations

$$K'_{A'} = \Gamma'_A(K) \tag{4.2}$$

and an infinite set of equations connecting the correlation functions g_A.

Since (4.1a) has the scalar product of P with the symmetric function $\exp[H]$ there is no loss of generality assuming that P itself is symmetric in S. Furthermore, restricting the RSRT to the class of $G-$invariant hamiltonians (homogeneous hamiltonians), then H' too has to be symmetric. Requiring that this be true for any given H implies that P is symmetric in S' as well. In other words, P belongs to the direct product space $V_{N'} \times V_N$.

Under these circumstances the most general weight functions which also satisfies (4.1b) is[13],

$$P(S', S) = 2^{-N'}[1 + \sum_{A'\mu} q_{A'\mu} W'_{A'}(S') W_\mu(S)]. \tag{4.3}$$

Although this is a non local way to write the weight function, any local weight function can be written in this form.

This is the most general and most economic way to write the weight function appearing in the RSRT. The $q_{A'\mu}$ are so far unknown parameters that quantify the arbitrariness in choosing the weight function. Assuming that $q_{A'0} = 0$ two equations follow:

$$\exp[G(K)]Z_0'(K') = Z_0(K) \tag{4.4}$$

and

$$\vec{g}(K') = b^{d/2}\,\mathcal{Q}\vec{g}(K) \tag{4.5}$$

where $\mathcal{Q}$ is the matrix of the coefficients q_{AB}. This is a stringent condition on the matrix $\mathcal{Q}$, showing that the parameters cannot be chosen with careless freedom. If the matrix $T_{AB} = \partial K_A'/\partial K_B$ is diagonalizable by means of a matrix R, $T = R\Lambda R^{-1}$ then Λ has in its diagonal elements $\lambda = b^{y_P}$, where b is the length scale associated to the RSRT, and the y_P are critical exponents that determine the scaling laws. In such case it follows immediately that

$$\mathcal{Q} = d^{d/2} \lim_{K \to K^*} [D' R\Lambda (DR)^{-1}] \tag{4.6}$$

where

$$D_{AB} = \frac{\sqrt{\gamma_A}}{\gamma}\frac{\partial^2 f(K)}{\partial K_A \partial K_B} \tag{4.7}$$

and K^* is the fixed point solution of the renormalization equations.

Eq. (4.6) explicitly shows that there is a delicate relation between the matrix $\mathcal{Q}$ defining the RSRT, the matrix Λ essencially defined by the (observable) critical exponents and the matrix of second derivatives of the free energy. Full scaling would require all the critical exponents to be preserved, not only the relevant ones.

References.

1) Th. Niemeijer and J.M.J. van Leeuwen, Phys. Rev. Letters **31** (1973) 1411.
2) Th. Niemeijer and J.M.J. van Leeuwen, Physica **71** (1974) 17.
3) Th. Niemeijer and J.M.J van Leeuwen in: Phase Transitions and Critical Behaviour, C. Domb and M.S. Green, eds., vol. 6 (Academic Press, New York, 1976).
4) L.P. Kadanoff and A. Houghton, Phys. Rev. B **11** (1975) 377.
5) R.B. Griffiths, Physica **106A** (1981) 59.
6) H.J. Hilhorst, M. Schick and J.M.J. van Leeuwen, Phys. Rev. Letters **40** (1988) 160.
7) H.J. Hilhorst, M. Schick and J.M.J. van Leeuwen, Phys. Rev. **B 19** (1979) 2149.
8) K. G. Wilson, Rev. Mod. Phys. **55** (1983) 583.
9) J.M.J. van Leeuwen, Phys. Rev. Lett. **34** (1975) 1056.
10) P. Cordero and M. Molina, Physica **151A** (1988) 139.
11) S.H. Swendsen, Phys. Rev. Lett. **52** (1984) 2321.
12) P. Cordero in: "Nonlinear Phenomena in Complex Systems", A.N. Proto, ed. (Elsevier, North-Holland, 1989).
13 P. Cordero in: "Proceedings of the Workshop on Nonequilibrium Statistical Phenomena", E.S. Hernández, ed. (in press).

Sand Piles, Combinatorial Games and Cellular Automata

Eric Goles*
Universidad de Chile
Facultad de Ciencias Físicas y Matemáticas
Departamento de Ingeniería Matemática
Casilla 170/3, Correo 3
Santiago, Chile
e-mail: egoles@uchcecvm

Abstract. We study some dynamical properties of a class of cellular automata which modelizes sand piles and combinatorial games on graphs. More precisely, we analize its periodic behavior on a tree, the one dimensional automaton and complexity properties of the high dimensional case.

1. Introduction

Let $G = (V, E)$ be a graph, where V is the set of sites and $E \subseteq V \times V$ is the set of bonds. We suppose that the neighborhood of a site i, $V_i = \{j \in V/(i,j) \in E\}$ is finite, i.e. $d_i = |V_i| < \infty$. At each site i, there is a discrete variable $x_i \in \mathbb{N}$ and a fixed critical threshold $z_i \geq d_i$. We define the transition rule:

If $x_i \geq z_i$ then:

$$\begin{aligned} x_i &\longleftarrow x_i - z_i \\ x_j &\longleftarrow x_j + 1 \quad \forall j \in V_i \end{aligned} \tag{1}$$

Obviously, if for any $i \in V$, $x_i < z_i$, the configuration $x \in \mathbb{N}^{|V|}$ is stable (also called a fixed point). A site i such that $x_i \geq z_i$ is called a firing site.

The set of sites can be updated sequentially (each site is updated one at a time, randomly or in a prescribed order) or parallel (all the sites are updated synchronously). The parallel update is the usual one in the context of cellular automata. In this framework a shigly similar local rule can be seen in [8]. In this paper we will studied some dynamical properties of both update modes.

This model has been introduced in several domains: combinatorics, computer sciences and physics. It was introduced first in the framework of an n-round perfect information game proposed by Spencer in [11]. The author determined the following local rule in a one dimensional lattice with nearest interactions:

* Partially supported by FONDECYT, Chile.

E. Tirapegui and W. Zeller (eds.), Instabilities and Nonequilibrium Structures III, 101–121.

$$x_i \longleftarrow x_i - 2\lfloor \frac{x_i}{2} \rfloor$$
$$\text{for } x_i \in I\!N,\ i \in \mathbb{Z} \qquad (2)$$
$$x_{i\pm1} \longleftarrow x_{i\pm1} + \lfloor \frac{x_i}{2} \rfloor$$

where $\lfloor\ \rfloor$ is the usual floor function.

The parallel update of the configuration $x_i(0) = 0\ \forall i \neq 0$, $x_0(0) = 11$, in the graph with nearest neighbors, $\mathbb{Z}$, is given in figure 1.

					11					
				5	1	5				
			2	1	5	1	2			
		1	0	1	1	1	0	1		
		1	2	0	5	0	2	1		
	1	2	0	3	1	3	0	2		
	1	0	2	1	3	1	2	0	1	
	1	1	0	3	1	3	0	1	1	
	1	1	1	1	3	1	1	1	1	
	1	1	1	2	1	2	1	1	1	
	1	1	2	0	3	0	2	1	1	
	1	2	0	2	1	2	0	2	1	
	2	0	2	0	3	0	2	0	2	
1	0	2	0	2	1	2	0	2	0	1
1	1	0	2	0	3	0	2	0	1	1
1	1	1	0	2	1	2	0	1	1	1
1	1	1	1	0	3	0	1	1	1	1
1	1	1	1	1	1	1	1	1	1	1

Figure 1. Parallel update of the Spencer game. After a finite transient time the initial configuration converges to a fixed point.

After the Spencer's paper, some combinatorial and dynamical aspects of this game were analized in [1]. In this reference it was established that the fixed points of the local rules (1) and (2) are the same. In fact, they proved that the application of the local rule:

$$\begin{aligned} x_i &\longleftarrow x_i - 2 \quad \text{if } x_i \geq 2 \\ x_{i\pm1} &\longleftarrow x_{i\pm1} + 1 \end{aligned} \qquad (3)$$

for the initial condition $x_0(0) = p \in I\!N$, $x_i(0) = 0\ \forall i \neq 0$, converges, in both cases, to the fixed point $(1, ..., 1) \in I\!N^p$ for p odd, and $(1, ..., 1, 0, 1, ..., 1, 1)$ if p is even. Moreover, they gave a bound of the transient time in $0(n^2)$ for $p = 2n + 1$.

Furthermore, an optimal bound for the transient time to reach an stable configuration on an arbitrary undirected graph, $G = (V, E)$, in $0(|V|^4)$ is given in [12].

Previous iteration schemes are particular cases of Petri Nets. A Petri Net is a model of parallel computing which has been extensively studied and applied to several domains in last years [7]. Roughly, it corresponds to a finite graph, with a configuration of integers in each site. A site may be updated when its value is bigger than its degree. In this case its diminishes its value in d_i units and it sends one unit to each of its neighbors. Usually, a Petri Net is updated at random and the graph admits non-symmetric bonds. An important property of Petri Nets is that they may simulate any computing machine [7]. So, it can be seen as an universal model of computation. Clearly, in this last case the bonds are non-symmetric. Our model, given by the local rule (1), is a Petri Net with symmetric bonds.

In physics, the iteration (1) was proposed by Bak,Tang and Wiesenfeld [2,3] to modelize the dynamical behavior of spatially extended dynamical systems which have the property of self-organized critically. In this context, they stablished, as a paradigm, a simple sand pile model in one and two dimensions. More precisely, a one dimensional sand pile is defined in the lattice $\mathbb{N} = \{1, 2, ...\}$ with nearest interactions. The grains of sand are represented by a non increasing sequence of integers: $h_1 \geq h_2 \geq ... \geq h_n \geq 0$. Each integer, h_i, corresponds to the number of grains in the i-th position (see figure 2)

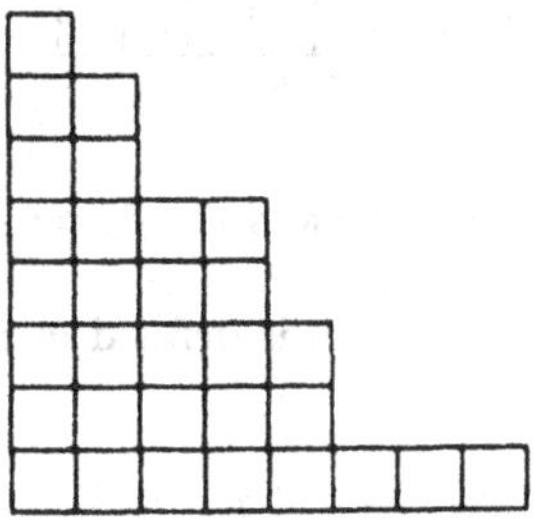

Figure 2. The sand pile configuration $h = (8, 7, 5, 5, 3, 1, 1, 1)$

Let $x_i = h_i - h_{i+1}$ be the high difference between consecutive piles, and $z \geq 2$ a critical threshold. The dynamics of the sand pile is given by the local rule:

$$\begin{aligned} h_i &\longleftarrow h_i - 1 \\ h_{i+1} &\longleftarrow h_{i+1} + 1 \end{aligned} \qquad \text{if } x_i \geq z \tag{4}$$

That is to say, when the high difference becomes higher than the critical threshold z, one grain tumbles to the lower level. The threshold z represents the maximum slope permitted without producing an avalanche.

It is direct to see that an equivalent dynamic can be done taking only the high differences (see figure 3), i.e.

$$
\begin{aligned}
x_i &\longleftarrow x_i - 2 \quad \text{if } x_i \geq z \\
x_{i\pm 1} &\longleftarrow x_{i\pm 1} + 1 \quad \text{for } i \geq 2 \\
x_1 &\longleftarrow x_1 - 2 \\
x_2 &\longleftarrow x_2 + 1
\end{aligned} \tag{5}
$$

It is direct from the transition rule (5) that the sand pile model is a particular case of iteration (1).

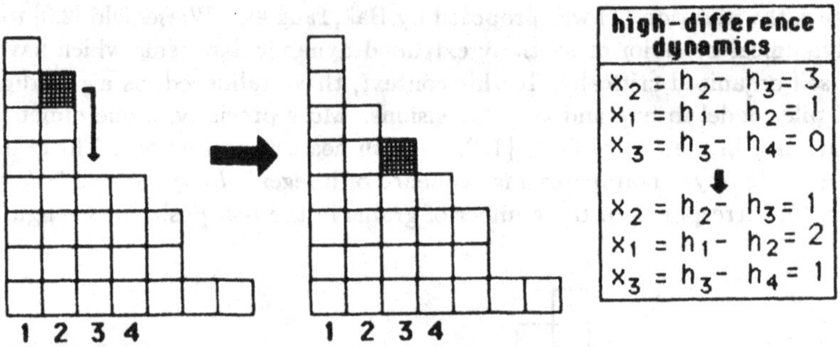

Figure 3. Equivalence between a sand pile and the dynamics of high differences.

The general sand pile model given in [3] is defined in the uniform two dimensional lattice accordingly to the local rule:

$$
\begin{aligned}
x_{ij} &\longleftarrow x_{ij} - 4 \quad \text{if } x_{ij} \geq z \\
x_{\ell\ell'} &\longleftarrow x_{\ell\ell'} + 1 \quad \text{for } (\ell, \ell') \in V_{ij}
\end{aligned} \tag{6}
$$

where V_{ij} is the set of nearest neighbors of (i, j).

In both cases, the one and the two dimensional model, the dynamics occurs in a confined region with some boundary conditions. For instance, in the one dimensional case, the boundary conditions of the high differences could be $x_0 = 0$, $x_{N+1} = 0$, i.e. the dynamics occurs in the lattice $\{1, 2, ..., N\}$ and when x_1 or x_N are bigger than the critical threshold z, a grain of sand is going out the system. Other boundary conditions can be seen in [3]. Figure 4 exhibits two examples of a particular sand pile. The initial configuration is a perfect stair (i.e. a configuration a, such trhat $a_i - a_{i+1} = 1 \; \forall i \in \mathbb{Z}$) where 1's have been distributed with probability p on each site

(i)

(ii)

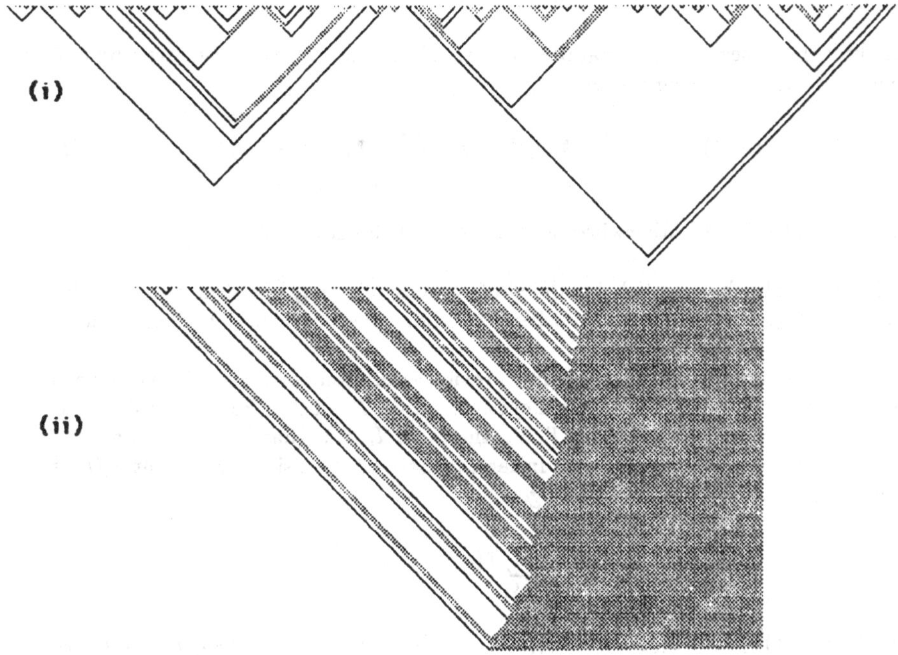

Figure 4. Parallel dynamics of sand piles, black dots code firing sites. **(i)** A random configuration. **(ii)** The initial configuration is $\cdots x_i, x_i, x_i - 1, x_i - 1, x_i - 2, x_i - 2, \cdots$ $\forall i \in \mathbb{Z}$, with a random perturbation for indexes $2i + 1$, $i < 0$.

In this paper we study some general aspects of the iteration (1). In particular we prove that, for the sequential update, the order in which sites are updated is irrelevant: the system converges to the same steady state independently of the choice of firing sites. Furthermore, we give for the one-dimensional sand pile, an exact bound of the transient time. Also, for high dimensional models we prove that on trees (graph without circuits) the iteration always converges to fixed points or two cycles. Moreover, for graphs with circuits (as the usual periodic one dimensional lattice or a rectangle of $\mathbb{Z}^2$) we exhibit large cycles.

To evaluate the complexity of such a model we determine particular networks which simulate registers. That is to say, the sand pile model has non-trivial computing capabilities, as the game of life proposed by Conway [4,8].

2. Parallel Dynamics of Sand Piles

It is not difficult to see that the parallel update of (1) (i.e. all the sites are updated synchronously) can be written as follows:

$$x_i(t+1) = x_i(t) - z_i \mathbb{1}(x_i(t) - z_i) + \sum_{j \in V_i} \mathbb{1}(x_j(t) - z_j) \tag{7}$$

where $\mathbb{1}(u) = 1$ iff $u \geq 0$ (0 otherwise), is the Heaveside function.

In this context, the parallel dynamics of a sand pile may be seen as a cellular automaton [8,9,13], with the local transition function given by the right hand side of equation (7).

Clearly, for a finite graph $G = (V, E)$, previous iteration converges to cycles with finite period. Let $(x(0), ..., x(T-1))$ be a cycle of period T. In this framework if $T > 1$, any site has a period T_i which divides T. Furthermore $T_i > 1$, that is to say, in a cycle all the sites are firing at least one time. In fact, if there exists a site i such that $x_i(t)$ is never fired in the cycle (i.e. $T_i = 1$), one gets:

$$x_i(t+1) = x_i(t) + \sum_{j \in V_i} \mathbb{1}(x_j(t) - z_j) \geq x_i(t)$$

hence, $x_i(0) \geq x_i(1) \geq ... \geq x_i(T-1) \geq x_i(0)$; i.e. $x_i(t) = a \leq z_i - 1$, for any $t \in [0, T-1]$. Now, let us take $k \in V_i$ such that there exits $t' \in [0, T+1]$ with $x_k(t') \geq z_k$, then

$$x_i(t'+1) \geq x_i(t') + 1 > x_i(t')$$

which is a contradiction. Then, any neighbor of the site i, is never fired. By recursive application of previous analysis and since G is finite, one concludes that all the sites are never fired. Hence, $T = 1$, which is a contradiction.

The same analysis holds when there exists a site which is always fired.

On the other hand, if there exists a firing site i with a critical threshold $z_i > d_i$, the network losts $z_i - d_i$ chips. Since in a cycle any site is fired almost one time and the number of chips must be constant, we conclude that, in this situation, any initial configuration converges to fixed points.

In the case $z_i = d_i$ for any $i \in V$, large cycles may appear. In figure 5 we exhibit a cycle in a 3×10 two dimensional torus. This example can be easily generalized to $3 \times m$ torus to exhibit cycles of perios m. Other families of cycles are given in [5].

$$
\begin{array}{ll}
 & 2\,1\,0\,4\,3\,2\,2\,2\,2\,2 \\
t=0 & 0\,4\,3\,2\,2\,1\,1\,1\,1\,1 \\
 & 3\,2\,2\,1\,0\,4\,3\,3\,3\,3
\end{array}
$$

$$
\begin{array}{lclr}
 & 2\,2\,1\,0\,4\,3\,2\,2\,2\,2 & & 2\,2\,2\,1\,0\,4\,3\,2\,2\,2 & \\
t=1 & 1\,0\,4\,3\,2\,2\,1\,1\,1\,1 & \longrightarrow & 1\,1\,0\,4\,3\,2\,2\,1\,1\,1 & t=2 \\
 & 3\,3\,2\,2\,1\,0\,4\,3\,3\,3 & & 3\,3\,3\,2\,2\,1\,0\,4\,3\,3 &
\end{array}
$$

$$
\begin{array}{lclr}
 & 2\,2\,2\,2\,1\,0\,4\,3\,2\,2 & & 2\,2\,2\,2\,2\,1\,0\,4\,3\,2 & \\
t=3 & 1\,1\,1\,0\,4\,3\,2\,2\,1\,1 & \longrightarrow & 1\,1\,1\,1\,0\,4\,3\,2\,2\,1 & t=4 \\
 & 3\,3\,3\,3\,2\,2\,1\,0\,4\,3 & & 3\,3\,3\,3\,3\,2\,2\,1\,0\,4 &
\end{array}
$$

$$
\begin{array}{lclr}
 & 2\,2\,2\,2\,2\,2\,1\,0\,4\,3 & & 3\,2\,2\,2\,2\,2\,2\,1\,0\,4 & \\
t=5 & 1\,1\,1\,1\,1\,0\,4\,3\,2\,2 & \longrightarrow & 2\,1\,1\,1\,1\,1\,0\,4\,3\,2 & t=6 \\
 & 4\,3\,3\,3\,3\,3\,2\,2\,1\,0 & & 0\,4\,3\,3\,3\,3\,3\,2\,2\,1 &
\end{array}
$$

$$
\begin{array}{lclr}
 & 4\,3\,2\,2\,2\,2\,2\,2\,1\,0 & & 0\,4\,3\,2\,2\,2\,2\,2\,2\,1 & \\
t=7 & 2\,2\,1\,1\,1\,1\,1\,0\,4\,3 & \longrightarrow & 3\,2\,2\,1\,1\,1\,1\,1\,0\,4 & t=8 \\
 & 1\,0\,4\,3\,3\,3\,3\,3\,2\,2 & & 2\,1\,0\,4\,3\,3\,3\,3\,3\,2 &
\end{array}
$$

$$
\begin{array}{lclr}
 & 1\,0\,4\,3\,2\,2\,2\,2\,2\,2 & & 2\,1\,0\,4\,3\,2\,2\,2\,2\,2 & \\
t=9 & 4\,3\,2\,2\,1\,1\,1\,1\,1\,0 & \longrightarrow & 0\,4\,3\,2\,2\,1\,1\,1\,1\,1 & t=10 \\
 & 2\,2\,1\,0\,4\,3\,3\,3\,3\,3 & & 3\,2\,2\,1\,0\,4\,3\,3\,3\,3 &
\end{array}
$$

Figure 5. Parallel iteration on a 3×10 two dimensional lattice with periodic boundary conditions.

Our main result, for the parallel update and critical threshold $z_i = d_i$, is the following:

When the graph G is a finite tree (i.e. a graph without circuits) the steady state associated to equation (7) admits only fixed points.

In order to prove this result, given a cycle $(x(0), ..., x(T-1))$, we define the local cycles and traces:

$$\forall i \in V \quad x_i = (x_i(0), ..., x_i(T-1))$$
$$\bar{x}_i = (\bar{x}_i(0), ..., \bar{x}_i(T-1))$$

where $\bar{x}_i = \mathbb{1}(x_i(t) - d_i)$

So, $\bar{x}(t) = 1$ means that at step t the i-th site is fired.

Also we define the support of a local trace $\bar{x}_i$ as follows:

$$supp(\bar{x}_i) = \{t \in [0, T-1] / \bar{x}_i(t) = 1\}$$

Clearly $supp(\bar{x}_i) = \bigcup_{k=1}^{p_i} C_k^i$, where C_k^i are maximal sets of $[0, T-1]$, coding blocks of 1's in the i-th trace, i.e.:

$$C_k^i = \{s, s+1, ..., s+r\}, \quad s, r \in [0, T-1]$$
$$x_i(s) = ... = x_i(s+r) = 1 \text{ and } x_i(s-1) = x_i(s+r+1) = 0$$

The same definition can be given for maximal sets coding blocks of 0's:

$$(supp(\bar{x}_i))^c = \bigcup_{k=1}^{q_i} D_k^i$$

Let M be the maximum length of consecutive 1's and N be the maximum length of consecutive 0's in the cycle $(x(0), ..., x((T-1))$, i.e.

$$M = \max_{i \in V} \max_{1 \le k \le p_i} |C_k^i|, \quad N = \max_{i \in V} \max_{1 \le k \le q_i} |D_k^i|$$

Clearly, $1 \le M, N \le T$. When M or N are equal to 1 or T, the steady state is a fixed point. So, the interesting case to analize is $2 \le M, N \le T-1$. First we prove the following property:

(H) *Given a maximal set* $[s-k, s] \subseteq supp(\bar{x}_i)$, *then there exists a site* $j \in V_i$ *such that* $[s-k-1, s-1] \subseteq supp(\bar{x}_j)$.

The proof follows by inductive application of equation (7) on the set $[s-k, s]$:

$$x_i(s) = x_i(s-k) - kd_i + \sum_{j \in V_i} \sum_{t=1}^{k+1} \bar{x}_j(s-t)$$

If the H-property does not hold, i.e. for any $j \in V_i$, there exists $t_j \in [s-k-1, s-1]$ with $\bar{x}_j(t_j) = 0$, one gets:

$$\sum_{j \in V_i} \sum_{t=1}^{k+1} \bar{x}_j(s-t) \le (k+1)d_i - d_i = kd_i$$

So, $x_i(s) \le d_i - 1$, then $\bar{x}_i(s) = 0$, which is a contradiction because $s \in supp(\bar{x}_i)$.

An analogous property holds for maximal sets of 0's.

The most important implication of the H-property is the following: when $G = (V, E)$ is a tree we have:

$$0 < M < T \Longrightarrow M = 1$$
$$0 < N < T \Longrightarrow N = 1$$

That is to say, when the limit cycle is not a fixed point (cases $M, N \neq 0$ or T), the trace is a cycle with period two: $\bar{x}_i = (101010...)$ or $\bar{x}_i = (010101...)$ for any $i \in V$.

We prove for M (for N the proof is similar). Let us suppose $M \geq 2$. Let $i_0 \in V$ be a site where a maximum set of 1's occurs, i.e.: $supp(\bar{x}_{i_0}) \supseteq C^0 = [t, t+M-1]$, where C^0 is a maximal set. From the H-property on maximal sets, there exists $i_1 \in V_{i_0}$ such that $supp(\bar{x}_{i_1}) \supseteq C^1 = [t-1, t+M-2]$. Furthermore, since M codes the maximum length of a set of 1's, C^1 is maximal. Also, in a similar way, there exists $i_2 \in V_{i_1}$ such that $supp(\bar{x}_{i_2}) \supseteq C^2 = [t-2, t+M-3]$ a maximal set and $i_0 \neq i_2$. In fact, since $t-1 \in supp(\bar{x}_2)$, $i_0 = i_2$ implies $t-1 \in supp(\bar{x}_{i_0})$ which is a contradiction. By recursive application of the H-property and since G is a finite tree, one determines a finite sequence of different sites $\{i_0, i_1, ..., i_s\}$ such that i_s is a leaf (i.e. $d_{i_s} = 1, V_{i_s} = \{i_{s-1}\}$) and:

$$C^0 = [t, t+M-1], C^1 = [t-1, t+M-2], ..., C^s = [t-s, t+M-(s+1)]$$

Finally, the H-property applied to the maximal set C^s implies that a maximal set $[t-s-1, t+M-(s+2)]$ must be contained in $supp(\bar{x}_{s-1})$, hence $\bar{x}_{i_{s-1}}(t-s) = 1$, which is a contradiction, because $i_{s-1} \notin C^{s-1}$. ■

Now we prove that in a tree the cycles have period two:

Let us suppose that for a tree the parallel update converges to a cycle $(x(0), ..., x(T-1))$ with period $T > 1$. Then, as we have proved, the traces are two cycles; i.e. for any $i \in V$, $\bar{x}_i(t+2) = \bar{x}_i(t)$.

Let $i \in V, t \in [0, T-1]$ such that $\bar{x}_i(t) = 1$, with exactly m firing neighbors, $\bar{x}_j(t) = 1$. From the two periodic property of the traces one gets:

$$x_i(t+1) = x_i(t) - d_i + m$$
$$x_i(t+2) = x_i(t+1) + d_i - m = x_i(t)$$

Similarly, if $\bar{x}_i(t) = 0$ we obtain $x_i(t+2) = x_i(t)$, then $\forall t \in [0, T-1]$, $x(t+2) = x(t)$, which proves that a tree admits only fixed points and/or two cycles. ■

Examples of two periodic behavior are given in figure 6.

(i)	1 − 1 − 3 − 0 − 2 − 0	(ii)	4	0
	↓↑			⇌
	0 − 3 − 1 − 2 − 0 − 1		0 0 0 0	1 1 1 1

Figure 6. **(i)** Two cycle in a finite one dimensional lattice. **(ii)** Two cycle in a star-tree.

For general graphs we have seen in figure 5 that there exist cycles with large periods. Furthermore, the complexity of the parallel update can be mesured by the computing capabilities of the iteration. In this context, one can simulates registers in the network, using gliders, 02, in a media of 1's (see figure 7(i)). A register R, which contains the value $n \in \mathbb{N}$, is simulated by a send pile network as in figure 7(ii).

```
(i)          t     ··· 1 1 1 0 2 1 1 ···
           t+1     ··· 1 1 1 1 0 2 1 ···

(ii)                   0 1  ···  ···  n
      1 1 1 ··· 1 2 1 1  ···  1 0 1 1 ···
                  1
                  1
                  ⋮
```

Figure 7. **(i)** A glider 02 in a one-dimensional media of 1's. **(ii)** The simulation of a register R. The bold numbers corresponds to relative positions. Position **(n)** = 0 codes the value n.

In order to substract one unit to R we send a glider from the left. This procedure is exhibited in figure 8.

```
1 1 1 0 2 2 1 1 0 1 1          1 1 1 1 0 3 1 1 0 1 1
          1              ⟶               1
          1                              1
          1                              1
                                         ↓

1 1 1 1 1 2 0 2 0 1 1          1 1 1 1 1 0 2 1 0 1 1
          0                              2
          2              ⟵               1
          1                              1
          ↓

1 1 1 1 1 2 1 0 1 1 1          1 1 1 1 1 2 1 0 1 1 1
          1              ⟶               1
          0                              1
          2                              0
```

Figure 8. Substraction of one unit from register R. The glider going out by the vertical branch, may be used as a signal to start the next instruction.

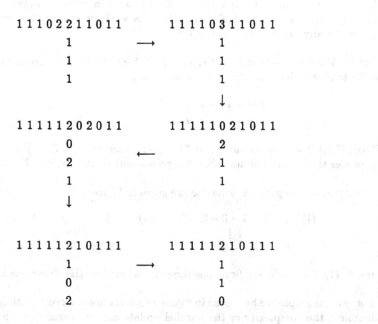

In order to add one unit to the content of register R, one sends a glider from the right. The register empty, $R = 0$, is coded by the configuration given in figure 9.

$$\begin{array}{c} \cdots\ 1\,1\,1\,1\,2\,0\,1\,1\,1\ \cdots \\ 1 \\ 1 \\ 1 \\ \vdots \end{array}$$

Figure 9. Simulation of a register empty. Position (0) = 0 and position $(k) = 1\ \forall k \geq 1$.

Let us take, for instance, the following program:

0 start
1 $R \longleftarrow 2$
2 $Q \longleftarrow 0$
3 $R \longleftarrow R - 1$
4 $Q \longleftarrow Q + 1$
5 If $R \neq 0$ goto 3
6 stop

This program uses two registers. Its work consists to copy the content of R in Q. When the program is finished, the content of R is 0. This procedure is simulayed by the network of figure 10. The horizontal branch simulates register R with two units. The vertical branch, from the bottom to the top, simulates register Q, which is inicialized empty. More sophisticated programs and the compute of logical functions can be coded similarily [5]. In this way, the parallel update of sand piles has powerful computing capabilities which are similar to those established for the Conway's game of life [4]. This fact proves that, in general graphs, the parallel update of sand pile belongs to the Wolfram's Class 4 [13]. Furthermore, as it was proved in [9], a Universal Turing Machine can be simulated by a finite set of instructions concerning only two registers. From our simulation it is very difficult to determine when a register is empty and to send this information to the right place in the network. In any case we have been able to simulate non trivial programs but we do not have a rigorous proof which asserts that any finite program can be simulated by sand piles. However, we conclude this paragraph, by saying that the parallel dynamics of sand piles has very powerful computing capabilities and it is almost universal.

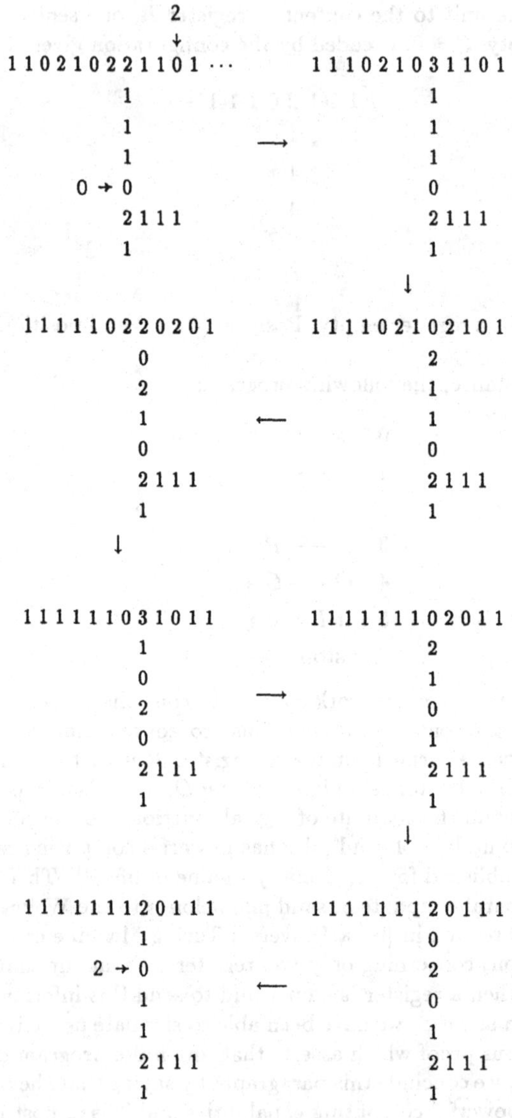

Figure 10. The network realizes the program which copies the context of register $R = 2$ in the register Q (initialized in 0). At the end, R is empty.

3. Sequential Update of Sand Piles

Let us take $G = (V, E)$, $|V| = n$, $|E| = m$. Given a configuration $x \in \mathbb{N}^n$, such that k is a firing site, (i.e. $x_k \geq z_k$) we define the local operator τ_k as follows:

$$(\tau_k(x))_i = \begin{cases} x_i & \text{for } i \notin V_k \cup \{k\} \\ x_i - z_i & \text{for } i = k \\ x_i + 1 & \forall i \in V_k \end{cases}$$

By using these operators, the sequential update can be viewed as the application of elements of the family $\{\tau_k\}_{k \in V}$ to a configuration $x \in \mathbb{N}^n$.

Let us suppose that sites i, j are firing. Directly from previous definition we have $y = \tau_i \tau_j(x)$, where:

$$y_k = \begin{cases} x_k & \text{for} \quad k \notin V_i \cap V_j \cup \{i, j\} \\ x_k - z_k & \text{for} \quad k = i, j \text{ and } (i, j) \notin V \\ x_k - z_k + 1 & \text{for} \quad k = i, j \text{ and } (i, j) \in V \\ x_k + 1 & \text{for} \quad k \in V_i \cup V_j \setminus V_i \cap V_j \\ x_k + 2 & \text{for} \quad k \in V_i \cap V_j \end{cases}$$

By interchanging the indexes i and j, and since $x_i \geq z_i$, $x_j \geq z_j$, we conclude that $\tau_i \tau_j x = \tau_j \tau_i x$. This commutative property is very important as we will see.

Similarily to the parallel update, if there exists a firing site i with critical threshold $z_i > d_i$, the number of chips decreases

$$\sum_{j=1}^{n} [T_i u(x)]_j = \sum_{j=1}^{n} x_j - (z_i - d_i)$$

hence, for a periodic configuration, i.e. $T_{j_1} T_{j_2}, ..., T_{j_s} x = x$, we have:

$$\sum_{j=1}^{n} [T_{j_1}, ..., T_{j_s} x]_j = \sum_{j=1}^{n} x_j$$

So, the site i (with $z_i > d_i$) can not be fired in the steady state. Furthermore, if there exists a neighbor of i which is fired, it will be fired infinitely often, so also i will be fired, which is a contradiction. Then, all the neighbor of the site i are never fired. Since G is finite, we conclude that all the sites are never fired. Hence, if there exists $z_i > d_i$, any initial configuration converges to fixed points.

In the case $z_j = d_j$, $\forall j \in V$, a sufficient condition to reach fixed points (with each component smaller than d_i) is $\sum_{j=1}^{n} x_j < m$, where m is the number of edges in the graph G [12].

Now, let us suppose that, for the initial configuration $x \in I\!N^n$ one gets $\tau_{j_1}...\tau_{j_s}x = y$, which is stable (i.e. $y_i < z_i$, $\forall i \in V$). Let us suppose that, given x, we update the firing sites in a different way, such that the new trajectory converges to other fixed point $y' \neq y$. Hence, there exists a configuration where the two trajectories different images, c and c', such that they do not have a common successor:

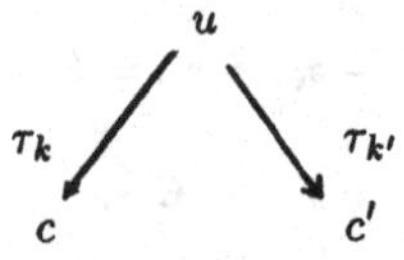

$k \neq k'$

Since, for firing sites k, k': $\tau_k \tau_{k'} = \tau_{k'} \tau_k$, one gets $\tau_k(c') = \tau_{k'}(c)$. Which is a contradiction, because c and c' do not have a common successor. Hence, the final orbit of the system is independent of the initial condition. This fact will be also used, in a shigly different context, in the next paragraph.

Finally, it is not difficult to see that the parallel update can be simulated by the sequential one. In fact, let $x \in I\!N^n$ such that the set of firing sites is $X = \{i_1, ..., i_s\}$. Clearly, if y is the image of x for the parallel update, we have $y = \tau_{i_1}, ..., \tau_{i_s}(x)$. For that it is sufficient to remark that a firing site remains firing independently of the update of other firing sites. This simulation implies that the fixed points are the same for both update modes.

4. One-Dimensional Sand Piles

In this paragraph we study the one dimensional sand pile model. We prove that any initial configuration converges to fixed points. Furthermore, we prove, for the sequential update, that the order to fire sand piles is irrelevant; i.e. convergence to a unique fixed point for a given initial condition.

Also we give an exact expression for the transient time. For that we introduce an energy associated to a sand pile, which is strictly increasing.

Let $S_n = \{(h_1, ..., h_i, ...)/ \sum_{i \geq 1} h_i = n, h_i \geq h_{i+1} \geq 0, h_i \in I\!N\}$ be the set of sand pile configurations associated to an integer $n \in I\!N$.

Given $w, w' \in S_n$, there exists a legal transition from w to w' iff there exists $j \in I\!N, j \geq 1$, such that $w_j - w_{j+1} \geq z$. In this case we note, similarly to previous paragraph, $w' = \tau_j(w) = (w_1, ..., w_j - 1, w_j + 1, ...)$. Clearly, if $z \geq z'$ and there exist a legal transition for the threshold z, then it also exists for z'. From previous remark, given a dynamical sequence of sand piles for $z \geq 2$, $\{w^t\}_{t \geq 0}$, such that $w^{t+1} = \tau_{j_t} w^t$, $j_t \geq 1$, i.e. $w^t_{j_t} - w^t_{j_t+1} \geq z$, then $w^t_{j_t} - w^t_{j_t+1} \geq 2$. Hence, since $z \geq 2$, the same sequence is produced for the threshold $z = 2$. From this fact one concludes that the largest transient sequences appear for the critical threshold $z = 2$. Throughout this paragraph we will focus our atention in this case.

A fixed point or a stable configuration is a sand pile where no legal moves may be done, i.e. $h \in S_n$ such that $\forall i \geq 1,\ h_i - h_{i+1} \leq 1$.

It is direct that any $n \in \mathbb{N}$ can be written as $n = \frac{k(k+1)}{2} + k'$, $k, k' \in \mathbb{N}$, $0 \leq k' \leq k$. For instance:

$$
\begin{aligned}
6 &= \frac{3 \cdot 4}{2}; \qquad k = 3, \quad k' = 0 \\
8 &= \frac{3 \cdot 4}{2} + 2 \quad k = 3, \quad k' = 2 \\
9 &= \frac{3 \cdot 4}{2} + 3 \quad k = 3, \quad k' = 3
\end{aligned}
$$

In this context we define the following configurations, $s^{(k,k')} \in S_n$:

$$s^{(k,0)} = (k, k-1, k-2, ...4, 3, 2, 1, 0....) \text{ for } k' = 0$$

and, for $k' \geq 1$:

$$s_i^{(k,k')} = \begin{cases} s_i^{(k,0)} & \text{for } 1 \leq i \leq k - k' + 1 \\ s_i^{(k,0)} + 1 & k - k' + 2 \leq i \leq k + 1. \end{cases}$$

It is easy to see that the configurations $s^{(k,k')}$ are fixed points; i.e. $s_i^{(k,k')} - s_{i+1}^{(k,k')} \leq 1$, $\forall i \geq 1$. For instance, given $k = 4, n = \frac{4 \cdot 5}{2} + 0 = 10$ we get the following fixed points:

$$
\begin{aligned}
s^{(4,0)} &= (4, 3, 2, 1, 0...) \in S_{10} \\
s^{(4,4)} &= (4, 4, 3, 2, 1, 0...) \in S_{14}
\end{aligned}
$$

The main result of this paragraph is the following:

For the critical threshold $z = 2$, and $n = \frac{k(k+1)}{2} + k'$, any sequential trajectory, from the initial sand pile $(n, 0, ...) \in S_n$, converges, in exactly $T(n) = \binom{k+1}{3} + kk' - \binom{k'}{2}$ steps, to the fixed point $s^{(k,k')}$.

To prove this result we introduce the energy of a sand pile as follows:

$$E(h) = \sum_{i \geq 1} h_i i, \qquad \text{for } h \in S_n$$

Clearly, for any $w \in S_n$ which accepts a legal move for an index $j \geq 1$ one gets:

$$E(\tau_j w) - E(w) = 1.$$

In fact, $\tau_j w = (w_1, ..., w_j - 1, w_{j+1} + 1, w_{j+2} ...)$, hence

$$E(\tau_j w) - E(w) = \sum_{i \le j-1} w_i i + (w_j - 1)j + (w_j + 1)(j+1) + \sum_{i \ge j+2} w_i i - \sum_{i \ge 1} w_i i = 1. \blacksquare$$

Now, let us suppose that, in steady state, there exists a cycle of period $T > 1$; i.e. a sequence of sand piles:

$$w^0 \to w^1 \to \dots \to w^{T-1} \to w^0$$

Since the energy is increasing, $E(w^0) < E(w') < \dots < E(w^0)$, which is a contradiction, then $T = 1$. Than is to say, for any initial sand pile the sequential update converges to a fixed points.

Moreover, given the initial sand pile $w \in S_n$, any sequential trajectory converges to the same fixed point.

In fact, we know that any trajectory from w, converges to a fixed point. Similarly to previous paragraph, let us suppose that there exist two fixed points $x, y \in S_n$ that are atteinted from w. That is to say, we have two trayectories which depend on the updated legal moves:

$$w \to w^1 \to w^2 \to \dots \to w^q = x$$

$$w \to h^1 \to h^2 \to \dots \to h^p = y$$

since $x \neq y$, there exists a common sand pile to both trajectories; $w^s = h^m$, such that, w^{s+1} and h^{m+1} do not have a common immediate successor:

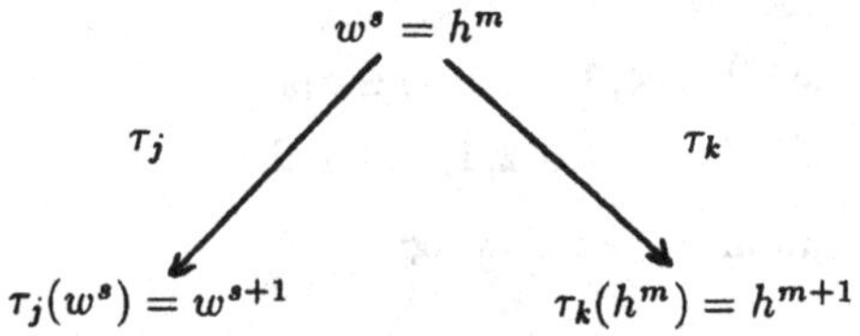

where $j, k \geq 1, j \neq k$, are indexes associated to legal moves.

If $k \geq j + 2$ one gets:

$$w^s = (\dots, w_j^s, w_{j+1}^s, \dots, w_k^s, w_{k+1}^s, \dots)$$

and

$$\tau_j(w^s) = (\dots, w_j^s - 1, w_{j+1}^s + 1, \dots, w_k^s, w_{k+1}^s \dots) = w^{s+1}$$

so, we can apply the legal move τ_k to w^{s+1}:

$$\tau_k(w^{s+1}) = \tau_k \tau_j(w^s) = (\dots, w_j^s - 1, w_{j+1}^s + 1, \dots, w_k^s - 1, w_{k+1}^s + 1, \dots)$$

In a similar way, for $h^m = w^s$:

$$\tau_k(h^m) = \tau_k(w^s) = (..., w_j^s, w_{j+1}^s, ..., w_k^s - 1, w_{k+1}^s + 1, ...) = h^{m+1}$$

and $\tau_j(h^{m+1}) = \tau_j\tau_k(h^m) = (..., w_j^s - 1, w_{j+1}^s + 1, ..., w_k^s - 1, w_{k+1}^s + 1, ...) = \tau_k\tau_j(w^s)$.

That is to say, the sand pile $\tau_j\tau_k(w^s)$ is a common successor of h^{m+1} and w^{s+1}, which is a contradiction.

The other case is $k = j + 1$; i.e. τ_j and τ_{j+1} are legal moves. So:

$$\tau_j(w^s) = (..., w_j^s - 1, w_{j+1}^s + 1, w_{j+2}, ...) = w^{s+1}$$

$$\tau_{j+1}(w^s) = (..., w_j^s, w_{j+1}^s - 1, w_{j+2}^s + 1, ...) = h^{m+1}$$

since τ_j, τ_{j+1} are legal moves: $w_j^s - w_{j+1}^s \geq 2$ and $w_{j+1}^s - w_{j+2}^s \geq 2$ we get $h_j^{m+1} - h_{j+1}^{m+1} = w_j^s - w_{j+1}^s + 1 \geq 3$ and $w_{j+1}^{s+1} - w_{j+2}^{s+1} = w_{j+1}^s + 1 - w_{j+2}^s \geq 3$. So one can applied the legal moves τ_{j+1}, τ_j to w^{s+1}, h^{m+1} respectively:

$$v = T_{j+1}T_j(w^s) = (..., w_j - 1, w_{j+1}, w_{j+2} + 1, ...)$$

which is a common successor of w^{s+1} and h^{m+1}, which is a contradiction. ■

From this result it is easy to see that

Given $n = \frac{k(k+1)}{2}$, there exists a sequential trajectory which converges to the fixed point $s^{(k,0)} = (k, k-1, ..., 4, 3, 2, 1, 0, ...)$

The idea of the proof consists to transport each grain of sand from the rightmost active sand pile of the configuration $(k, k-1, ...3, 2, 1, 0...)$ to the first one, until to obtain $(n, 0, ...)$. First, one traslates the unique grain of the k-th pile as follows:

$$(k, k-1, ..., 3, 2, 1, 0...) \leftarrow (k, k-1, ...4, 3, 3, 0, 0...) \leftarrow (k, k-1, ...4, 4, 2, 0, ...)$$

$$\leftarrow ... \leftarrow (k, k-1, ..., \ell+1, \ell, \ell, \ell-2, \ell-3, ..., 4, 3, 2, 0, 0, ... \leftarrow (k+1, k-1, ..., 3, 2, 0...)$$

After that, one traslates in the same way the grains of other piles. It is direct that any transition in the sequence corresponds to a legal move. So, we have generated a sequential trajectory from $(n, 0, ...)$ to $s^{(k,0)}$. ■

In figure 11 we give an example of previous algorithm for $n = 6$.

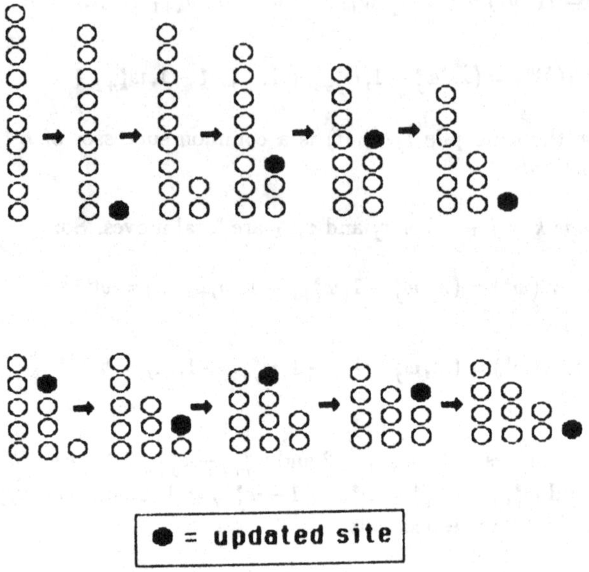

Figure 11. Legal trajectory between the fixed point (4,3,2,1) and the initial sand pile (10,0,0).

In a similar way we have, for $n = \frac{k(k+1)}{2} + k', 1 \leq k' < k$, that there exists a sequential trajectory which converges to the fixed point $s^{(k,k')}$.

To prove that, we use the same previous procedure to build a sequential trajectory, C_1, between $(n, 0, ...)$ and the sand pile $w = (k + k', k-1, k-2, ..., 4, 3, 2, 1, 0...)$, where $w_j = 0 \quad \forall j \geq k+1$.

From w to the fixed point $s^{(k,k')}$, we built the following trajectory, C_2:

$$w \rightarrow (k + k' - 1, k, k-2, ..., 3, 2, 1, 0...) \rightarrow (k + k' - 1, k-1, k-1, k-3, ..., 3, 2, 1, 0)$$
$$... \rightarrow (k + k' - 1, k-1, k-2, ..., \ell+1, \ell, \ell, \ell-2, ..., 3, 2, 1, 0...) \rightarrow$$
$$\rightarrow (k + k' - 1, k-1, k-2, ..., \ell+1, \ell, \ell-1, \ell-1, \ell-3, ..., 3, 2, 1, 0...) \rightarrow ...$$
$$\rightarrow (k + k' - 1, k-1, k-2, ..., 3, 2, 2, 0...) \rightarrow (k + k' - 1, k-1, ..., 3, 2, 1, 1, 0...)$$

After, we generate, for $s \in \{1, ..., k'-1\}$:

$$\rightarrow (k+k'-s, k-1, k-2, ..., s+1, s, s, s-1, s-2, ..., 3, 2, 1, 1, 0, ...) \rightarrow$$
$$\rightarrow (k+k'-s-1, k, k-2, ..., 3, 2, 1, 1, 0...) \rightarrow$$
$$\rightarrow (k+k'-s-1, k-1, k-2, ..., \ell+1, \ell, \ell, \ell-2, ..., s+2, s+1, s, s, s-1, ..., 3, 2, 1, 1, 0..)$$
$$... \rightarrow (k+k'-s-1, k-1, k-2, ..., ..., s+2, s+1, s+1, s, s-1, ..., 3, 2, 1, 1, 0...)$$

So the trajectory $C_1 \cup C_2$ goes from $(n, 0...)$ to $s^{(k,k')}$ ■

From the uniqueness of fixed point, any sequential trajectory from $(n, 0, ..., 0) \in S_n$ converges to the fixed point $s^{(k,k')}$, where $n = \frac{k(k+1)}{2} + k'$; $0 \leq k' < k$. ■

Since the energy increases exactly one unit, step by step, we can calculate the sequential transient time $T(n)$ by the difference

$$T(n) = E(s^{(k,k')}) - E((n, 0, ...)) = \binom{n+1}{3} + kk' - \binom{k'}{2}$$

It is important to point out that, in all the cases, the relaxation transient time is $0(n\sqrt{n})$, i.e. the sand pile has a fast dynamics. Moreover, for any other critical threshold $z > 2$ the convergence to fixed point is faster, and the same analysis can be done.

Clearly, since the parallel dynamics can be simulated by the sequential one, $T(n)$ is a bound for the parallel transient time.

Taking into account the equivalence between one dimensional sand piles and the high difference dynamics (given by equations (4) and (5)), we can also conclude that the sequential trajectories of the automaton defined by rule (5):

if $x_i \geq 2$; $x_i \leftarrow x_i - 2$, $\quad x_{i\pm1} \leftarrow x_{i\pm1} + 1 \quad$ for $i \geq 2$

or $x_1 \leftarrow x_1 - 2$, $\quad x_2 \leftarrow x_2 + 1 \quad$ for $i = 1$.

converges in exactly $T(n)$ steps to the fixed point $(1, 1, ..., 1)$ for $n = \frac{k(k+1)}{2}$ or $(1...101...1)$ if $n = \frac{k(k+1)}{2} + k', 1 \leq k' < k$. In the second case, the component zero appears in position $k - k' + 1$.

In figure 12 we give, for $n = 18$, an example of a sequential and the parallel dynamics of the sand pile and its high differences. It is important to point out that $T(n)$ is not a good bound for the parallel update. By using the energy one can give better bounds of the parallel transient time.

(i) sand pile	high differences
1 8	1 8 0
1 7 1	1 6 1
1 6 2	1 4 2
1 5 3	1 2 3
1 4 4	1 0 4
1 3 5	8 5
1 2 6	6 6
1 1 7	4 7
1 0 8	2 8
9 9	0 9
9 8 1	1 7 1
9 7 2	2 5 2
8 8 2	0 6 2
8 7 3	1 4 3
8 6 4	2 2 4
8 5 5	3 0 5
8 5 4 1	3 1 3 1
8 5 3 2	3 2 1 2
8 4 4 2	4 0 2 2
8 4 3 3	4 1 0 3
8 4 3 2 1	4 1 1 1 1
7 5 3 2 1	2 2 1 1 1
7 4 4 2 1	3 0 2 1 1
7 4 3 3 1	3 1 0 2 1
7 4 3 2 2	3 1 1 0 2
7 4 3 2 1 1	3 1 1 1 0 1
6 5 3 2 1 1	1 2 1 1 0 1
6 4 4 2 1 1	2 0 2 1 0 1
6 4 3 3 1 1	2 1 0 2 0 1
6 4 3 2 2 1	2 1 1 0 1 1
5 5 3 2 2 1	0 2 1 0 1 1
5 4 4 2 2 1	1 0 2 0 1 1
5 4 3 3 2 1	1 1 0 1 1 1

(ii) sand pile	high differences
1 8	1 8
1 7 1	1 6 1
1 6 2	1 4 2
1 5 2 1	1 3 1 1
1 4 3 1	1 1 2 1
1 3 3 2	1 0 1 2
1 2 4 1 1	8 3 0 1
1 1 4 2 1	7 2 1 1
1 0 4 3 1	6 1 2 1
9 5 2 2	4 3 0 2
8 5 3 1 1	3 2 2 0 1
7 5 3 2 1	2 2 1 1 1
6 5 4 2 1	1 1 2 1 1
6 5 3 3 1	1 2 0 2 1
6 4 4 2 2	2 0 2 0 2
5 5 3 3 1 1	0 2 0 2 0 1
5 4 4 2 2 1	1 0 2 0 1 1
5 4 3 3 2 1	1 1 0 1 1 1

Figure 12. Sand pile dynamics for $n = 18 = \frac{5 \cdot 6}{2} + 3, k = 5, k' = 3$. **(i)** A sequential update with transient time $T(18) = 32$. **(ii)** Parallel dynamics.

References.

[1] Anderson R.J., Lovász L., Shor P., Spencer J., Tardos E., Winograd S., Disk, Balls and Walls: Analysis of a Combinatorial Game, Preprint 1988, Amer. Math. Monthly, **96**(6), 481-493, 1989.

[2] Bak P., Tang Ch., Wiesenfeld K., Self-Organized Critically: An Explanation of $1/f$ Noise, Phys. Rev. Lett. **59**(4), 381-384, 1987.

[3] Bak P., Tang Ch., Wiesenfeld K., Self-Organized Criticaly, Physics Review A, **38**(1), 364-373, 1988.

[4] Berlekamp E.R., Conway J.H., Guy R.K., Winning Ways for your Mathematical Plays, **2**, Chapter 25, Academic Press, Third printing, 1985.

[5] Bitar J., Juegos Combinatoriales en Redes de Autómatas, Tesis Ingeniero Matemático, Departamento de Ingeniería Matemática, Universidad de Chile, 1989.

[6] Björner A., Lovasz L., Short P., Chips Firing Games on Graphs, Preprint, 1988.

[7] Brams G.W., Réseaux de Petri: Théorie et Practique, **1**(2), Masson, 1983.

[8] Goles E., Martínez S., Neural and Automata Networks: Dynamical Behaviour and Applications, Kluwer Acad. Pub., Collection Math. and its Applications, **58**, 1990.

[9] Herrmann H., Cellular Automata, in Nonlinear Phenomena in Complex Systems, Proto A.N. ed., Elsevier, Amsterdam, 1989.

[10] Minsky L.M., Computation: Finite and Infinite Machines, Prentice-Hall series in Automatic Computation, 1967.

[11] Spencer J., Balancing Vectors in the Max Norm, Combinatorica, **6**, 55-65, 1986.

[12] Tardós G., Polynomial Bounds for a Chip Firing Game on Graphs, SIAM J. on Disc. Maths., **1**(3), 397-398, 1990.

[13] Wolfram S., Theory and Applications of Cellular Automata, World Scientific, 1986.

PART II.

STOCHASTIC EFFECTS ON DYNAMICAL SYSTEMS

NONEQUILIBRIUM POTENTIALS IN SPATIALLY EXTENDED PATTERN FORMING SYSTEMS

R. GRAHAM[1] and T. TÉL[2]
[1] *Fachbereich Physik, Universität Essen-GHS*
D-4300 Essen 1, FRG
[2] *IFF, Research Center Jülich,*
Pf. 1913, D-5170 Jülich, FRG
(On leave from Eötvös University, Budapest)

ABSTRACT. We point out that nonequilibrium potentials playing the role of Lyapunov functionals do exist in spatially extended pattern forming systems even in the generic case when the equations of motion cannot be written in a gradient from. Such potentials reflect global stability properties not derivable from a linear analysis. Potential barriers among coexisting attractors provide useful tools for characterizing their relative stability. Explicit results are given for the one-dimensional complex Ginzburg-Landau equation in the limit of weak spatial diffusion.

1. Introduction: Nonequilibrium Potentials

Stable macroscopic states in thermal equilibrium are selected by the condition of minimal free energy [1]. To be more specific, let q^ν, $\nu = 1, 2, ..., n$ denote thermodynamic variables in a system not necessary in equilibrium but having contact with a heat bath of temperature T. A *coarse grained* free energy $F(q)$ is defined, as usual, via the partition function which is, however, evaluated by keeping the q^ν's fixed. In terms of $F(q)$, a stable equilibrium state q_0 is characterized by the condition

$$F(q_0) = \text{minimum}. \tag{1}$$

This free energy can also be constructed from the equation of state $p_\nu = p_\nu(q)$, which connects conjugate thermodynamic forces p_ν and variables q^ν, by simple integration:

$$F(q) = \min\{\int_{q_0}^{q} p_\nu(q)dq^\nu + F(q_0)\}. \tag{2}$$

The minimum plays a role only if the integral is multivalued. Here, and in the following, repeated lower and upper indices imply summation.

Coarse grained free energies exist, of course, in spatially extended systems, too. Perhaps the most prominent application of this concept is provided by symmetry breaking phase transitions [2], where around a critical point the Ginzburg-Landau functional

$$F_{GL}(\{\psi\}) = 2\int dx\{-a|\psi(x)|^2 + \frac{b_r}{2}|\psi(x)|^4 + D_r|\nabla\psi(x)|^2\} + \text{const.} \tag{3}$$

governs the behaviour of the order parameter field $\psi(x)$. A complex ψ corresponds to a two-component order parameter. The quantities a, b_r, and D_r are real numbers ($b_r, D_r > 0$).

E. Tirapegui and W. Zeller (eds.), Instabilities and Nonequilibrium Structures III, 125–142.

By minimizing the Ginzburg-Landau free energy one obtains below the critical point, i.e., for $a < 0$:

$$\psi \equiv 0, \tag{4}$$

while in the symmetry breaking region, $a > 0$

$$\psi \equiv \left(\frac{a}{b_r}\right)^{1/2} e^{i\varphi_0} \tag{5}$$

where φ_0 is an arbitrary phase. These solutions are homogeneous and describe the equilibrium state in the framework of the mean-field theory. As well known, the general treatment requires the inclusion of order parameter fluctuations [2]. This can be done by using the canonical distribution with F_{GL} which yields for the probability W of observing a given field $\psi(x)$

$$W(\{\psi\}) \sim \frac{1}{Z} e^{-F_{GL}(\{\psi\})/k_B T}. \tag{6}$$

Here k_B stands for Boltzmann's constant, and $1/Z$ ensures normalization. The inverse relation can be used as an additional definition of the coarse grained free energy which turns out to be essentially the logarithm of $1/W$ in the formal limit $k_B \to 0$:

$$F_{GL}(\{\psi\}) = - \lim_{k_B \to 0} k_B T \ln W(\{\psi\}). \tag{7}$$

It is worth emphasizing that F_{GL} is completely different from the total free energy (the latter is proportional to $-\ln Z$).

The aim of this paper is to show that the notion of coarse grained free energies can be generalized to any nonequilibrium system. In such cases, the central quantity is, instead of an *equation of state*, a kind of *equation of motion*. Let us, first, return to systems which are not extended in space. The dynamics of macroscopic variables $q^\nu(t)$ is then governed by a set of nonlinear ordinary differential equations which we write in the form

$$\dot{q}^\nu = K^\nu(q) \tag{8}$$

where K^ν are given functions.

Thermodynamic fluctuations are often replaced in nonequilibrium systems by *Langevin forces*. Consequently, the equations of motion become stochastic ones of type

$$\dot{q}^\nu = K^\nu(q) + \eta g_i^\nu(q)\xi^i(t) \tag{9}$$

where $\xi^i(t)$ stands for delta correlated Gaussian noise:

$$< \xi^i(t) >= 0, \quad < \xi^i(t)\xi^j(t') >= \delta^{ij}\delta(t-t'). \tag{10}$$

Functions g_i^ν characterize the coupling to the noise, and η is the noise intensity measuring the overall strength of this coupling.

The role of the *canonical ensemble* is then played by a *steady state distribution*, the existence of which we assume in what follows. Let $W_\eta(q)$ denote the stationary probability density for process (9) which can be computed by evaluating path integrals over different realizations of noise. Using the saddle point approximation in the weak noise limit ($\eta \to 0$) one can show that the distribution is always of type

$$W_\eta(q) \sim \exp(-\phi(q)/\eta) \tag{11}$$

where ϕ is independent of the noise intensity. $\phi(q)$ is called the *nonequilibrium potential* [3] - [20] and is the direct analogue of the coarse grained free energy $F(q)$ (for a review and references to early works, see [15]). Inverting equation (11) we obtain

$$\phi(q) = -\lim_{\eta\to 0} \eta W_\eta(q) \tag{12}$$

generalizing relation (7). Notice, η plays the same role as the thermal noise intensity k_BT in equilibrium.

It follows from the path integral solution [4] that $\phi(q)$ can always be constructed as an integral of a certain Lagrangian over a fictitious time. In fact, one finds that there exists a conservative deterministic system associated with (9) defined by the Hamiltonian

$$H = \frac{1}{2}Q^{\nu,\mu}(q)p_\nu p_\mu + K^\nu(q)p_\nu \tag{13}$$

where p_ν are conjugate mommenta to q^ν, and the symmetric matrix $Q^{\nu,\mu}$ is obtained from the couplings g_i^ν as

$$Q^{\nu,\mu}(q) = g_i^\nu(q)g_i^\mu(q). \tag{14}$$

The Legendre transform of $H(q,p)$ is a Lagrangian $L(q,\dot{q})$ (dot denotes here the derivative with respect to the time variable τ of the conservative system)

$$L = \sum p_\nu \dot{q}^\nu - H \tag{15}$$

from which

$$L(q,\dot{q}) = \frac{1}{2}(Q^{-1}(q))_{\nu,\mu}(\dot{q}^\nu - K^\nu(q))(\dot{q}^\mu - K^\mu(q)) \tag{16}$$

follows. The nonequilibrium potential was shown [4] to appear as

$$\phi(q) = \min\{\int_{-\infty}^{0} d\tau L(q,\dot{q}) + \phi(A)\}. \tag{17}$$

The integral is to be taken over path $q(\tau)$ satisfying Lagrange's or Hamilton's equations *on the zero energy surface* $H \equiv 0$, and connecting the attractor A of the deterministic system (8) with an endpoint $q(0) = q$. For simplicity, we assume there is one attractor only.

As a natural consequence of (17) ϕ is minimal in the attractor:

$$\phi(q \in A) = \text{minimum} \tag{18}$$

generalizing property (1) of the free energy. One can also show that ϕ is maximal and extremal in the repellers and saddles of (8), respectively, and that $\dot{\phi}(q) \leq 0$ holds along solutions of (8). Thus, we find that the nonequilibrium potential not only governs the stationary distribution in the presence of a weak noise, but simultaneously it is a *Lyapunov function for the deterministic system.*

Inserting (15) into (17) and using that the total energy is zero, we find the potential in the form of a phase space integral:

$$\phi(q) = \min\{\int_{A}^{q} p_\nu(q)dq^\nu + \phi(A)\} \tag{19}$$

where $p_\nu(q)$ describes the separatrix, i.e. the unstable manifold of the hyperbolic object created by embedding the attractor A in the Hamiltonian system (13). A comparison with eq. (2) shows that the separatrix equation plays a role similar to that of the equation of state in equilibrium. Relations (12), (18) and (19) complete the analogy between coarse grained free energy and nonequilibrium potential.

The concept of nonequilibrium potential can naturally be generalized to spatially extended systems [21] - [24]. The macroscopic variables are then (often complex) fields $\psi(x,t)$ evolving according to partial differential equations. The effect of noise can in many cases be described by adding a spatial Gaussian force $\eta\xi(x,t)$ with local correlations

$$< \xi(x,t)\xi^*(x',t') >= Q\delta(x-x')\delta(t-t') \tag{20}$$

and zero mean value. The stationary distribution and the nonequilibrium potential become functionals of ψ. Relation (17) is then extended in the form

$$\phi(\{\psi\}) = \min\{\int_{-\infty}^{0} d\tau L(\psi,\dot{\psi}) + \phi(A)\} \tag{21}$$

where $L(\{\psi,\dot{\psi}\})$ is a Lagrangian which can uniquely be constructed from the equation of motion and the parameter Q.

Before treating a nontrivial example, it is worth listing some spatially extended systems possessing an obvious potential which can be read off by inspection.

2. Some pattern forming systems with obvious potentials

Certain partial differential equations describing nonequilibrium structures have the special property that their right hand side is the derivative of some functional V. The equation of motion of such *gradient systems* is

$$\dot{\psi}(x) = -\frac{1}{2}\frac{\delta V(\{\psi\})}{\delta\psi^*}. \tag{22}$$

By adding Gaussian noise $\eta\xi(x,t)$ with correlation as given in (20), one easily sees that the logarithm of the stationary distribution is $-V/(\eta Q)$, consequently, the *functional V is essentially the nonequlibrium potential* ϕ for such systems. More precisely,

$$\phi(\{\psi\}) = \frac{1}{Q}V(\{\psi\}). \tag{23}$$

(In the case of a real field variable, the derivative in (22) is to be taken with respect to ψ.) Below we give some examples with gradient type dynamics.

2.1. BÉNARD CONVECTION

After convection sets in in a fluid layer heated from below, a typical pattern is a regular lattice of rolls. Let the x_1 and x_2 axis be chosen orthogonal and parallel to the rolls, respectively. The velocity in the third (vertical) direction can then be written as

$$v_3 = (\psi(x_1, x_2)e^{ik_c x_1} + \text{c.c.})\sin(\pi x_3). \tag{24}$$

The thickness of the layer is taken to be unity, $2\pi/k_c$ is the wavelength of the roll pattern at the onset, and ψ is a slowly varying complex amplitude, the order parameter. Its dynamics is governed in this case by the Newell-Whitehead-Segel equation [25]

$$\dot{\psi} = (a - b_r|\psi|^2)\psi + D_r(\partial_1 - \frac{i}{2k_c}\partial_2^2)^2\psi \tag{25}$$

where a, b_r, D_r (all real) can be expressed by hydrodynamic parameters. The corresponding potential has been found [26] in the form

$$\phi(\{\psi\}) = \frac{2}{Q}\int dx\{-a|\psi|^2 + \frac{b_r}{2}|\psi|^4 + D_r\,|(\partial_1 - \frac{i}{2k_c}\partial_2^2)\psi|^2\}. \tag{26}$$

When the symmetry breaking phase is isotropic, the order parameter ψ is defined via the relation

$$v_3 = (\psi(x_1, x_2)e^{iS(x_1,x_2)} + \text{c.c.})\sin(\pi x_3) \tag{27}$$

where $|\nabla S| = k_c$. The dynamics of ψ is then described by the Swift-Hohenberg equation [27] which in dimensionless form reads

$$\dot{\psi} = (\alpha - (1 + \nabla^2)^2)\psi - |\psi|^2\psi. \tag{28}$$

Here α is a real control parameter and ∇^2 denotes the two-dimensional Laplacian. The nonequilibrium potential is

$$\phi(\{\psi\}) = \frac{2}{Q}\int dx\{-\alpha|\psi|^2 + \frac{1}{2}|\psi|^4 + (\frac{1}{2}\psi^*(1+\nabla^2)^2\psi + c.c.)\} \tag{29}$$

as can be checked easily.

2.2. CONVECTION IN NEARLY INSULATED LIQUID LAYERS

In a horizontal layer of liquid enclosed between two rigid plates the convection problem is augmented by a new effect, namely by the heat exchange with the boundary. In this case the local temperature disturbance can be considered as a (real) slow variable ψ. After appropriate rescaling it fulfills the equation [28]:

$$\dot{\psi} = -\beta\psi - \nabla^4\psi - \nabla[(1 - (\nabla\psi)^2)\nabla\psi] \tag{30}$$

where β is a coefficient measuring the heat exchange with the boundary. Eq.(30) is also of gradient type with nonequilibrium potential

$$\phi(\{\psi\}) = \frac{1}{Q}\int dx\{\beta\psi^2 - (\nabla\psi)^2 + \frac{1}{2}(\nabla\psi)^4 + (\nabla^2\psi)^2\}. \tag{31}$$

2.3. PHASE VARIATION OF STATIONARY PATTERNS

In geometrics which suppress instbilities along Bénard rolls the pattern has only one phase variable ψ. Its spatial and temporal variation is of importance since different local wavelengths might coexist. The dynamics of the phase is described by the dimensionless equation [29]

$$\dot{\psi} = (1 + E\nabla\psi + (\nabla\psi)^2)\nabla^2\psi - \nabla^4\psi. \tag{32}$$

E denotes here a nonlinearity parameter. The corresponding nonequilibrium potential is of the from:

$$\phi(\{\psi\}) = \frac{1}{Q}\int dx\{(\nabla\psi)^2 + \frac{E}{3}(\nabla\psi)^3 + \frac{1}{6}(\nabla\psi)^4 + (\nabla^2\psi)^2\}. \tag{33}$$

The knowledge of these potentials has turned out to be very useful in identifying different attractors, i.e., stable macroscopic states, and in specifying their stability border in the parameter space. Instabilities, e.g. the Eckhaus instability, are reflected by the disappearance of the potential's local minimum in the given state.

Unfortunately *gradient systems are exceptional*, therefore, in the generic case one must construct the potential via the recipies discussed in the first section.

3. Example without obvious potential: the complex Ginzburg-Landau equation

The complex Ginzburg-Landau equation is the normal form for spatio-temporal Hopf bifurcations and plays a relevant role in different phenomena ranging from fluid dynamics to chemical reactions [30]. Its canonical form is

$$\dot{\psi} = (a - b|\psi|^2)\psi + D\nabla^2\psi \tag{34}$$

where ψ is a complex field, a is a real parameter, and $b \equiv b_r + ib_i, D \equiv D_r + iD_i$. The equation is valid around the bifurcation point ($|a|$ small) on time and length scales long compared to microscopic scales. Notice that the right hand side of (34) cannot be written as the derivative of any functional. The noisy equation is defined by adding a complex Langevin force $\eta\xi(x,t)$ with correlation (20).

Interestingly, in the special case when the combination

$$D_- \equiv D_r b_i - D_i b_r \tag{35}$$

vanishes, the nonequilibrium potential for the equation is just the Ginzburg-Landau functional given by (3) ([22]):

$$\phi(\{\psi\}) = \frac{1}{Q}F_{GL}(\{\psi\}) \quad \text{for} \quad D_- = 0. \tag{36}$$

Note, however, that even in this case the system is not of gradient type since

$$\dot{\psi}(x) = -\frac{1}{2}\frac{\delta F_{GL}(\{\psi\})}{\delta\psi^*} + ib_i(-|\psi|^2 + \frac{D_r}{b_r}\nabla^2)\psi.$$

In the general case $D_- \neq 0$ the potential is to be constructed via equation (21). The Lagrangian turns out to have the form

$$L = \frac{1}{Q} \int dx |\dot{\psi} - a\psi + b|\psi|^2\psi - D\nabla^2\psi|^2 \tag{37}$$

as the generalization of (16) for a spatially extended case. The integration in (21) is taken along a solution $\psi(x, \tau)$ of Lagrange's equations starting with $\psi(x, -\infty)$ on an attractor A of (34) in the infinite past, and ending on $\psi(x, 0) = \psi(x)$ at time zero.

An analytic solution of this problem is not known. Therefore, our strategy is to evaluate the potential in a perturbation expansion. As a convenient parameter we choose the diffusion coefficient D. Since the calculation can be performed up to low orders only, our results will be valid in the limit of a weak spatial diffusion or, equivalently, for large length scales. In the next section the perturbation scheme will briefly be presented.

4. General expansion scheme

For simplicity, we illustrate the method for a case with a finite number of variables. The extension to fields is straightforward. Let ϵ be a small parameter of the system. First, split the Lagrangian in terms proportional to different powers of this parameter, i.e., write

$$L = L_0 + \epsilon L_1 + \epsilon^2 L_2 + \dots \; . \tag{38}$$

Note that all the L_i's are explicitly given since the form of the Lagrangian is determined by the stochastic process (9). Next, expand also the attractor of the deterministic system according to a similar scheme :

$$A = A_0 + \epsilon A_1 + \epsilon^2 A_2 + \dots \tag{39}$$

where A_0 is the attractor corresponding to the Lagrangian in lowest order, etc. The quantity to be computed is the nonequilibrium potential represented by the series

$$\phi = \phi_0 + \epsilon\phi_1 + \epsilon^2\phi_2 + \dots \; . \tag{40}$$

In order to find ϕ_i we use relation (17).

In zeroth order, one solves the Euler-Lagrange equations with L_0:

$$\frac{d}{d\tau}\frac{\partial L_0}{\partial \dot{q}} = \frac{\partial L_0}{\partial q}. \tag{41}$$

Let $q_0(\tau)$ denote the solution fulfilling the boundary conditions: $q_0(-\infty) \in A_0$, $q(0) = q$. The corresponding potential is then given by the integral

$$\phi_0(q) = \int_{-\infty}^{0} d\tau L_0(q_0, \dot{q}_0) + \phi_0(A_0). \tag{42}$$

For clarity, we assume there is one attractor only, and the integral is single valued so that the min sign of (17) can be discarded.

In first order, the Euler-Lagrange equations are taken with $L^{(1)} \equiv L_0 + \epsilon L_1$. We seek for a solution $q^{(1)}(\tau)$ with the boundary condition $q^{(1)}(-\infty) \in A_0 + \epsilon A_1$, $q^{(1)}(0) = q$. Write

$q^{(1)} \equiv q_0 + \epsilon q_1$ from which $q_1(-\infty) \in A_1, q_1(0) = 0$ follows. Performing a Taylor expansion in ϵ, one finds an explicit equation for q_1 :

$$\frac{d}{d\tau}\Big(\frac{\partial^2 L_0}{\partial \dot{q}_0^2}\dot{q}_1\Big) - q_1\Big[\frac{\partial^2 L_0}{\partial q_0{}^2} - \frac{d}{d\tau}\Big(\frac{\partial^2 L_0}{\partial q_0 \partial \dot{q}_0}\Big)\Big] = \frac{\partial L_1}{\partial q_0} - \frac{d}{d\tau}\frac{\partial L_1}{\partial \dot{q}_0}. \tag{43}$$

Relation (17) now reads

$$\begin{aligned}\phi^{(1)} \equiv \phi_0 + \epsilon\phi_1 &= \int_{-\infty}^{0} d\tau[L_0(q_0 + \epsilon q_1, \dot{q}_0 + \epsilon\dot{q}_1) + \epsilon L_1(q_0, \dot{q}_0)] + \\ &\quad \phi_0(A_0 + \epsilon A_1) + \epsilon\phi_1(A_0).\end{aligned} \tag{44}$$

According to Hamilton's principle, the correction in the argument of L_0, ϵq_1, does not change the integral to first order in ϵ. Furthermore, since the potential is minimal on the attractor, ϵA_1 does not contribute to this order. Thus, we obtain [17]

$$\phi_1(q) = \int_{-\infty}^{0} d\tau L_1(q_0, \dot{q}_0) + \phi_1(A_0) \tag{45}$$

telling us that the integral is nothing but the action belonging to L_1 evaluated along the zeroth order path q_0.

In second order, let us write the solution $q^{(2)}$ fulfilling the required boundary conditions as $q^{(2)} \equiv q_0 + \epsilon q_1 + \epsilon^2 q_2$. Thus, $q_2(-\infty) \in A_2$ and $q_2(0) = 0$. Equation (17) can then be written in the form

$$\begin{aligned}\phi^{(2)} &\equiv \phi_0 + \epsilon\phi_1 + \epsilon^2\phi_2 \\ &= \int_{-\infty}^{0} d\tau[L_0(q_0 + \epsilon q_1 + \epsilon^2 q_2, \dot{q}_0 + \epsilon\dot{q}_1 + \epsilon^2\dot{q}_2) \\ &\qquad + \epsilon L_1(q_0 + \epsilon q_1, \dot{q}_0 + \epsilon\dot{q}_1) + \epsilon^2 L_2(q_0, \dot{q}_0)] + \\ &\qquad \phi_0(A_0 + \epsilon A_1 + \epsilon^2 A_2) + \epsilon\phi_1(A_0 + \epsilon A_1) + \epsilon^2\phi_2(A_0).\end{aligned} \tag{46}$$

Expanding the integrands and the contributions from the attractor up to second order, one finds an equation for ϕ_2. Using identity (43) of the first order approach, and making use of partial integration, L_0 can completely be eliminated from the integrand. Contributions from the lower boundary $\tau = -\infty$ cancel those from ϕ_0, ϕ_1 on the attractor by applying the relation $\partial\phi/\partial q = \partial L/\partial\dot{q}$ which expresses two ways of obtaining the momentum along Euler-Lagrange trajectories (see (19)). Finally, we have

$$\phi_2(q) = \int_{-\infty}^{0} d\tau\Big[\frac{1}{2}q_1\Big(\frac{\partial L_1}{\partial q_0} - \frac{d}{d\tau}\frac{\partial L_1}{\partial \dot{q}_0}\Big) + L_2(q_0, \dot{q}_0)\Big] + \phi_2(A_0). \tag{47}$$

The second order contribution to the potential can, thus, uniquely be evaluated along the path of the first order solution.

5. Results for the complex Ginzburg-Landau equation

The perturbation procedure has been applied [20] to the complex Ginzburg-Landau equation taking the diffusion coefficient as small parameter: $D, D^* \sim \epsilon$. Here we summarize the results obtained for the one-dimensional model with periodic boundary conditions in the symmetry breaking region: $a > 0$.

5.1. ZEROTH ORDER APPROXIMATION

In this leading order the Lagrangian does not contain any spatial derivative and the problem performing the integral in (42) is formally reduced to one with two degrees of freedom:

$$L_0 = \frac{1}{Q}\int dx |\dot{\psi} - a\psi + b|\psi|^2\psi|^2. \tag{48}$$

Note, however, that ψ *depends parametrically on the space coordinate* x: $\psi = \psi(x,\tau)$. After separating a trivial temporal oscillation via a simple transformation the attractor is found, in this order, to be the same as the minimum (5) of the Ginzburg-Landau functional, therefore, the Euler-Lagrange equations (41) are solved with the boundary conditions

$$\psi_0(x,-\infty) = (\frac{a}{b_r})^{1/2}e^{i\varphi(x)}, \quad \psi_0(x,0) = \psi(x). \tag{49}$$

The result is easily found to be

$$\psi_0(x,\tau) = \psi[\frac{b_r|\psi|^2}{a} + (1 - \frac{b_r|\psi|^2}{a})e^{2a\tau}]^{-1/2+ib_i/(2b_r)}e^{-ia(b_i/b_r)\tau} \tag{50}$$

(the argument of ψ has not been written out here explicitly) from which the potential ϕ_0 follows as

$$\phi_0 = \frac{2}{Q}\int dx\{-a|\psi|^2 + \frac{b_r}{2}|\psi|^4\} + C(\{\varphi_0\}). \tag{51}$$

The last term C is the contribution on the attractor. An essentially new feature of our case, when compared with models with a finite number of variables, is that *the attractor is highly degenerate* in zeroth order: the function $\varphi(x)$ is arbitrary. The form of $C(\{\varphi\})$ has to be calculated independently. We found this can always be done by using a local (quadratic) ansatz for C around the attractor .

5.2. THE POTENTIAL IN SECOND ORDER

Proceeding along the lines sketched in the previous section, and determining the contribution from the attractor order by order,one can construct the nonequilibrium potential $\phi^{(2)}$ up to second order in the diffusion coefficient. It is convenient to give the result in polar coordinates of the field, i.e., to write:

$$\psi(x) = r(x)e^{i\varphi(x)}. \tag{52}$$

In terms of the amplitude r, the phase φ, and their spatial derivatives the potential was found [20] to have the form

$$\phi^{(2)} = \frac{1}{Q}\int dx\mathcal{L}(\nabla^2\varphi, \nabla\varphi, \nabla^2 r, \nabla r, r). \tag{53}$$

The density $\mathcal{L}$ of the potential splits into a part containing the amplitude r only, and a rest containing all spatial inhomogeneities:

$$\mathcal{L} = K(a, \nabla^2\varphi, \nabla\varphi, \nabla^2 r, \nabla r, r) - U(a,r) \tag{54}$$

where

$$K = \alpha(\nabla\varphi)^4 + \frac{\alpha_1}{r^2}(\nabla\varphi)^2(\nabla r)^2 + \frac{\alpha_2}{r}(\nabla\varphi)^3(\nabla r) + \beta(\nabla^2\varphi)^2 + 2f_1(a,r)(\nabla\varphi)^2 + 4f_2(a,r)\nabla\varphi\nabla r + 2f_3(a,r)(\nabla r)^2, \tag{55}$$

and

$$U(a,r) = 2ar^2 - b_r r^4. \tag{56}$$

Terms of type $(\nabla^2 r)^2, (\nabla r)^4$, and $(\nabla r)^3\nabla\varphi$, which will not be relevant in what follows, have been dropped in K. The coefficients are given as

$$\begin{aligned}
\alpha &= \frac{1}{3}\frac{D_- D_r b_i}{b_r |b|^2}, \qquad \beta = -\frac{2D_- \mathrm{Re}(iD^* b^2)}{|b|^4}, \\
\alpha_1 &= \frac{2}{3}\frac{D_- D_r b_i}{b_r^3 |b|^2}(b_i^2 - 2b_r^2), \quad \alpha_2 = \frac{2}{3}\frac{D_- D_r}{b_r^2|b|^2}(b_i^2 - b_r^2), \\
f_1(a,r) &= D_r r^2 - \frac{D_- b_i}{|b|^2}\frac{a}{b_r}, \\
f_2(a,r) &= \frac{D_- a}{r|b|^2}[1 - \frac{|b|^2}{2b_r^2}(1 - \frac{b_r r^2}{a})], \\
f_3(a,r) &= D_r + \frac{D_- b_i}{r^2 b_r}(\frac{a}{|b|^2} - \frac{1}{3}\frac{a - b_r r^2}{b_r^2})
\end{aligned} \tag{57}$$

where the parameter D_- has been defined in (35). Notice that in the limit $D_- = 0$ the majority of terms vanishes in K, and what survives corresponds just to the polar coordinate representation of the Gizburg-Landau functional. The form (53) for $D_- \neq 0$, thus, provides a considerable generalization of F_{GL}. For global stability to this order we have to require $\alpha > 0$, i.e., $D_- b_i > 0$, which we assume in the following.

5.3. EXTREMUM CONDITIONS FOR THE POTENTIAL

Minima of the potential correspond to attractors of (34). One might be interested more generally in extrema of $\phi^{(2)}$ which correspond to invariant objects in the phase space of the complex Ginzburg-Landau equation. Extrema of $\phi^{(2)}$ follow from the variational principle

$$\delta\phi^{(2)} = 0 \tag{58}$$

with $\phi^{(2)}$ as given by (53-57). This problem can be considered as the analogue of Hamilton's principle with x as a kind of time and $\mathcal{L}$ as a Lagrangian. In fact, K and U play the role of a kinetic and a potential energy, respectively. It is worth emphasizing that this new Lagrangian is something completely different from L defined by (37) and (21).

The Euler-Lagrange equations of the variational problem (58) are

$$\nabla(\frac{\partial\mathcal{L}}{\partial\nabla\varphi} - \nabla\frac{\partial\mathcal{L}}{\partial\nabla^2\varphi}) = 0, \tag{59}$$

and

$$\nabla\frac{\partial\mathcal{L}}{\partial\nabla r} = \frac{\partial\mathcal{L}}{\partial r}. \tag{60}$$

Since the phase is a cyclic variable, the first equation expresses the conservation of an "angular momentum" J:

$$J \equiv \frac{\partial \mathcal{L}}{\partial \nabla \varphi} - \nabla \frac{\partial \mathcal{L}}{\partial \nabla^2 \varphi} = \text{const.} \tag{61}$$

Another conservation law is implied by the fact that $\mathcal{L}$ does not explicitly depend on x. We, thus, find in the manner familiar from analytical mechanics a conserved "energy" E:

$$E = \nabla^2 \varphi \frac{\partial \mathcal{L}}{\partial \nabla^2 \varphi} + J \nabla \varphi + \nabla r \frac{\partial \mathcal{L}}{\partial \nabla r} - \mathcal{L} = \text{const.} \tag{62}$$

The two conservation laws provide conditions for the fields $r(x), \varphi(x)$ belonging to extrema of the potential $\phi^{(2)}$.

It is worth emphasizing that the Euler-Lagrange equations (59,60) define, in general, a *nonintegrable* Hamiltonian system. Because of the term $\nabla^2\varphi$ in the potential, equation (59) is of *fourth* order, thus, two conservation laws are not sufficient to determine the motion completely. Therefore, it must be expected that *chaotic* extrema (in x) of $\phi^{(2)}$ exist. They must be degenerate because of the translation invariance, consequently, $\phi^{(2)}$ can be constant on a manifold of degenerate minima, which must appear already for a single chaotic attractor. This should correspond to well-known turbulent solutions of (34) [30]. In order to check this hypothesis a numerical simulation is neccessary which has not yet been carried out.

In what follows we restrict our attention to nonchaotic extrema and assume, therefore, that the term $\nabla^2\varphi$ is negligible in the potential. Consequently, $\partial\mathcal{L}/\partial\nabla^2\varphi$ can be neglected in (59). The Euler-Lagrange equations define then an integrable system of two degrees of freedom determined by the conserved quantities:

$$J \equiv \frac{\partial \mathcal{L}}{\partial \nabla \varphi} = \text{const.}, \quad E = J \nabla \varphi + \nabla r \frac{\partial \mathcal{L}}{\partial \nabla r} - \mathcal{L} = \text{const.}. \tag{63}$$

By eliminating $\nabla\varphi$ via the conserved angular momentum, in an analogous way as in Kepler's problem, one finds the energy conservation in the form

$$E = \frac{1}{2} M(r, J)(\nabla r)^2 + U_{eff}(r, J) \tag{64}$$

where M stands for the "mass" of a particle moving in the "effective potential" U_{eff}. The qualitative form of U_{eff} can be read off from Figure 2a below.

5.4. PLANE WAVE ATTRACTORS AND THEIR STABILITY

For plane wave solutions we have

$$\varphi(x) = kx, \quad r(x) = \text{const.} , \tag{65}$$

where k is a parameter, the wave number. The conservation laws express then relations between the amplitude and wave number, and can be written as $J \equiv J_k = J_k(r)$ and $E \equiv E_k = E_k(r)$. With energy and angular momentum conservation the action, i.e., the potential $\phi^{(2)}$, has the form [20]

$$\phi^{(2)} = \frac{1}{Q} \int dx \{kJ_k + E_k - 2U_{eff}(r, J_k)\}. \tag{66}$$

This can be an extremum if r takes the special value r_k defined by

$$\frac{\partial U_{eff}(r, J_k)}{\partial r}_{|r=r_k} = 0, \tag{67}$$

from which

$$r_k{}^2 = \frac{a - D_r k^2}{b_r} \tag{68}$$

is obtained. For a minimum of $\phi^{(2)}$, one must require stability against phase, amplitude, and amplitude gradient fluctuations, which are expressed by the relations

$$\frac{\partial J_k(r)}{\partial k}_{|r=r_k} > 0, \tag{69}$$

$$\frac{\partial^2 U_{eff}(r, J_k)}{\partial r^2}_{|r=r_k} < 0, \tag{70}$$

and

$$M(r_k, J_k) > 0, \tag{71}$$

respectively. Notice that the first and third conditions require the positivity of "moment of inertia" and "mass", respectively, while the second one implies that a stable plane wave belongs to a local *maximum* of the effective potential at r_k. From (69),(70) the inequalities

$$D_+ \equiv D_r b_r + D_i b_i > 0, \tag{72}$$

$$k^2 < k_c{}^2 \equiv \frac{D_+}{3D_+ b_r + 2D_- b_i} \frac{ab_r}{D_r}. \tag{73}$$

follow. If the first or second one is not valid, the system undergoes a Newel-Kuramoto [31] or an Eckhaus-Benjamin-Feir [32] type instability, respectively. From relation (71) we obtain

$$D_r D_+ + D_i D_- > 0 \tag{74}$$

which seems to be a novel stability condition.

The value of the potential for stable plane waves in a system of length l is found to be

$$\phi_k^{(2)} = \frac{2lD_+ a}{Q|b|^2} k^2 (1 - \frac{k^2}{6k_c{}^2}). \tag{75}$$

This shows that the most stable plane wave attractor is the homogenous state $k = 0$. A schematic plot of $\phi_k^{(2)}$ will be given in the lower part of Fig.4.

5.5. POTENTIAL BARRIERS

Although, the potential $\phi_k^{(2)}$ depends smoothly on k, there exist, in the space of all possible solutions, potential barriers among plane wave attractors. We investigate here the barrier between two neighbouring plane waves. By neighbouring we mean that the wave numbers are as close to each other as possible. In a system of length l the smallest difference is $2\pi/l$.

Thus, we consider a wave with wave number k and another one with $k' = k - 2\pi/l$. The schematic shape of such waves are depicted in Figs. 1a, 1c. These minimizing solutions of the potential are separeted by a *saddle* solution characterized by a local maximum of the potential. The saddle is expected [33] to behave asymptotically (for $x \to \pm\infty$) as a plane wave with some other wave number k_s. Its amplitude is, however, not constant but possesses a well defined minimum around the origin (Fig. 1b).

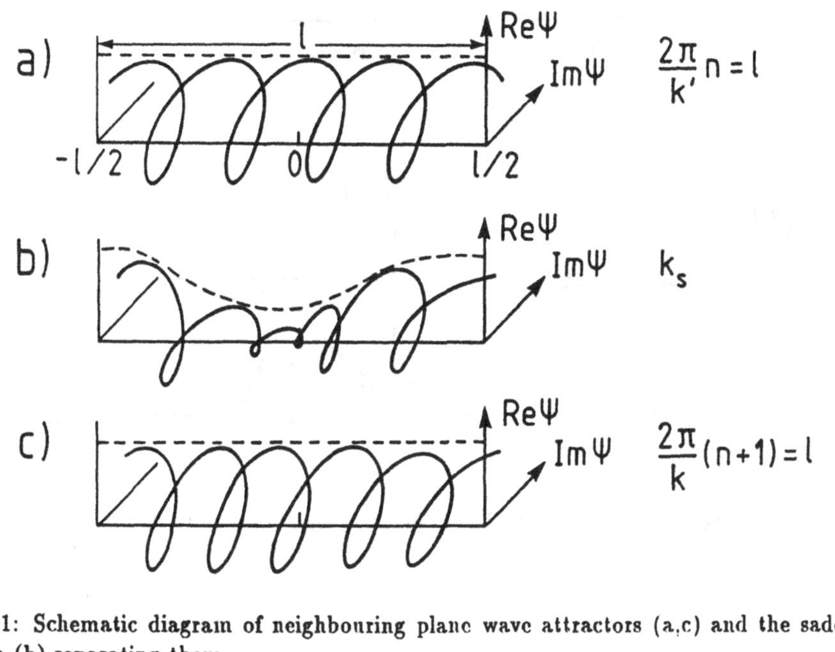

Figure 1: Schematic diagram of neighbouring plane wave attractors (a,c) and the saddle solution (b) separating them

In the potential picture, the saddle solution corresponds to a motion with an energy E_{k_s} which is just the local maximum of the effective potential $E_{k_s} = U_{eff}(r_{k_s}, J_{k_s})$ where r_{k_s} is the asymptotic amplitude of the saddle, as illustrated in Fig.2a. Consequently, the amplitude of the saddle behaves as shown in Fig.2b.

The nonequilibrium potential belonging to the saddle is just the action (53) evaluated along the saddle solution. To estimate this value, one can take a characteristic energy of the motion and multiply it with the time T_{eff} of harmonic oscillations around the minimum of the effective potential. The potential barrier $\Delta\phi_k$ between plane wave k and the saddle solution, $\Delta\phi_k \equiv \phi_k^{(2)} - \phi_s^{(2)}$, (see Fig.3) is then guessed as the difference between local extrema of $U_{eff}(r, J_{k_s})$ multiplied by T_{eff}. When doing this, one might approximate U_{eff}

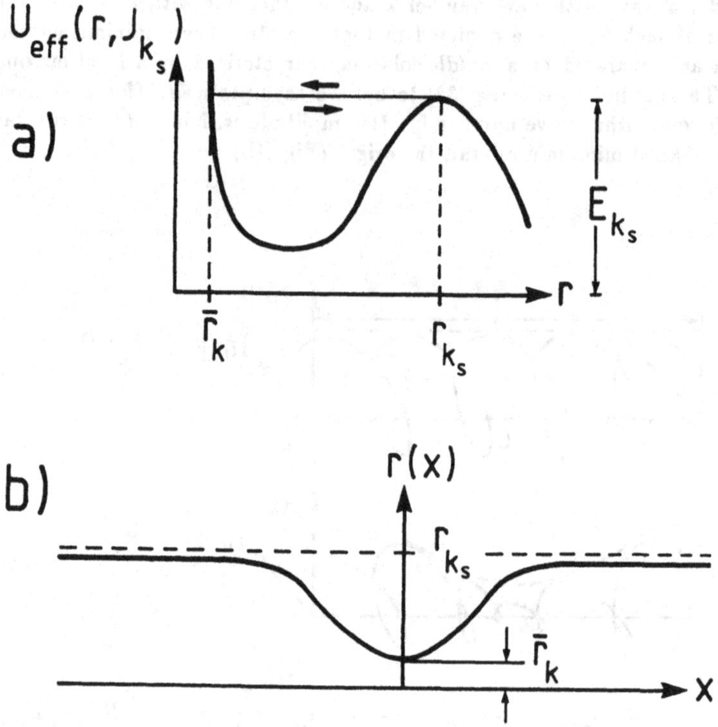

Figure 2: Energy diagram (a) and the amplitude (b) of the saddle shown in Fig. 1b

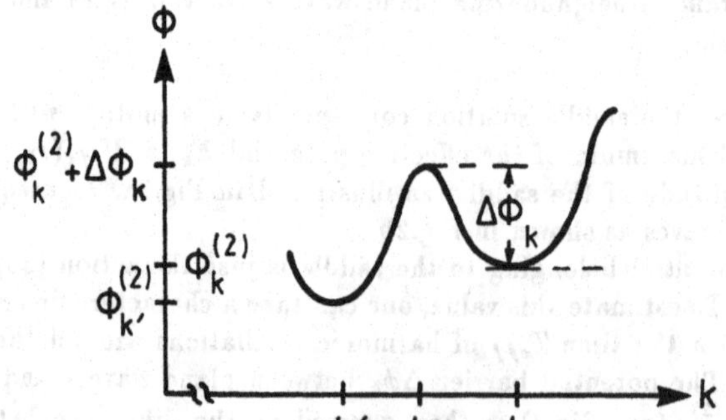

Figure 3: Schematic plot of the nonequilibrium potential for different spatially extended states. k denotes here the asymptotic wave number characterizing the state for $x \to \pm\infty$. $\Delta\phi_k$ represents the potential barrier between two neighbouring plane waves

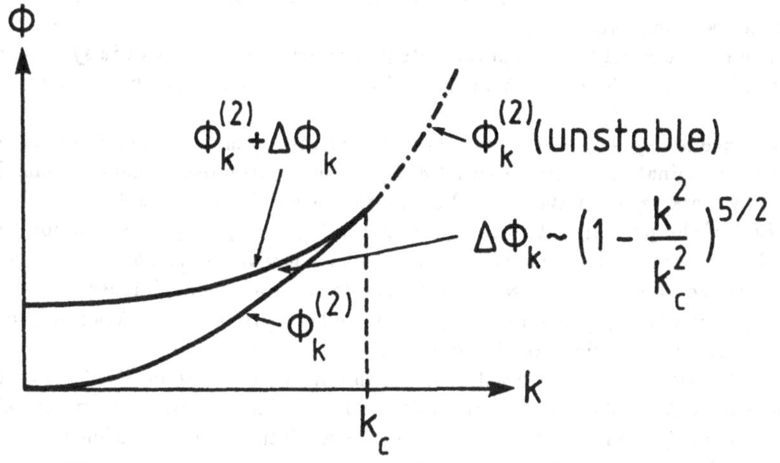

Figure 4: Nonequilibrium potential $\phi_k^{(2)}$ for a plane wave attractor and the barrier separating it from a neighbouring plane wave as a function of wave number k

by a cubic parabola. The result is then obtained in the form

$$\Delta\phi_k \sim \frac{b_r(3D_+b_r + 2D_-b_i)}{QD_r{}^3 |\, b\,|^6}(D_+D_r + D_-D_i)^{1/2}(aD_+)^{3/2}(1 - \frac{k^2}{k_c{}^2})^{5/2} \tag{76}$$

The proportionality factor not written out in (76) is found in a more detailed calculation [20] to be $4\sqrt{2}/15$.A schematic diagram of the k-dependence of the barrier and of $\phi_k^{(2)}$ is given in Fig.4.

Note that $\Delta\phi_k$ is independent of the system's size l. It is zero both at the Hopf, Newel-Kuramoto, and Eckhaus-Benjamin-Feir instabilities. Thus, we conclude that the vanishing of potential barriers is simultaneously a sign of local instabilities.

6. Conclusions

The aim of this paper has been to show that nonequilibrium potentials do exist for spatially extended systems including the generic case when the equation of motion is *not* of gradient type. Such a potential is a Lyapunov functional of the system and governs simultaneously the stationary distribution in the presence of a weak external noise. It reflects global properties not obtainable via a deterministic analysis. Nonequilibrium potentials provide useful tools for studying instabilities leading to pattern formation, and potential barriers among coexisting attractors. The latter characterize the relative stability of coexisting nonequilibrium states.

A general algorithm was presented to construct nonequilibrium potentials. For analytic purposes a perturbation scheme has been worked out and applied to a particular spatially extended system, to the complex Ginzburg-Landau equation. Results up to second order in the diffusion coefficient were given.

Here we summarize novel features of the potential which are not present in systems with finite degrees of freedom. Some of them have already been mentioned in the text, some others not.

The coexistence of *degenerate* attractors seems to be quite common in spatially extended cases, and the determination of the potential $\phi(A)$ on these attractors is unavoidable. This can be solved by working out suitable local expansions around such attractors.

Singularities of the potential at certain special states (like e.g. repellers) cannot be excluded. Results for the complex Ginzburg-Landau equation with the $1/r$ behaviour provide a hint only since the approximation we used breaks down around singularities.

Boundary conditions may play an important role and can properly be taken into account by means of surface contributions to the potential.

The form of the potential might drastically depend on the *dimensionality* of the system. We have found indications [20] that in higher than one spatial dimension the potential for the complex Ginzburg-Landau equation is no longer analytic in the diffusion coefficient: for $d \geq 2$ the perturbation scheme does not work with D as a small parameter. What a suitable expansion coefficient could be and how the potential looks like in two or higher dimensions remains to be answered by further studies.

Acknowledgment

This work was supported by the Deutsche Forschungsgemeinschaft through Sonderforschungsbereich 237 "Unordnung und grosse Fluktuationen".

References

[1] H. B. Callen, *Thermodynamics* (Wiley, New York, 1960)

[2] S. K. Ma, *Modern Theory of Critical Phenomena* (Benjamin, Reading, 1976)

[3] M. I. Freidlin and A. D. Wentzell, *Random Perturbations of Dynamical Systems* (Springer, Berlin, 1984)

[4] R. Graham and T. Tél, Phys. Rev. Lett. **52**, 9 (1984); J. Stat. Phys. **35**, 729 (1984); Phys. Rev. A **31**, 1109 (1985)

[5] R. Graham et al., Phys. Rev. A **31**, 3364 (1985)

[6] L. Schimansky-Geier et al., Phys. Lett. **108 A**, 329 (1985)

[7] H. R. Jauslin, J. Stat. Phys. **40**, 147 (1985); **42** 573 (1986); Physica A **144**, 179 (1987)

[8] R. Graham and T. Tél, Phys. Rev. A **33**, 1322 (1986); **35**, 1328 (1987);
R. Graham, Europhys. Lett. **2**, 901 (1986)

[9] D. Roekaerts and F.Schwartz, J. Phys. A **20**, L217 (1987);
D. Roekaerts and H. Yoshida, J. Phys. A **21**, 3547 (1988)

[10] E. Sulpice et al., Phys. Lett. **121 A**, 67 (1987);
A. Lemarchand et al., J. Stat. Phys. **53**, 613 (1988)

[11] G. Hu, Phys. Rev. **A 36**, 5782 (1987); **38**, 3693 (1988); **39**, 1286 (1989)

[12] P. Talkner et al., J. Stat. Phys. **48**, 231 (1987);
P. Talkner and P. Hänggi, in *Noise in Nonlinear Dynamical Systems*, Vol. 2, ed.: F. Moss and P. McClintock (Cambridge Univ. Press, 1989) pp 87- 99.

[13] T. Tél, J. Stat. Phys. **50**, 897 (1988)

[14] R. L. Kautz, Phys. Rev. **A 38**, 3693 (1988);
P. Grassberger, J. Phys. **A 22**, 3283 (1989)

[15] R. Graham, in *Noise in Nonlinear Dynamical Systems*, Vol. 1, ed.: F. Moss and P. McClintock (Cambridge Univ. Press, 1989) pp 225- 278.

[16] O. Descalzi and E. Tirapegui, J. Stat. Phys. **57**, 993 (1989)

[17] T. Tél, R. Graham and G. Hu, Phys. Rev. **A 40**, 4065 (1989)

[18] J. Grasman, Math. and Comp. in Simulation **31**, 41 (1989);
H. Roozen, Thesis, Wageningen, 1990

[19] G. Hu and H. Haken, Phys. Rev. **A 40**, 5966 (1989); Phys. Rev. **A 41**, 2231; 7078 (1990)

[20] R. Graham and T. Tél, Potential for the complex Ginzburg-Landau equation, preprint, 1990;
Steady-state ensemble for the complex Ginzburg-Landau equation with weak noise, to appear in Phys. Rev. A, 1990

[21] D. Walgraef, G. Dewel and P, Borckmans, Adv. Chem. Phys. **49**, 311 (1982);
J. Chem. Phys. **78**, 3043 (1983)

[22] P. Szépfalusy and T. Tél, Physica **112A**, 146 (1982)

[23] W. G. Faris and G. Jona-Lasinio, J. Phys. **A 15**, 3025 (1982)

[24] H. Lemarchand and G. Nicolis, J. Stat. Phys. **37**, 609 (1984);
H. Lemarchand, Bull. Sci. Acad. R. Belg. **70**, 40 (1984);
A. Fraikin and H. Lemarchand, J. Stat. Phys. **41**, 531 (1985)

[25] A. Newel and J. Whitehead, J. Fluid. Mech **38**, 279 (1969);
L. A. Segel, J. Fluid. Mech **38**, 203 (1969)

[26] R. Graham, Phys. Rev. **A 10**, 1762 (1974)

[27] J. Swift and P. C. Hohenberg, Phys. Rev. **A 15**, 319 (1977)

[28] V. L. Gertsberg and G. I. Sivashinsky, Prog. Theor. Phys. **66**, 1219 (1981)

[29] H. R. Brand and R. J. Deissler, Phys. Rev. Lett. **63**, 508 (1989);
Phys. Rev. **A 41**, 5478 (1990)

[30] K. Stewartson and J. S. Stuart, J. Fluid. Mech. **48**, 529 (1971);
Y. Kuramoto and T. Tsuzuki, Prog. Theor. Phys. **52**, 1399 (1974);
A. Wunderlin and H. Haken, Z. Phys. **B 21**, 393 (1975);
P. J. Blenerhassett, Philos. Trans. Roy. Soc. London Ser. **A1441**, 43 (1980);
H. Moon, P. Huerre and L. Redekopp, Phys. Rev. Lett. **49**, 458 (1982);
K. Nozaki and N. Bekki, Phys. Rev. Lett. **51**, 2171 (1983);
L. R. Keefe, Stud. Appl. Math. **73**, 91 (1985);
H. R. Brand et al., Physica **D 23**, 345 (1986); Phys. Lett. **118 A**, 67 (1986);
C. R. Doering et al., Nonlinearity **1**, 279 (1988);
P. Coullet et al., Phys. Rev. Lett. **59**, 884 (1987); **62**, 1619 (1989);
S. Rica and E. Tirapegui, Phys. Rev. Lett. **64**, 878 (1990)

[31] A. C. Newell, Lect. Appl. Math. **15**, 157 (1974);
Y. Kuramoto and T. Tsuzuki, Prog. Theor. Phys. **55**, 356 (1976)

[32] W. Eckhaus, *Studies in Nonlinear Stability Theory* (Springer, Berlin, 1965);
T. B. Benjamin and J. E. Feir, J. Fluid. Mech. **27**, 417 (1967);
J. T. Stuart and R. C. Di Prima, Proc. Roy. Soc London **A362**, 27, (1978)

[33] J. S. Langer and V. Ambegaokar, Phys. Rev. **164**, 498 (1967)

PASSAGE TIME DESCRIPTION OF DYNAMICAL PROCESSES

M. SAN MIGUEL, E. HERNÁNDEZ-GARCÍA, P. COLET, M.O. CÁCERES*, F. DE PASQUALE+
Departament de Física
Universitat de les Illes Balears
E-07071 Palma de Mallorca, Spain

ABSTRACT. The description of dynamical processes in terms of passage time distributions is generally advocated. Two specific situations are considered, transient relaxation at a pitchfork bifurcation and transport in disordered chains. The scaling description at a supercritical bifurcation is reviewed. For the subcritical case an approximation for the stochastic paths allowing the calculation of the passage time distribution is presented. In the case of disordered chains, a characterization of strong disorder is given in terms of the long tail of the passage time distribution to escape from a finite interval.

1. Introduction.

Problems in which fluctuations are important in some dynamical sense are described in terms of a stochastic process[1] $x(t)$. A standard way of characterizing the process $x(t)$ is by the time dependent moments $< x^n(t) >$. This characterization gives de statistics of the random variable x at a fixed given time t. An alternative characterization is given by considering t as a function of x. One then looks for the statistics of the random variable t at which the process reaches for the first time a fixed given value x. The distribution of such times is the First Passage Time Distribution (FPTD). This alternative characterization emphasizes the role of the individual realizations of the process $x(t)$. It is particularly well suited to answer questions related to time scales of evolution. For example, the lifetime of a given state can be defined as the Mean FPT(MFPT) to leave the vicinity of that state. The associated variance of the PTD identifies if that lifetime is a meanigful quantity.

An example of situations in which a Passage Time description is useful is the study of transient relaxation processes triggered by noise in which a system leaves a state which is not globally stable. A first well known case is that of metastability, in which the lifetime of the metastable state, identified as the MFPT, is $T \sim exp(\Delta V/\epsilon)$, where ΔV is an activation energy and ϵ measures noise intensity[1]. A second case is the relaxation from the state that becomes unstable in a supercritical pitchfork bifurcation[2]. In a deterministic treatment, $x(t)$ starts to grow exponentially as $x^2 \sim e^{2at}$. However, the lifetime of the state is not a^{-1} because such lifetime is determined by fluctuations. It turns out to be $T \sim \frac{1}{2a} ln\epsilon^{-1}$. The relaxation from an unstable state in a supercritical pitchfork bifurcation follows, initially, Gaussian statistics for x. More complicated is the description of the relaxation from the state that looses its stability in a saddle-node bifurcation[3] or in a supercritical pitchfork bifurcation[4]. Those are states of marginal stability for which Gaussian statistics, or equivalently, linear theory, do not hold at any time of the relaxation process.

A second example of situations amenable of a PT description is given by transport in a random chain[5]. It is well known that Random Walk in a chain leads to $< x^2 > \sim t$, so that the time scale

* Permanent address: Centro Atómico Bariloche, 8400, Bariloche, Argentina.
+ Permanent address: Dipartimento di Fisica, Universitá di L'Aquila, I-67100 L'Aquila, Italy.

E. Tirapegui and W. Zeller (eds.), Instabilities and Nonequilibrium Structures III, 143–155.

for the walker to leave an interval $[-L, L]$ is $T \sim L^2$. In a chain with strong disorder one finds subdiffusive behavior $< x^2 > \sim t^\delta, \delta < 1$. In this case the lifetime of the interval $[-L, L]$ is not $T \sim L^{\frac{2}{\delta}}$. It happens that although the walker leaves the interval with probability one, the mean time to leave the interval diverges, $T = \infty$. The divergence law can be used to characterize the degree of disorder[6].

In this paper we deal with the two examples mentioned above. In Section 2 we will discuss relaxation at a pitchfork bifurcation, reviewing first the supercritical case from a new point of view and considering next more recent developments for the subcritical case[4]. In Section 3 we give a characterization of disorder in random chains in terms of FPT techniques[6].

The calculation of PT statistics is standard[1] for a Markov process $x(t)$ whose probability density $P(x,t)$ obeys an equation of the form

$$\partial_t P(x,t) = HP(x,t) \tag{1.1}$$

where H is the operator defining the dynamics. The probability that at time $t, x(t)$ is still in an interval (a, b) given an initial condition x_0 at $t = 0$ is known as the survival probability $F(x_0, t)$. It obeys the equation

$$\partial_t F(x_0, t) = H_D^+ F(x_0, t) \tag{1.2}$$

where H_D^+ is the adjoint of H_D. The latter operator differs from H in that it avoids the probability of reentering the interval (a, b) after having left it. The FPTD is given by $f(x_0, t) = -d_t F(x_0, t)$ and as a consequence the MFPT T_{x_0} obeys the Dynkin equation

$$H_D^+ T_{x_0} = -1 \tag{1.3}$$

Eqs. (1.2)-(1.3) completed with adequate boundary conditions give a solution of the problem of calculating the PT statistics for a Markov process. However, it is particularly interesting in cases of transient processes to have approximations for the individual realizations of the process $x(t)$. Such approximations give a physical picture of the process. In Sect. 2 we will discuss how to obtain these approximations and how to extract PT statistics from them. To calculation of PT statistics for nonMarkovian processes is a complicated problem[7]. The random walk in a disordered medium can be represented[8] by an effective nonMarkov problem given by the average of (1.1), but the PT statistics of the effective process do not coincide with the PT statistics of the posed problem. To calculate the PT statistics it is then convenient to average (1.2) over the distributions of disorder. We address the problem of transport in disordered medium from this point of view in Sect.3.

2. Relaxation at a Pitchfork Bifurcation.

The normal form of the dynamical equation associated with a pitchfork bifurcation is

$$d_t x = ax + bx^3 - cx^5 + \sqrt{\epsilon}\xi(t), \qquad c > 0 \tag{2.1}$$

where we have added a noise term $\xi(t)$ with noise intensity ϵ. The noise $\xi(t)$ is assumed to be Gaussian white noise of zero mean and correlation

$$< \xi(t)\xi(t') >= 2\delta(t - t') \tag{2.2}$$

We are here interested in describing the relaxation triggered by noise from the state $x = 0$. In the supercritical bifurcation $b < 0$ and we consider the relaxation for $a > 0$, so that $x = 0$ is an unstable state. In this case the quintic term in (2.1) is irrelevant and we can set $c = 0$. In the subcritical bifurcation $b > 0$. When $-\frac{3b^2}{16c} < a < 0$, $x = 0$ is a metastable state and it becomes unstable for $a > 0$. Changing the control parameter a from negative to positive values there is a crossover in the relaxation mechanism from relaxation via activation to relaxation from an unstable state. We will consider the relaxation from $x = 0$ in the critical case $a = 0$. In this case the state $x = 0$ has marginal stability.

In both cases of subcritical and supercritical bifurcations the MFPT can be calculated by a straightforward application of (1.3). We will follow here the alternative route of first finding an approximation for the individual stochastic paths of the process and then extracting the PT statistics from this description of the paths. This approach allows a complete description of the relaxation process and also to address questions as the existence of dynamical scaling. In addition, it allows the calculation of the PT statistics in situations in which (1.2)-(1.3) do not hold, as for example problems with time dependent parameters[9] or processes driven by non-white noise[10-12]. It might also be useful in many-variable problems in which a formulation like (1.2)-(1.3) does not lead easily to explicit results.

2.1 SUPERCRITICAL BIFURCATION: RELAXATION FROM AN UNSTABLE STATE.

This case corresponds to (2.1) with $c = 0$, $b < 0$. The operator H^+ in (1.2) is here the adjoint of the Fokker-Planck operator H associated with the Langevin equation (2.1). Rewritting the equation in terms of a potential $V(x)$ as

$$d_t x = -\partial_x V + \sqrt{\epsilon}\xi(t) \tag{2.3}$$

H^+ becomes

$$H^+ = -(\partial_x V)\partial_x + \epsilon {\partial_x}^2 \tag{2.4}$$

where $H^+ = H_D^+$ if $x \in (a, b)$. The operator H_D^+ vanishes for $x \notin (a, b)$. The potential is given by $V(x) = -\frac{a}{2}x^2 - \frac{b}{4}x^4$ and $x = 0$ is a local maximum of $V(x)$. The lifetime of the state $x = 0$ is given by the MFPT for x^2 to reach a value R_0^2 starting at $x = 0$. From (1.3) we obtain

$$T = \frac{2}{\epsilon}\int_0^{R_0} dx_1 e^{V(x_1)/\epsilon}\int_0^{x_0} e^{V(x_2)/\epsilon} dx_2 \tag{2.5}$$

which for asymptotically small ϵ gives

$$T \sim \frac{1}{2a} ln \frac{aR_0^2}{\epsilon} - \frac{\psi(1)}{2a} \tag{2.6}$$

where $\psi(1)$ is the digamma function. We wish now to reobtain this result taking a closer look at the relaxation process. In order to approximate the individual paths of the relaxation process defined by (2.3) and $x(0) = 0$, we write $x(t)$ as the ratio of two stochastic processes

$$x(t) = z(t)y^{-1/2}(t) \tag{2.7}$$

Then (2.3) is equivalent to the set of equations

$$d_t z(t) = az(t) + \sqrt{\epsilon}y^{1/2}(t)\xi(t) \tag{2.8}$$

$$d_t y(t) = -2bz^2(t) \tag{2.9}$$

with $z(0) = x(0) = 0$, $y(0) = 1$. Eqs. (2.8)-(2.9) can be solved iteratively from the initial conditions. In the zeroth order iteration

$$x(t) \sim z(t) = h(t)e^{at} \tag{2.10}$$

where

$$h(t) = \sqrt{\epsilon}\int_0^t e^{-as}\xi(s)ds \tag{2.11}$$

is a Gaussian stochastic process. In the first order iteration

$$x(t) = \frac{e^{at}h(t)}{[1 - 2b\int_0^t e^{2as}h^2(s)ds]^{1/2}} \tag{2.12}$$

Eq. (2.12) gives an accurate representation of the solution of (2.3) except for the small fluctuations around the final steady state $x_f^2 = -a/b$. In this approximation the decomposition (2.7) is interpreted as follows. The process $z(t)$ coincides with the linear approximation ($b = 0$) to (2.3). In the linear regime initial fluctuations are amplified exponentially and $h(t)$ in (2.10) plays the role of an effective stochastic initial condition. The process $y(t)$ introduces saturation effects killing the exponential growth of $z(t)$.

It is important to note that, beyond criticality, the time regime in which $at >> 1$ is rapidly achieved. In this regime (2.12) is safely approximated by

$$x(t) = \frac{e^{at}h(t)}{[1 - \frac{b}{a}h^2(t)(e^{2at} - 1)]^{1/2}} \tag{2.13}$$

Eq. (2.13) implies a dynamical scaling result for the process $x(t)$, in the sense that $x(t)$ is given by a time dependent nonlinear transformation of another stochastic process, namely the Gaussian process $h(t)$. This transformation, or mapping, is the mapping that gives the deterministic solution (2.3) ($\epsilon = 0$) from an initial condition $x(0) \neq 0$. The scaling transformation of (2.13) is then the deterministic mapping of the effective stochastic initial condition $h(t)$. This is the contents of the original scaling theory for this problem[2]. The relaxation process has two important time scales. A first one corresponds to the linear regime described by (2.10). A second one is given by the nonlinear deterministic evolution of the initial fluctuations as described by (2.12) or (2.13). In this regime initial fluctuations are amplified and give rise to transient anomalous fluctuations[2] of order ϵ^0 as compared with the initial or final fluctuations of order ϵ.

The calculation of PT statistics from the representation (2.12) can be done realizing that in this problem the escape from $x = 0$ occurs during the linear regime. In fact T sets the upper limit of validity of the linear approximation. The PT statistics can then be calculated from (2.10). Considering times of interest $at >> 1$ we can approximate $h(t)$ by $h(\infty)$ so that (2.10) can be solved for t as a function of a reference value $x^2 = R_0^2$:

$$t = \frac{1}{2a} ln \frac{R_0^2}{h^2(\infty)} \tag{2.14}$$

This results gives the statistics of the Passage Time as a transformation of the statistics of the Gaussian random variable $h(\infty)$. The statistics of t are determined by the generating function $W(\lambda)$:

$$W(\lambda) \equiv < e^{-\lambda t} >= \int dh P(h) e^{-\lambda t(h)} \qquad (2.15)$$

where $P(h)$ is the Gaussian probability distribution of h and $t(h)$ is given by (2.14). We find

$$W(\lambda) = e^{-\lambda < t >} \qquad (2.16)$$

with $< t >= T$ given by (2.6)

Besides the calculation of PT statistics, (2.13), allows also the calculation of the transient moments $< x^n(t) >$ by averaging over the Gaussian distribution of $h(t)$. The calculation of PT statistics along the lines described above for situations in which the parameter a is time dependent and $\xi(t)$ has a finite correlation time is discussed elsewhere[9,12].

2.2 SUBCRITICAL BIFURCATION: RELAXATION FROM A MARGINAL STATE

We now consider the relaxation from $x = 0$ as described by (2.1) with $b > 0$. The MFPT can be again calculated from (1.3). The same equations (2.3)-(2.5) hold for this case with the potential $V(x)$ being now $V(x) = -\frac{a}{2}x^2 - \frac{b}{4}x^4 + \frac{c}{6}x^6$. The state $x = 0$ is a local minimum of $V(x)$ for $a < 0$ and a local maximum for $a > 0$. For the relaxation process from $x = 0$ the saturation term $\frac{c}{6}x^6$ is irrelevant, so that $V(x)$ can be replaced in (2.5) by a potential in which $c = 0$ is for $|x| < R_0$. The asymptotic evaluation of (2.5) for small ϵ gives

$$T = (b\epsilon)^{-1/2}\phi(k) \qquad (2.17)$$

$$\phi(k) = \sum_{n=0}^{\infty} B_n(-1)^n \frac{k^n}{n!} \qquad (2.18)$$

$$B_n = \frac{\sqrt{2}}{16} 2^{n/2} \left[\Gamma\left(\frac{n+1}{4}\right)\right]^2 \qquad (2.19)$$

where the parameter k

$$k \equiv a/(b\epsilon)^{1/2} \qquad (2.20)$$

measures the distance to the situation of marginality $a = 0$. The result (2.17) interpolates between the MFPT for relaxation from a metastable state and the MFPT for relaxation from an unstable state. Indeed, for $a < 0$ and $|k| >> 1$, (2.18) is a series of positive terms which converges to the well known Kramers result for relaxation by activation through a barrier

$$T = \frac{\pi}{\sqrt{2}|k|} e^{k^2/4} \qquad (2.20)$$

On the other hand, for $a > 0$ and $|k| >> 1, \phi(k)$ becomes an alternating series which converges to

$$\phi(k) \sim \frac{1}{2k} ln k^2 \qquad (2.21)$$

This value of $\phi(k)$ replaced in (2.17) reproduces the dominant contribution in (2.6).

A delicate relaxation mechanism occurs at the marginal situation $a = 0$. We now consider this case looking for a representation of the individual stochastic paths. In the same way that in the previous section we look for a decomposition of the form (2.7). Since we are only interested in the escape process we set $c = 0$ in (2.1). Eq. (2.1) with $a = c = 0$ is equivalent to the set (2.8)-(2.9) with $a = 0$. In the zeroth order iteration

$$x(t) \sim z(t) = \sqrt{\epsilon}W(t), \quad W(t) = \int_0^t \xi(s)ds \tag{2.22}$$

the process $x(t)$ coincides with the Wiener process giving diffusion in the locally flat potential. In the first order iteration we find

$$x(t) = \frac{\sqrt{\epsilon}W(t)}{[1 - 2\epsilon b \int_0^t W^2(s)ds]^{1/2}} \tag{2.23}$$

Two important differences between (2.23) and (2.12) should be stressed. The first one is that in the first stage of evolution given by (2.22) there is no escape from $x = 0$. The nonlinearities introduced by the process $y(t)$ are essential for the escape process to occur. This means that contrary to (2.12) there is no regime of the escape process in which Gaussian statistics hold. The second difference is that $x(t)$ given by (2.23) has not a scaling form since it is not a transformation of a single stochastic process. Indeed, $x(t)$ in (2.23) depends on two non independent processes, namely $W(t)$ and $\int_0^t W^2(s)ds$. A naive approach to the problem would be to assume, by analogy with (2.13), a scaling form in which $x(t)$ is given by the deterministic nonlinear mapping of the fluctuating initial process $\sqrt{\epsilon}W(t)$, i.e., $x(t)$ is given by the deterministic solution of (2.3) with $x(0)$ replaced by $\sqrt{\epsilon}W(t)$. This would amount to take $y(t) = 1 - 2\epsilon btW^2(t)$. This scaling representation is qualitatively incorrect since it leads to a diverging MPT, $T = \infty$.

The accuracy of the representation (2.23) is shown in Fig.1 where an individual trajectory given by (2.23) is compared with the corresponding one of the exact process given by (2.1). In this figure we observe that $x(t)$ coincides initially with the Wiener process, it later departs from it and at a time rather sharply defined it departs from the vicinity of $x = 0$. In fact, the strong nonlinearity implies that the solution of (2.1) with $c = 0$ and $x(0) \neq 0$ reaches $|x| = \infty$ in a finite time. It is then natural to identify the PT as the escape time from $x = 0$. This is a random time t^* in which $x(t^*) = \infty$, or equivalently $y(t^*) = 0$. From (2.23) we find

$$1 = 2b\epsilon \int_0^{t^*} W^2(s)ds \tag{2.24}$$

which can be solved for t^* as

$$t^* = \left[\frac{1}{2b\epsilon\Omega}\right]^{1/2} \tag{2.25}$$

with

$$\Omega \equiv \int_0^1 W^2(s)ds \tag{2.26}$$

Eq. (2.25) gives the statistics of t^* as a transformation of the statistics of another random variable Ω. This scaling result for t^* has the same basic contents than (2.14). In (2.14) the result appeared for $at >> 1$ as a consequence of Gaussian statistics while here the transformation (2.25) appears as an exceptional scaling at the critical point $a = 0$ and Ω has non-Gaussian statistics.

The result (2.25) can be adjusted making use of an exact result for the process $y(t)$. It follows from (2.8)-(2.9), with $a = 0$, that $< y(t) >$ satisfies

$$d_t^2 < y(t) >= cos2(b\epsilon)^{1/2}t \tag{2.27}$$

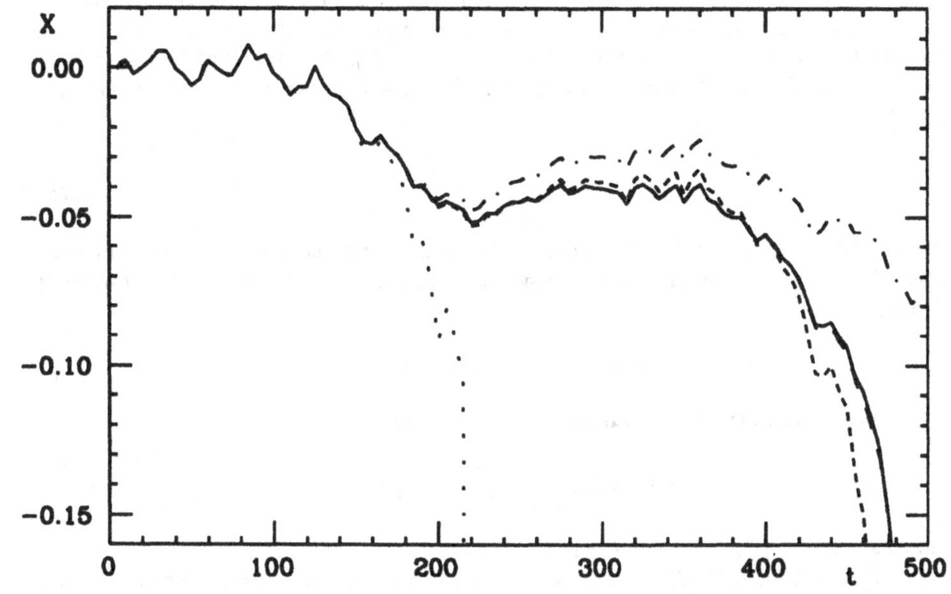

Fig. 1. Different stochastic paths for a given realization of the noise. Solid line corresponds to a simulation of the exact process (2.1). Dotdashed line corresponds to a Wienner process. Dots corresponds to the scaling approach analogous to (2.13) (see text), small dashed line corresponds to the approximation (2.23) and long dashed line to a second order approximation. We take b=c=1 and $\varepsilon=10^{-6}$. The time step of integration is $\eta=0.01$. In Fig. 2 averages are over 10000 trajectories.

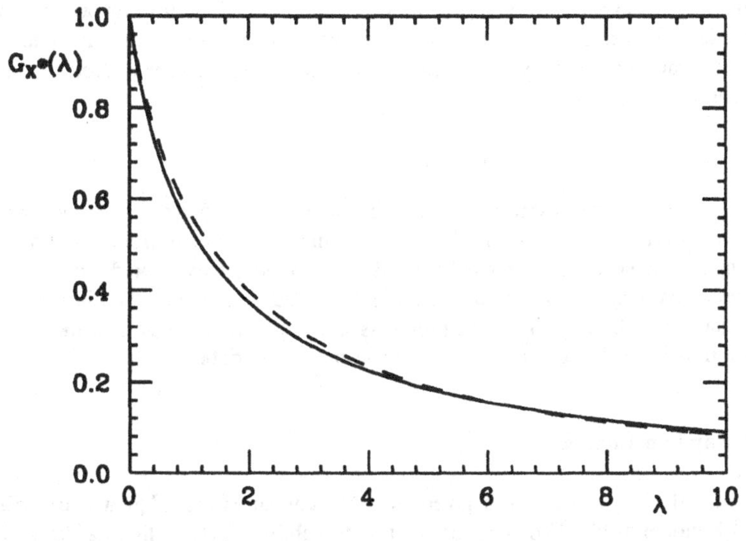

Fig. 2. Generating function $G_{X^*}(\lambda)$. Solid line has been calculated from simulation of the exact process (1.1) with x*=0.3. Long dashed line corresponds to the analytical result $\left(\cosh\left(2\,\delta\sqrt{\lambda}\,\right)\right)^{-1/2}$ (see text).

so that $< y(t_m) >= 0$ for $t_m = \pi/4(b\epsilon)^{1/2}$. The average of the approximate process $y(t)$ used in (2.23) vanishes at $t'_m = \delta t_m$ with $\delta = 2\sqrt{2}/\pi$. We can improve our approximation including an ad hoc factor δ^2 so that $y(t)$ in (2.23) becomes $y(t) = 1 - 2\epsilon b\delta^2 \int_0^t W^2(s)ds$. With this adjustment the correct value of t_m is obtained in our approximation. The inclusion of the factor δ^2 changes (2.15) into

$$t^* = \left[\frac{1}{2b\epsilon\delta^2\Omega}\right]^{1/2} \tag{2.28}$$

The calculation of the statistics of t^* from (2.28) requires the knowledge of the statistics of Ω. The latter is completely determined by the generating function $G(\lambda)$ for which an exact result is available[4],

$$G(\lambda) \equiv < e^{-\lambda\Omega} >= (cosh 2\lambda^{1/2})^{-1/2} \tag{2.29}$$

The moments of the PTD are obtained from (2.28) in terms of $G(\lambda)$ as

$$< t^{*2n} >= \frac{(2b\epsilon\delta^2)^{-n}}{\Gamma(n)} \int_0^\infty d\lambda \quad \lambda^{n-1} G(\lambda) \tag{2.30}$$

In particular for $n = 1/2$, (2.30) reproduces (2.17) for $a = 0$.

It should be noted that (2.28) identifies a PT which is independent of the reference value R_0. In fact the reference value has been taken $R_0 = \infty$ when setting $y(t^*) = 0$. Since the process $x(t)$ achieves its final value very fast once it has left the vicinity of $x = 0$, our result for the PT is reliable for any reference value away from the vicinity of the marginal state $x = 0$. A stringent test of the accuracy of (2.28) is shown in Fig. 2 in which the generating function $cosh(2\delta\lambda^{1/2})^{-1/2}$ is compared with $< e^{-\lambda/2b\epsilon t^*} >$, where the distribution of t^* is obtained from a numerical simulation of the exact process (2.1).

Up to here we have a good representation of the paths of the escape process (2.23) and a way of calculating the PT statistics (2.30) obtained from this approximation. It turns out that for this problem the transient moments $< x^n(t) >$ can be simply obtained in terms of the statistics of the PT. Given the strong nonlinearity of the escape process a good approximation to calculate ensemble averages is to represent $x(t)$ by[13,4]

$$x^2(t) = x_0^2\theta(t - t^*) \tag{2.31}$$

where x_0 is the final stable state (local minima of the potential V) and $\theta(t - t^*)$ is the Heaviside step function. The individual path is approximated by a jump from $x = 0$ to $x = x_0$ at a random time t^*. The transient moments are then easily calculated as averages over the distribution of the times t^*. The numerical validity of this procedure is discussed elsewhere[4]. The important final result is that we have a scaling result for the moments $< x^n(t) >$ in the sense that their statistics are obtained as a transformation of the statistics of the random variable t^*.

3. Diffusion in Random Chains.

A usual model of diffusion in a chain is given by a Markovian Master Equation describing a one-dimensional Random Walk. The equation for the probability $P_n(t)$ of finding the walker at site n at time t is

	$\rho(\omega)$	$<x^2>$	PTD
MODEL A	$<\omega_n^{-M}> \neq \infty$	$\sim 2\langle \omega^{-1}\rangle^{-1} t$	$\langle T_{n_0}^L \rangle \sim \Omega^A(L, n_0) = \frac{(L+1)^2 - n_0^2}{2}\langle \omega_n^{-1}\rangle$
MODEL B	$\rho(\omega) = 1$ $\omega \in (0, 1)$	$\sim t / \ln t$	$< T_{n_0}^L > = \infty$ $< f_{n_0}^L(t) > \sim \Omega^B(L, n_0)\, t^{-2}$ $\Omega^B = ((L+1)^2 - n_0^2)/2$
MODEL C	$\rho = (1-\alpha)\,\omega^{-\alpha}$ $0 < \alpha < 1$ $\omega \in (0, 1)$	t^δ $\delta = \frac{2(1-\alpha)}{2-\alpha}$	$< T_{n_0}^L > = \infty$ $< f_{n_0}^L(t) > \sim \Omega^C(L, n_0)\, \frac{1-\alpha}{\Gamma(\alpha)}\, t^{\alpha-2}$ $\Omega^C = \Omega^B 2^\alpha \pi (1-\alpha) / (L+1)^\alpha \sin(\pi\alpha)$
SITE-PERCOLATION	$\rho(\omega_n = 0) = 1/2$ $\rho(\omega_n = 1) = 1/2$	t^δ $\delta = 0$	$< T_{n_0}^L > = \infty$ $\int_0^\infty < f_{n_0}^L(t) > dt = p < 1$

TABLE 1

$$\partial_t P_n(t) = \omega_{n+1}^- P_{n+1}(t) + \omega_{n-1}^+ P_{n-1}(t) - (\omega_n^- + \omega_n^+) P_n(t) \tag{3.1}$$

where $\omega_n^\pm$ are the transition probabilities per unit time to jump from site n to site $n \pm 1$. A disordered chain can now be modeled by considering the ω_n as random variables. A general model of a disordered chain, named Random Trap model, considers that the sites are wells of random depth. The walker jumps from the well n to the well $n+1$ or $n-1$ with equal probabilities so that $\omega_n^+ = \omega_n^- = \omega_n$. The transition probabilities ω_n are taken as independent random variables with the same probability distribution $\rho(\omega_n)$. Specific random trap models are defined by specific distributions $\rho(\omega_n)$. It is usual to distinguish between strong and weak disorder by the long time behavior of the mean square displacement[5]. Diffusive motion $< x^2 > \sim t$ is associated with weak disorder while subdiffusive motion $< x^2 > \sim t^\delta, \delta < 1$ is associated with strong disorder. A model of weak disorder is given by model A of Alexander et al[5] in which $\rho(\omega_n)$ is chosen so that the inverse moments $< \omega_n^{-M} >$ are finite. (See Table 1). Examples of models of strong disorder are model C of Alexander et al[5] and a site-percolation model both defined in Table 1. In the site percolation model traps of infinite depth ($\omega_n = 0$) are randomly placed in the chain. An intermediate situation between diffusive and subdiffusive behavior is given by model B obtained as the limit $\alpha = 0$ of model C. Another situation with subdiffusive behavior is given by Sinais model[20]. This differs from Random Trap models in that $\omega_n^+ \neq \omega_n^-$. For this model $< x^2 > \sim (lnt)^4$. The strategy followed here is to characterize diffusion in a Random Chain by the statistics of the FPT to leave a finite interval $[-L, L]$, instead of characterizing it by the long time behavior of $< x^2 >$.The MFPT, $< T_{n_0}^L >$ involves here an average over realizations of the process and an extra average over configurations of disorder. The index $n_0 \epsilon [-L, L]$ indicates the initial condition. It happens that the FPTD might have long time tails so that configurations with small statistical weight dominate the value of $< T_{n_0}^L >$. This leads to situations in which $< T_{n_0}^L > = \infty$ even if the walker leaves the interval with probability one. We can then give an alternative definition of strong disorderd, namely, situations in which $< T_{n_0}^L > = \infty$ for any finite L. This gives a clear cut difference between weak and strong disorder. For example, it is known[14] that $< T_{n_0}^L >$ is finite for Sinai's model while we find $< T_{n_0}^L > = \infty$ for model B. Given this definition we can also characterize the degree of disorder by the form of the divergence. This is given by the tail of the FPTD, $< f_{n_0}^L(t) > \sim t^\nu, t \to \infty$, or alternatively by the small-z behavior of the Laplace transform of the survival probability, $< \hat{F}_{n_0}^L(z) > \sim z^{-\bar{\nu}}$. The averages of $f_{n_0}^L(t)$ and $\hat{F}_{n_0}^L(z)$ are here meant to be taken over the ensemble of configurations defined by $\rho(\omega_n)$.

The calculation of passage time statistics starts from (1.2) written for a given configuration of the ω_n

$$\partial_t F_{n_0}^L(t) = \omega_{n_0}(E^+ + E^- - 2) F_{n_0}^L(t) \tag{3.2}$$

$$F_{n_0}^L = 0 \quad \forall n_0 \notin [-L, L]$$

where $E^\pm$ are shifting operators, $E^\pm F_{n_0}^L = F_{n_0 \pm 1}^L$. For finite L, (3.1) is a linear algebraic equation for the Laplace transform $< \hat{F}_{n_0}^L(z) >$ involving a $(2L+1) \times (2L+1)$ matrix. This can be easily solved and then averaged with $\rho(\omega_n)$. The normalization $F_{n_0}^L(t = 0) = 1$ gives the general asymptotic dependence $\hat{F}_{n_0}^L(z) \sim z^{-1}$ as $z \to \infty$. We are here interested in the opposite limiting behavior $z \to 0$. From this behavior the MFPT, $< T_{n_0}^L >$ and the probability p of escape of the interval are given by

$$< T_{n_0}^L > = < \hat{F}_{n_0}^L(z = 0) > \tag{3.3}$$

$$p = 1 - \lim_{t\to\infty} < F^L_{n_0}(t) >= 1 - \lim_{z\to 0} z < \hat{F}^L_{n_0}(z) > \tag{3.4}$$

For the site-percolation model the small z-behavior of $< \hat{F}^L_{n_0}(z) >$ is easily expressed in terms of p. It is obvious that there are a finite number of configurations in which the walker never leaves the interval, so that $p < 1$ and $< T^L_{n_0} >= \infty$. From (3.3) we then have $< \hat{F}^L_{n_0}(z) >\sim (1-p)z^{-1}$. By numerical exact enumeration of configurations the average with $\rho(\omega_n)$ can be computed exactly. The numerical solution evidentiates this behavior[6] with $p = 5/24, 7/80, 503/13440...$ for $L = 1, 2, 3...$ respectively.

For models A, B and C the small z-behavior is more delicate because in these cases $p = 1$. Again for finite L the problem can be exactly solved from (3.2), and numerical averaging by Monte Carlo sampling of the configurations given by $\rho(\omega_n)$. Results for $< T^L_{n_0} >$ and $< f^L_{n_0}(t) >$ for models A, B, C obtained from the solution for $< \hat{F}^L_{n_0}(z) >$ are given in Table 1. Results for $< \hat{F}^L_{n_0}(z) >$ for models B and C and different values of L and n_0 are shown in Figs 3 and 4. For model A one obtains a finite value of $< \hat{F}^L_{n_0}(z = 0) >= \Omega^A(L, n_0)$ which indicates a finite MFPT and a clear situation of weak disorder. For model B there is a weak divergence of $< \hat{F}^L_{n_0}(z) >\sim \Omega^B(L, n_0)|lnz|$ which indicates a situation of strong disorder within our definition. For model C we also find a divergence given by $< \hat{F}^L_{n_0}(z) >\sim \Omega^C(L, n_0)z^{-\alpha}$. These results identify the type of divergence with exponents independent of the size of the interval and of the initial condition. Such dependence is entirely included in the amplitudes $\Omega(n_0, L)$.This fact implies a scaling form for the small z-behavior of $< \hat{F}_{n_0}(z) >$ in the sense that it is given by a unique function of z when the amplitudes $\Omega(n_0, L)$ are scaled out. This is shown in the inserts of Figs. 3 and 4.

The characteristic divergence and the value of the amplitudes given in Table 1 can be obtained within a calculational scheme devised to solve the average of (3.2). The scheme parallels the one used[8] to calculate the mean-square displacement $< x^2(t) >$ by a perturbation around an Effective Medium. The lowest order calculation in this scheme is also given by a type of Effective Medium Approximation (EMA) in which $< \hat{F}^L_{n_0}(z) >$ satisfies the equation

$$z < \hat{F}^L_{n_0}(z) > -1 = \Gamma(z)(E^+ + E^- - 2) < \hat{F}^L_{n_0}(z) > \tag{3.5}$$

where the kernel $\Gamma(z)$ is defined in a selfconsistent way by

$$\left\langle \frac{\omega_n - \Gamma(z)}{1 - (\omega_n - \Gamma(z))(E^+ + E^- - 2)G_{00}(\Gamma(z), z)} \right\rangle = 0 \tag{3.6}$$

Here $G_{00}(\Gamma(z), z)$ is the Green' s function for a Random Walk with effective rates Γ and absorbing boundary conditions at $n = \pm L$. It is necessary to stress that $\Gamma(z)$ is not the effective rate that appears in the standard EMA[5] for the calculation of the mean square displacement. From (3.5) and (3.6) the results in Table 1 are obtained. The accuracy of these results is clearly established in Figs. 3 and 4. We note that for weak disorder (model A), we obtain the same result than in the absence of disorder with ω_n^{-1} replaced by $< \omega_n^{-1} >$. Within a perturbation scheme around the result given by (3.5) and (3.6) one can check that higher order terms do not modify the asymptotic dependences given in Table 1. The amplitudes Ω^A and Ω^B are also exactly obtained from (3.5)-(3.6), while there is a small modification of Ω^C by higher order contributions which, as seen from Fig.4, is of very little numerical significance.

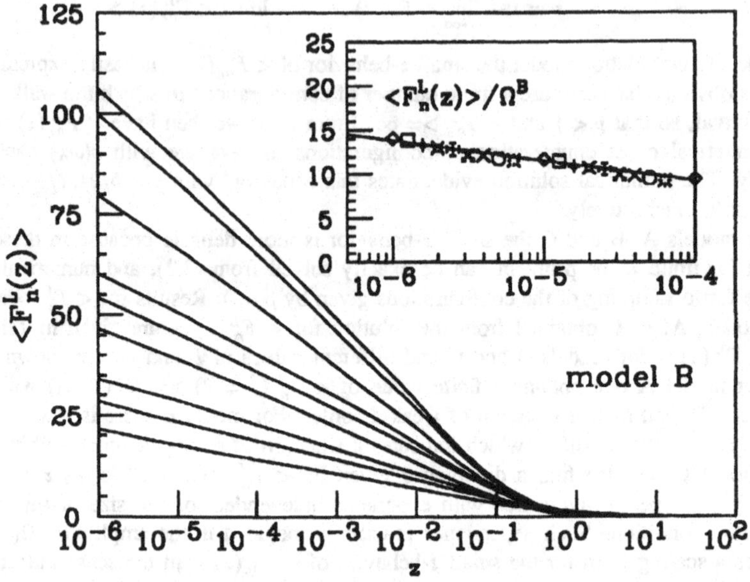

Fig. 3. The Laplace-Transformed survival probability for model B of strong disorder obtained by Monte Carlo averaging over 10^7 configurations the solution of (3.2). Different continuous lines result from different initial conditions n_0 and system sizes L. The insert shows the comparison of the Monte Carlo results (different symbols corresponding to different values of L and n_0) with the leading behavior at small z obtained from Eqs. (3.4)-(3.5) (solid line).

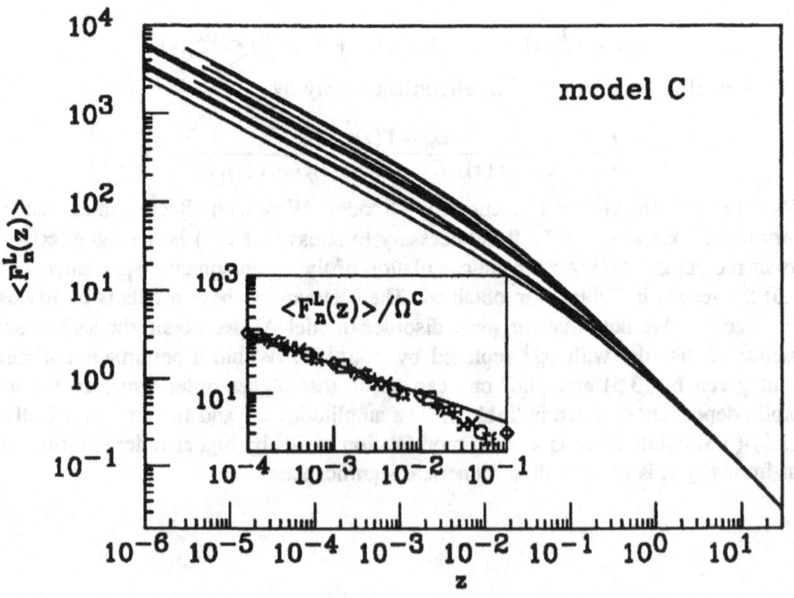

Fig. 4. Same as Fig.3, for model C of strong disorder (see Table 1) with $\alpha = 0.5$. $<F^L_{n_0}(z)>$ diverges as $z^{-\alpha}$ for small z.

ACKNOWLEDGMENTS: Financial support from Dirección General de Investigación Científica y Técnica (DGICYT) Spain, Project No. PB-86-0534 is acknowledged. F.D.P. and M.O.C. also acknowledge financial support from DGICYT during their stay in the Universitat de les Illes Balears.

References.

1- N.G. Van Kampen, *Stochastic Processes in Physics and Chemistry*, North-Holland, Amsterdam, 1981.
R.L. Stratonovich *Topics in the Theory of Random Noise*, Gordon and Breach, New York, 1967.
C.W. Gardiner *Handbook of Stochastic Methods*, Springer-Verlag, Berlin, 1985.
2- M. Suzuki, Adv. Chem. Phys. 46, 195 (1981).
F. de Pasquale and P. Tombesi, Phys. Lett. 72A, 7 (1979).
F. De Pasquale, P. Tartaglia and P. Tombesi, Z. Phys. B43, 353 (1981); Phys. Rev. A25, 466 (1982).
3- P. Colet, M. San Miguel, J. Casademunt and J.M. Sancho, Phys. Rev. A39, 149 (1989).
F.T. Arecchi, A. Politi and L. Ulivi, Nuovo Cimento, 71B, 119 (1982).
D. Sigetti and W. Horsthemke. J. Stat. Phys. 54, 1217 (1989).
4- P. Colet, F. De Pasquale, M.O. Cáceres and M. San Miguel, Phys. Rev. A41, (1990).
5- S. Alexander, J. Bernasconi, W.R. Schneider and R. Orbach, Rev. Mod. Phys. 53, 175 (1981).
J.W. Haus and K.W. Kehr, Phys. Rep. 150, 263 (1987).
S. Havlin and D. Ben-Avraham, Adv. Phys. 36, 695 (1987).
6- E. Hernández-García, M.O. Cáceres and M. San Miguel, Phys. Rev. A-Rapid Comm. 41, (1990).
E. Hernández-García and M.O. Cáceres (unpublished)
7- P. Hänggi and P. Talkner, Phys. Rev. Lett. 51, 2242 (1983).
8- M.A. Rodríguez, E. Hernández-García, L. Pesquera and M. San Miguel, Phys. Rev. B40, 4212 (1989).
9- M.C. Torrent and M. San Miguel, Phys. Rev. A38, 245 (1988).
10- F. De Pasquale, J.M. Sancho, M. San Miguel and P. Tartaglia, Phys. Rev. Lett. 56, 2473 (1986).
11- J.M. Sancho and M. San Miguel, Phys. Rev. A39, 2722 (1989).
12- M.C. Torrent, F. Sagués and M. San Miguel, Phys. Rev. A40, 6662 (1989).
13- A. Valle, L. Pesquera and M.A. Rodríguez, Optics Comm. (1990).
14- Ya G. Sinai, Theory Prob. Its Appl. 27, 247 (1982).

STOCHASTIC PROCESSES DRIVEN BY COLORED NOISE: A PATH INTEGRAL POINT OF VIEW

H.S. WIO*
Centro Atómico Bariloche
8400 Bariloche, Argentina

P. COLET, M. SAN MIGUEL
Department de Física, Universitat de les Illes Balears
E-07071 Palma de Mallorca, Spain

L. PESQUERA and M.A. RODRIGUEZ
Departamento de Física Moderna, Universidad de Cantabria
E-39005 Santander, Spain

1. INTRODUCTION

The study of stochastic differential equations driven by colored noise is a recurrent topic in many contexts of physics and other sciences. In its simplest form the problem can be defined considering a relevant variable q(t) which satisfies a stochastic differential equation of the form

$$\partial_t q(t) = f(q(t)) + \xi(t) \tag{1}$$

where f is a general function of q. The name of colored noise indicates that the stochastic process $\xi(t)$ has not a flat white spectrum. Here we will restrict ourselves to the case in which $\xi(t)$ is an Ornstein-Uhlenbeck process, that is, a Gaussian process of zero mean and correlation

$$C(t,t') \equiv \langle\xi(t)\xi(t')\rangle = (D/\tau)\exp(-|t-t'|/\tau) \tag{2}$$

The parameter D measures the noise intensity and τ its correlation time. The problem is the calculation of the statistical properties of the process q(t) defined by (1)-(2) and specified initial conditions. The difficulty of the colored noise problem is due to the nonvanishing value of τ. This implies that the process q(t) is nonMarkovian, so that well known techniques to deal with continuous Markow processes cannot be applied. In the limit $\tau \to 0$, $\xi(t)$ becomes a white noise

* Member of CONICET, Argentina.

E. Tirapegui and W. Zeller (eds.), Instabilities and Nonequilibrium Structures III, 157–169.

$$\langle\xi(t)\xi(t')\rangle \to 2D\delta(t-t') \tag{3}$$

and the transition probability $P(q,t|q_o,t_o)$ obeys the Markovian Fokker-Planck equation

$$\partial_t P(q,t|q_o t_o) = -\partial_q f(q)\, P(q,t|q_o,t_o) + D\, \partial_q^2 P(q,t|q_o,t_o) \tag{4}$$

As the contributions of Stratonovich (1962) and Van Kampen (1974) show, the problem of dealing with nonMarkovian processes driven by a colored noise is by no means new. However, it is in the last decade that this field has become more fashionable, mainly in connection with external noise problems, and at the same time rather controversial, particularly in connection with the calculation of escape times in bistable systems. The recent literature in this problem is very extense and it usually contains statements about confusion and controversy (Doering *et al.*, 1988; Fox, 1988; Grigolini *et al.*, 1988; Hänggi *et al.*, 1988; Jung and Hänggi, 1988; Leiber *et al.*, 1988; Peacock-López *et al.*, 1988; Tsironis and Grigolini, 1988),however a clear and useful theoretical understanding is yet to be achieved. Many proposed theories are based on expansions in the parameters D and τ. A basic difficulty is to establish the validity of different truncations or resummations of such expansions. Early analogical (Sancho *et al.*, 1982a) and numerical (Sancho *et al.*, 1982b) simulations were useful to check general qualitative predictions of these expansions. However, it is now clear (Jung and Hänggi, 1988) that the results of more recent simulations have often added to the confusion, either because of lack of accuracy or because the values of the parameters used were unappropriate. In view of this general situation it is clear the need of exact results and methods which do not use τ as an expansion parameter. One of such methods is the path integral formulation. The exact result for the action integral in a path integral formulation was obtained some years ago (Pesquera *et al.*, 1983), but it passed rather unnoticed in the literature. A remarkable fact of this formulation is that the action integral does not contain any terms involving powers of τ larger than τ^2. As a consequence, the difficulties associated with τ-expansions are, in principle, bypassed in this formulation. The path integral formulation proposed by Pesquera *et al.*, (1983) is in the configuration q-space, while other very recent formulations (Förster and Mikhailov, 1988; Luciani and Verga, 1988; Hänggi, 1989) introduce a phase-space representation. An exception is the formulation of Bray and McKane (1989). The advantage of a configuration space representation is that the process is discussed in the space of physical interest without the introduction of additional unphysical variables.

We review here the path integral formulation of the stochastic differential equation (1). We discuss the origin and contents of the action integral featuring in such formulation with emphasis on aspects related to the nonMarkovian nature of the process. We also consider the calculation of the stationary probability density of the process q(t) (Colet *et al.*, 1989) as an example of the posibilities offered by this formulation to perform practical calculations. Other statistical properties of the process $q(t)$, such as steady-state correlation functions, relaxation times and transient dynamics properties are discussed by more conventional methods by Sancho and San Miguel (1989). A path integral calculation of such properties is also of clear interest. We start discussing several routes to the path integral formula for the transition probability. Alternative derivations are helpful to understand the origen of the simple τ-dependence of the action integral and the role of initial conditions and preparation of the process. Afterwards we show how a Markovian Fokker Planck approximation for the process $q(t)$ can be obtained from the path integral formulation. A more detailed discussion of the derivation, from the path integral formulation, of approximate equations for the probability density is given elsewhere (Wio *et al.*, 1989). Finally we consider the direct calculation of the stationary probability density from the path integral formulation. The variational problem involved is studied from two points of view. The first one is the direct calculation of the minimizing path and the second a Hamilton-Jacobi type formulation.

2. ROUTES TO THE PATH INTEGRAL

There are several equivalent and alternative derivations of the path integral expression for the transition probability $P(q,t|t_o,t_o)$. The different possibilities are discussed in more detail elsewhere (Wio *et al.*, 1989). A first possible starting points is the original stochastic differential equation (1). This gives the process $q(t)$ as a functional of $\xi(t)$. Inverting this relation and using the known characteristic functional of the Gaussian process $\xi(t)$ the desired result follows (Pesquera *et al.*, 1983). However, the role of initial conditions and preparation effects is not particularly transparent in this approach. This is more easily discussed taking as a starting point the path integral representation of a Markovian process defined in the enlarged space (q,ξ). Indeed, $\xi(t)$ obeys the stochastic differential equation

$$\partial_t \xi(t) = -\tau^{-1}\xi(t) + \tau^{-1}\eta(t) \tag{5}$$

where $\eta(t)$ is a Gaussian white noise with correlation $\langle\eta(t)\eta(t')\rangle = 2D\,\delta(t-t')$. The transition probability $P(q,\xi,t|q_o,\xi_o,t_o)$ of the two-variable Markovian process (q,ξ) obeys the Fokker-Planck equation

$$\partial_t P(q,\xi,t|q_o,\xi_o,t_o) = \left\{-\partial_q(f(q)+\xi)+\tau^{-1}\partial_\xi \xi+D\partial_\xi^2\right\}P(q,\xi,t|q_o,\xi_o,t_o) \tag{6}$$

The path integral representation of (6) is standard (Langouche *et al.*, 1982). The only difficulty is that the diffusion matrix in (6) is singular since $q(t)$ obeys a deterministic equation in the (q,ξ)-space. The path integral is then defined in a phase obtained by introducing additional conjugate variables $\hat{q}$ and $\hat{\xi}$. In the prepoint discretization:

$$P(q,\xi,t|q_o,\xi_o,t_o)= \int_{(q_o,\xi_o)}^{(q,\xi)} D[q(s)]D[\hat{q}(s)]D[\xi(s)]D[\hat{\xi}(s)]\exp(-S) \quad (7)$$

where the integral is over paths defined in the interval (t,t_o) with fixed initial (q_o,ξ_o) and final (q,ξ) values. The action integral is given by

$$S = \int_{t_o}^{t} ds\left\{i\ \dot{q}(s)\hat{q}(s) + i\dot{\xi}(s)\ \hat{\xi}(s) - H\right\} \quad (8)$$

and the Hamiltonian-like function H is,

$$H = i\hat{q}(f(q)+\xi) - i\tau^{-1}\hat{\xi}\xi - D\tau^{-2}(i\hat{\xi})^2 \quad (9)$$

At this point it is interesting to note that a Legendre transformation of H gives

$$L = iq\hat{q} + i\dot{\xi}\hat{\xi} - H = (\tau^2/4D)(\dot{\xi} + (\xi/\tau))^2 \quad (10)$$

L plays the role of a Lagrangian and if ξ and $\dot{\xi}$ are substituted in (10), as obtained from (1), one obtains the correct Lagrangian (see (17)), except for important boundary terms which appear in the action integral.

To obtain the path integral representation in configuration space one has to carry out explicitly the integrals over $\hat{q}(s)$, $\xi(s)$ and $\hat{\xi}(s)$ in (7). This can be done in six different but equivalent ways depending on the order of integration. We consider here two particular ways. In the first one the origin of the simple dependence on τ of the action integral becomes clear, while the second permits a simpler discussion of preparation effects. In the first case we integrate in the following order $\hat{\xi}(s)$, $\xi(s)$, $\hat{q}(s)$. Integrating over $\hat{\xi}(s)$, $\xi(s)$, over all possible final values $\xi(t) = \xi$ and also over $\xi(t_o) = \xi_o$ with the stationary distribution $P_{st}(\xi_o)$ of the process (5) we obtain:

$$P(q,t|q_o,t_o) = \int_{q_o}^{q} D[q(s)]D[\hat{q}(s)]\exp(-A) \quad (11)$$

where

$$A = -i\int_{t_o}^{t} ds\ \hat{q}(s)[\dot{q}(s)-f(q(s))]+ \frac{1}{2}\int ds\ ds'\hat{q}(s)\ C(s,s')\hat{q}\ (s') \quad (12)$$

Equations (11)-(12) give a phase-space path integral representation.

They reproduce a result by Phythian (1977) for general Gaussian processes $\xi(t)$. The crucial point to obtain the configuration-space path integral is the Gaussian integral over $\hat{q}(s)$ which involves the inversion of the matrix C(s,s'). The origin of the familiar τ-expansion approximations (Sancho *et al.*, 1982b) is the expansion

$$C(s,s') = (D/\tau)\exp - (|s-s'|/\tau) = 2D \sum_{n=0}^{\infty} \tau^n \delta^n (s-s'), \quad s>s' \tag{13}$$

The inverse R(s,s') of C(s,s') is defined by

$$\int_{t_o}^{t} ds' C(s,s') R(s',s'') = \delta(s-s') \tag{14}$$

It turns out that (Wio *et al.*, 1989)

$$\begin{aligned} R(s,s') = {} & (2D)^{-1}\left[\delta(s-s') - \tau^2\delta''(s-s')\right] + \\ & + (\tau/D)\left[\delta(s'-t_o)\delta(s''-t_o)+\delta(s'-t)\delta(s''-t)\right]+ \\ & + (\tau^2/D)\left[\delta'(s'-t_o)\delta(s''-t_o)-\delta'(s'-t)\delta(s''-t)\right]+ \end{aligned} \tag{15}$$

so that, while C(s,s') has a power series expansion in τ, the associated expansion for R(s,s') is cut at terms of order τ^2. The final result is then

$$P(q,t|q_o,t_o)=\int_{q_o}^{q} D[q(s)]\exp\left[-\int_{t_o}^{t} ds\ L-(\tau/2D)(\dot{q}_o-f(q_o))^2\right] \tag{16}$$

where

$$L = [\tau\ddot{q} + (1-\tau f'(q))\dot{q}-f(q)]^2/4D \tag{17}$$

The nonMarkovian nature of the process q(t) is reflected on two main aspects of (16)-(17). First is the dependence of the Lagrangian-like function on $\ddot{q}$. Second is the appearance of a boundary term in (16) which depends explicitly on the initial condition. This term, as well as the dependence on $\ddot{q}$, dissappear in the Markovian limit $\tau \to 0$. Equations (16)-(17) are an exact result containing all the required information on statistical properties of the process q(t) which depend on a single time t. They are a good starting point for explicit calculations in which τ-expansions can be avoided.

For an explicit discussion of initial conditions and preparation effects we rewrite (7) as

$$P(q,\xi,t|q_o,\xi_o,t_o) = \int D[q(s)]D\left[\hat{q}(s)\right]D[\xi(s)]D\left[\hat{\xi}(s)\right]\delta(q(t)-q) \times$$
$$\times\ \delta(\xi(t)-\xi)\delta(t_o)-q_o)\delta(\xi(t_o)-\xi_o)\exp(-S) \tag{18}$$

where now the integral is over all the paths. We now consider integration in the following order. $\hat{q}(s)$, $\xi(s)$, $\hat{\xi}(s)$. The integration over q(s) yields a delta functional $\delta[\dot{q}(s)-f(q(s))-\xi(s)]$ which makes the integral over $\xi(s)$ immediate. The remaining integral over $\hat{\xi}(s)$ is Gaussian. We finally arrive at

$$P(q,\xi,t|q_o,\ \xi_o,\ t_o)=\int D[q(s)]\delta(\dot{q}(t)-f(q(t))-\xi(t)) \times$$
$$\times\ \delta(\dot{q}(t_o)-f(q(t_o)-\xi_o)\delta(q(t)-q)\delta(q(t_o)-q_o)\left\{\exp - \int ds\ L\right\} \tag{19}$$

where L is given by (17). This is a general result valid for arbitrary preparation of the process at time t_o. Our previous result (16)-(17) is reobtained if one assumes decoupled initial conditions at time to for q(t) and $\xi(t)$, $P(q_o,\xi_o)=P(q_o)P(\xi_o)$, and in addition that $P(\xi_o)$ is the stationary probability density $P_{st}(\xi_o)$ associated with (5). In this case integration of (19) over all possible values of ξ and over initial conditions ξ_o weighted with $P_{st}(\xi_o)$ reproduce (16)-(17).

We have seen that the explicit boundary terms in (16) reflect preparation effects which are crucial in nonMarkovian processes. In this connection it is worth noticing that our formulation is based on the standard technique of considering paths defined in the interval (t_o,t). For Markovian processes an equivalent formulation in terms of paths defined for the time interval $(-\infty,\infty)$ can be given. Such a procedure is questionable for nonMarkovian processes and neglectes boundary terms which reflect preparation effects. In particular, the recent configuration-space path integral formulation of Bray and McKane (1989), differs from (16)-(17) in boundary terms. It features an action integral that can be obtained as mentioned after (10). It seems to correspond to a probability distribution $P(q_o,\xi_o)$ which is the stationary joint probability distribution of the process (q,ξ). This is the stationary solution of (6). In this sense such formulation cannot describe the standar situation in which the initial condition q_o is chosen arbitrarily.

As a final remark we note that the simple form (17) does not hold in more general cases. For instance consider the case in which a Gaussian white noise $\eta(t)$ is added to the right hand side of (1). The new effective noise $\bar{\xi}(t) = \xi(t)+\eta(t)$ is still Gaussian but it is no longer an Ornstein-Uhlenbeck process. The results is that the inverse of the correlation function of $\bar{\xi}(t)$ has a series expansion cntaining all powers τ^n as in (13). As a consequence the Langrangian-like function contains terms of all orders in τ and time derivatives of q(t) of arbitrary order.

3. MARKOVIAN FOKKER PLANCK APPROXIMATION

The path integral formulation (16)-(17) is a useful starting point to obtain approximate evolution equations for the probability density of the process P(q,t). As particularly useful, it is possible to seek for Fokker-Planck approximations. This can be done from (16)-(17) without relying on τ-expansions. We have already remarked that the nonMarkovian nature of the process is reflected in the presence of terms involving $\ddot{q}$ in (17), as well as in the boundary terms in (16). A possible Markovian approximation consists in neglecting the boundary terms and setting $\ddot{q} = 0$. It turns out that this drastic approximation leads to a Lagrangian associated with a Markovian Fokker Planck equation. Indeed, the comparison of the resulting Lagrangian with the general structure (Langouche *et al.*., 1982) of the path integral formulation of Fokker-Planck equations identifies the following Markovian Fokker-Planck equation

$$\partial_t P(q,t) = -\partial_q\left[(f/(1-\tau f'(q)))+ \frac{1}{2} D'(q)-\partial_q D(q)\right]P(q,t) \tag{20}$$

where

$$D(q) = D/(1-\tau f'(q))^2 \tag{21}$$

Equations (20)-(21) are associated with a stochastic differential equation for the process q(t) driven by Gaussian multiplicative white noise. A most noticeable fact is that (20) coincides with the adiabatic approximation to (1) proposed by Jung and Hänggi (1987). This alternative derivation justifies its dynamical contents as a consistent Markovian approximation to the process. The same idea can be applied to more general cases, as for example when the starting equation (1) contains multiplicative noise or additional white noise. In these cases the derivation of Jung and Hänggi can not be directly applied. A Markovian Fokker-Planck approximation is obtained from the path integral formulation neglecting boundary terms and setting to zero terms of the Langrangian involving time derivatives of q of order higher than one, and also terms involving $\dot{q}^n$ with n>2.

4. STATIONARY PROBABILITY DENSITY

We now addres the question of calculating the stationary distribution $P_{st}(q)$ for the process q(t) starting from the path integral formulation (16)-(17) and without relying in the use of any approximate equation for the time dependent probability density P(q,t). The calculation that we develop here is in the weak noise limit and it is based in the idea of a nonequilibrium potential (Graham and Tel, 1984, 1985, 1986, 1987) defined as

$$\phi(q) \equiv - \lim_{D\to 0} D \ln P_{st} \tag{22}$$

It can be calculated by a minimum principle written in terms of the path integral representation of the stochastic process (Graham and Tel, 1985)

$$\phi(q) = \min \int_{q(-\infty)\in A}^{q(0)=q} dt\, L_o(q(t),\dot{q}(t),\ddot{q}(t)) \tag{23}$$

L_o is defined as the singular part of the Lagrangian in the limit D->0

$$L_o = \lim_{D\to 0} D\, L\, (q(t),\dot{q}(t),\ddot{q}(t) \tag{24}$$

In our case (17), $L_o \equiv DL$. We note that the boundary term coming from initial conditions in (16) does not contribute to $\phi(q)$. In addition, the formula (23) is explicitly written for the case in which the deterministic dynamics (eq.(1) with $D = 0$) has a single attractor. In this case the action integral is minimized over all the paths starting from the attractor **A** at $t = -\infty$ and reaching the point q at time $t = 0$. The case of several coexisting attractors is not explicitly considered here.

There are two possible approaches to calculate ϕ which we discuss separately below. The first is a direct evaluation of (23) with the explicit solution for the minimizing path. The second is to obtain a differential equation for ϕ to be solved a posteriori.

The path minimizing the action integral (16) satisfies the Euler-Lagrange like equations given by

$$\frac{d^2}{dt^2}\frac{\partial L}{\partial \ddot{q}} - \frac{d}{dt}\frac{\partial L}{\partial \dot{q}} + \frac{\partial L}{\partial q} = 0 \tag{25}$$

In our case they explicitly read

$$\tau^2 \dddot{q} - 3\tau^2 f'' \dot{q}\, \ddot{q} - \tau^2 f''' \dot{q}^3 - (1+\tau^2 f'^2)\ddot{q} - \tau^2 f'' f' \dot{q}^2 + ff' = 0 \tag{26}$$

We are interested in minimizing paths with fixed initial and final value for q and whatever initial and final values of $\dot{q}$. A particular enlightening case is the linear problem $f(q) = aq$, $a>0$, for which the exact stationary distribution is known (Sancho and San Miguel, 1989). The solution of (26) for this case is

$$q(t) = A_1 e^{t/\tau} + A_2 e^{-t/\tau} + B_1 e^{at} + B_2 e^{-at} \tag{27}$$

with four undetermined constants A_1, A_2, B_1 and B_2. The choice of a path starting at the attractor $q = 0$ at $t = -\infty$ requires $A_2=B_2=0$. This path reaches at $t = 0$ a point $q = A_1 + B_1$ with velocity $\dot{q} = A_1 + \tau a B_1$. Calculating the action integral along this path we find

$$\int_{-\infty}^{0} dt\ L_o(q,\dot{q},\ddot{q}) = \int_{-\infty}^{0} dt\ \frac{1}{4D}\ [\tau\ddot{q}+(1+\tau a)\dot{q}+aq]^2 = \frac{1}{D}(1+a\tau)(\frac{aq^2}{2}+\frac{\tau\dot{q}^2}{2}) \qquad (28)$$

We now choose the final free value $\dot{q}$ of the path by the minimum action requirement. This is $\dot{q} = 0$ in (28) which leads to the exact stationary distribution $P_{st} = N\ \exp-[((1+a\tau)/D)(aq^2/2)]$.

The general solution of (26) requires four boundary conditions. Two of them are given by the fixed initial and final values of q. The other two boundary conditions follow from the variational problem for the action S in (16) with free boundary values of $\dot{q}$.

$$S = -\int L\ dt - \frac{\tau}{2D}\ [\dot{q}-f(q)]^2 \qquad (29)$$

$\delta S = 0$ requires in addition of the Euler-Lagrange equations (26) the following boundary conditions. At $t = 0$

$$\left(\frac{\partial L}{\partial \ddot{q}}\right)_{t=0} = \tau\left[\tau\ddot{q} + (1-\tau f')\dot{q}-f)\dot{q}-f\right]_{t=0} = 0 \qquad (30)$$

and at $t = -\infty$

$$\left(\frac{\partial L}{\partial \ddot{q}}\right)_{t=0} -\frac{\tau}{D}[\dot{q}-f]_{t=-\infty}=0 \qquad (31)$$

so that

$$\tau\ddot{q}-(1+\tau f')\dot{q}+f\,\Big|_{t=-\infty} = 0 \qquad (32)$$

The boundary conditions used for the linear model are particular cases of (30) and (32). It can be shown that (30) and (32) coincide with the boundary conditions obtained from the equations for the minimizing path in the phase space representation (Luciani and Verga, 1988; Hänggi, 1989).

It is important to note that the differential equations (26) for the minimizing path are independent of boundary terms and preparation effects included in (16). However, the boundary conditions (30), (32) and the calculation of ϕ are not independent of them. A general solution of (26) with (30) and (32) for a general case in certainly nontrivial. Solutions for each particular function f(q) can be attemped. The possibility of using τ as an expansion parameter is not always well defined. For example, for the linear case above, an expansion in τ of the minimizing path is, at best, singular. An alternative to find the solution of (26) which permits to obtain general results is to look for the equation satisfied by ϕ starting from its definition (22).

In the white noise limit $\tau = 0$, it is well known that ϕ satisfies the Hamilton-Jacobi equation of the Hamiltonian dynamics associated with the Fokker-Planck Lagrangian (Graham and Tel, 1984). In our case L_o in (23) is not properly speaking a Lagrangian function. However, the variational problem solved by the Hamilton-Jacobi equation can be generalized to Lagrangian-like functions which depend on time derivatives of q(t) of order higher than one (Buchdahl, 1988). In this generalization one obtains Hamilton-Jacobi like equations for variational problems as the one posed by (23). The basic idea is the introduction of generalized conjugate momenta associated with the time derivatives of q. In our case one introduces momenta Π^0 conjugate to q and Π^1 conjugate to $\hat{q}$ as

$$\Pi^0 \equiv \frac{\partial L_o}{\partial \dot{q}} - \frac{d}{dt}\frac{\partial L_o}{\partial \ddot{q}} \tag{33}$$

$$\Pi^1 \equiv \frac{\partial L_o}{\partial \ddot{q}} \tag{34}$$

and the generalized Hamiltonian becomes

$$H(q,\dot{q}\Pi^0,\Pi^1) = \Pi^0\dot{q} + \Pi^1\ddot{q} - L_o \equiv \dot{q}\Pi^0 + \frac{1}{\tau^2}(\Pi^1)^2 - \frac{1}{\tau}\,[(1-\tau f')\,\dot{q}-f]\Pi^1 \tag{35}$$

The Hamilton-Jacobi-like equation associated with the variational problem (23) is now obtained in the usual way replacing Π^0 by $\partial_q\bar{\phi}$ and Π^1 by $\partial_{\dot{q}}\bar{\phi}$ in the Hamiltonian and equating it to zero. Explicitly we have

$$\dot{q}\,\frac{\partial\bar{\phi}}{\partial q} + \frac{1}{\tau^2}\left(\frac{\partial\bar{\phi}}{\partial\dot{q}}\right)^2 - \frac{1}{\tau}\,[(1-\tau f')\dot{q}-f]\,\frac{\partial\bar{\phi}}{\partial\dot{q}} = 0 \tag{36}$$

Eq.(36) is for a function $\bar{\phi}(q,\dot{q})$. The dependence on $\dot{q}$ comes from the fact that (36) is associated with a variational problem in which the final velocity $\dot{q}$ is fixed. As explicitly seen in the linear case we are interested in the function $\bar{\phi}(q,\dot{q})$ at the value $\dot{q}_0$ which makes $\bar{\phi}$ minimum

$$\phi(q) = \bar{\phi}(q,\dot{q}_0) \;;\; \left(\frac{\partial\bar{\phi}}{\partial\dot{q}}\right)_{\dot{q}=\dot{q}_0} = 0 \tag{37}$$

The important point to notice is that the structure of (36) identifies that precisely $\dot{q} = 0$ is the value $\dot{q}_o$ for which $\partial_{\dot{q}}\bar{\phi} = 0$. As a consequence, a natural approximation seems to be to look for a solution of (36) of a quadratic form

$$\bar{\phi}(q,\dot{q}) = \phi(q) + \dot{q}^2\phi_1(q) \tag{38}$$

Replacing (38) in (36) we obtain the following equation for ϕ and ϕ_1

$$\partial q \; \phi(q) = -2f(q)\phi_1(q)/\tau \tag{39}$$

$$\phi_1(q) = \tau(1-\tau f'(q))/2 \tag{40}$$

This yields the result

$$\phi(q) = -\int f(q)\; dq + \tau\; f^2(q)/2 \tag{41}$$

This form of the potential $\phi(q)$ is well known. It seems to be favored by empirical and numerical considerations (Grigolini *et al.*, 1988) and it has been justified in several ways: It was obtained by an ad hoc exponentiation proposed by Sancho *et al.*, (1982b) to solve a Fokker-Planck equation to first order in τ. The same result was obtained by an equivalent related technique (Horstemke and Lefever, 1984). A more refined justification of this procedure is to look for solutions of such approximate Fokker-Planck equation of the form $\mathbf{P}_{st} = e^{-\phi/D}$ (Schenzle and Tel, 1985). Another justification relies on the fact that this form of ϕ appears in the formal solution of a Fokker-Planck approximation (Fox, 1986) which is intermediate between the small-τ and small-D approximations (Sancho and San Miguel, 1989). In addition it also follows from a τ-expansion calculation in the Hamilton-Jacobi formulation of the two variable Markov process defined by (16) (Schimansky-Geier, 1988). These justifications invoke a small-τ assumption. It also happens that (41) becomes exact in the limit $\tau \to \infty$ (Sancho *et al.*, 1982b) and that it is the result which follows from the Fokker-Planck equation (20). It is rather intringuing to understand the wisdom of the general result (41). the novelty of our calculation above is to show that (41) can be obtained in a natural way without invoking either Markovian nor Fokker-Planck approximations, and without explicit use of τ-expansions. This derivation puts the result (41) on a much safer and clear grounds.

A question that, however, cannot be overlooked is the validity of the quadratic approximation (38). A first consistency requirement from (36) is that $\phi_1(q)>0$ so that $\dot{q} = 0$ is a minimum. This is fulfilled in the cases with a single attractor considered here ($f'< 0$). A detailed analysis of this question requires the consideration of higher order terms in the expansion in powers of $\dot{q}$ in (38). This is discussed elsewhere (Wio *et al.*, 1989). For a general problem the only parameter left is τ, and one finds that the quadratic approxiamtion (38) can be understood as a useful interpolation between the limit τ>>1 and τ<<1. The problem can be analyzed for each particular function f(q) without restoring to considerations on the value of τ. In any case (36) and (37) give an exact interesting formulation of the problem of calculating the potential $\phi(q)$.

ACKNOWLEDGMENTS

Financial support from Directión General de Investigación Científica y Técnica (Spain), Projects PB86-0534 and PB87-0014 is acknowledged.
One of us (H.S.W.) acknowledges the kind hospitality extended to him during his stay at the Department of Física of the Universitat de las Illes Balears, Spain.

REFERENCES

Bray, A.J. and McKane, A.J. 1989, *Phys.Rev.Lett.*62, 493.
Colet, P., Wio, H.S. and San Miguel, M. 1989. *Phys.Rev.A* 39, 6094 (1989).
Doering, C.R., Bagley, R.J., Hagan and Levermore, C.D. 1988. *Phys.Rev.Lett.*60, 2805.
Forster, A. and Mikhailov, A.S. 1988 *Phys.Lett.A* 126, 459.
Fox, R.F. 1986 *Phys.Rev.A* 34, 4525.
Fox, R.F. 1988 *Phys.Rev.A* 37, 911.
Graham, R. and Tel, T. 1984. *J.Stat.Phys.* 35, 729.
Graham, R. and Tel, T. 1985. *Phys.Rev.A* 31, 1109.
Graham, R. and Tel, T. 1986. *Phys.Rev.A* 33, 1322.
Graham, R. and Tel, T. 1987. *Phys.Rev.A* 35, 1328.
Grigolini, P., Lugiato, L., Mannella, R., McClintock, P.V.E., Merri, M. and Pernigo, M. 1988. *Phys.Rev.A* 38, 1966.
Hänggi, P. 1989. *Z.Phys.B* 75, 275.
Hänggi, P., Jung, P., and Talkner, P. 1988. *Phys.Rev.Lett*, 60, 2804.
Horsthemke, W. and Lefever, R. 1984. *Noise Induced Transitions*, Berlin: Springer.
Jung, P. and Hänggi, P. 1987. *Phys.Rev.A* 35, 4464.
Jung, P. and Hänggi, P. 1988. *Phys.Rev.Lett.*61, 11.
Langouche, F., Roekaerts, D. and Tirapegui, E. 1982. *Functional Integration and Semiclassical Expansions*. Dordrecht: Reichl Publ.Co.
Leiber, T., Marchesoni, F. and Risken, H. 1988. *Phys.Rev.A.*38, 983.
Luciani, J.F. and Verga, A.D. 1988 *J.Stat.Phys.*50, 567.
Peacock-Lopez, E., West, B.J. and Lindenberg, K. 1988 *Phys.Rev.A* 37, 3530.
Pesquera, L., Rodríguez, M. and Santos, E. 1983. *Phys.Lett.A* 94, 287.
Phytian, R. 1977. *J.Phys.A* 10, 777.
Sancho, J.M. and San Miguel, M. 1989. In *Noise in Nonlinear Dynamical Systems* (F.Moss and P.V.E. McClintock, eds.) vol.1, pp. 72-109. Cambridge: Cambridge University Press.
Sancho, J.M., San Miguel, M., Yamazaki, H. and Kawakubo, T. 1982a. *Physica A* 116, 560.
Sancho, J.M., San Miguel, M., Katz, S. and Gunton, J.D. 1982b. *Phys.Rev.A* 26, 1589.

Schenzle, A. and Tel, T. 1985. *Phys.Rev.A* 32, 596.
Stratonovich, R.L. 1963. *Introduction to the Theory of Random Noise*, vol.1. New York: Gordon and Breach.
Tsironis, G.P. and Grigolini, P. 1988. *Phys.Rev.Lett.*61, 7.
Van Kampen, N.G. 1976. *Phys.Reports* 24, 171.
Wio, H.S., Colet, P., San Miguel, M., Pesquera, L. and Rodriguez, M. 1989. *Phys.Rev.A* 40, 7312.

Sauermann, [illegible] and [illegible] 198[illegible] Phys. Rev. A 32, [illegible]
Stratonovich, R. L. 1963 Introduction to the Theory of Random Noise, vol. I, New York: Gordon and Breach.
Tsironis, G. and Grigolini, P. 1988 Phys. Rev. Lett. 61, 7[illegible]
Van Kampen, N. G. 1976 Phys. Reports 24, 171.
Zhu, S., [illegible] Cole, [illegible] San Miguel, M., [illegible] and [illegible] 1989 Phys. Rev. A 40, 7312.

EFFECT OF NOISE ON A HOPF BIFURCATION

R. C. BUCETA
Departamento de Física, Facultad de Ciencias Exactas y Naturales, Universidad Nacional de Mar del Plata.
Funes 3350, (7600) Mar del Plata, Argentina.

E. TIRAPEGUI
Departamento de Física, Facultad de Ciencias Físicas y Matemáticas, Universidad de Chile.
Casilla 487-3, Santiago, Chile.

ABSTRACT. The normal form of a system making a Hopf bifurcation generically breaks rotational invariance $z \to z\exp(i\delta)$. We study the consequences of this effect on the correlation function and show that observable new peaks appear in the spectral density $\Delta(k)$.

We consider here a stochastic differential equation in $[0, 2\pi]$ of the form

$$\dot{\theta} = \omega + \varepsilon^{1/2}\, g(\theta)\, F(t)\,, \tag{1}$$

where ω is a frequency, $g(\theta)$ a periodic function, $F(t)$ a given white or colored noise and ε a parameter measuring the intensity of the noise. The motivation is the problem of the effect of noise on a Hopf bifurcation. It has been shown that the normal form of a system making a Hopf bifurcation [1] [2] and forced by external noise is

$$\dot{z} = (\mu + i\omega_0)\, z - (\alpha + i\beta)\, |z|^2\, z + \varepsilon^{1/2} f(t; z, \bar{z})\,, \tag{2}$$

where the noise term $\varepsilon^{1/2} f(t; z, \bar{z})$ breaks in general the rotational invariance $z \to z\exp(i\delta)$ of the normal form without noise. Take for example $f(t; z, \bar{z}) = iF(t)$ in (2) where $F(t)$ is the given noise which we take real. Putting $z = r\exp(i\theta)$ one has

$$\begin{aligned} \dot{r} &= \mu\, r - \alpha\, r^3 + \varepsilon^{1/2} \sin\theta\, F(t) \\ \dot{\theta} &= (\omega_0 - \beta\, r^2) + \varepsilon^{1/2}\, \tfrac{1}{r} \cos\theta\, F(t) \end{aligned} \,. \tag{3}$$

If we neglect the fluctuations of r and put r equal to its mean value $(\mu/\alpha)^{1/2}$ we obtain for θ equation (1) with $\omega = \omega_0 - \beta\mu/\alpha$ and $g(\theta) = \cos\theta$. We see then that in general the absence of rotational invariance in (2) will lead to equations of type (1) which break the invariance $\theta \to \theta + \delta$. The case $g(\theta) =$ *constant* corresponds to rotational invariance of

E. Tirapegui and W. Zeller (eds.), Instabilities and Nonequilibrium Structures III, 171–176.

(2), for example $f(t; z, \bar{z}) = zF(t)$. The complete study of (2) will be reported elsewhere and we shall limit ourselves here to consider equation (1) with $g(\theta) = \cos\theta$ and to study there the correlation function $\langle \cos\theta(t)\cos\theta(0)\rangle$ in the stationary state which corresponds in (2) to the study $\langle x(t)x(0)\rangle$, $x = r\cos\theta$, $r = (\mu/\alpha)^{1/2}$. In order to study (1) we shall construct a perturbation expansion using functional integral methods introduced in [3] and extended here to stochastic process taking values in compact regions (here the interval $[0, 2\pi]$). Similar techniques have been used in [4] for a related problem. We consider then equation (1) with $g(\theta) = \cos\theta$ and we make there the change of variables $\theta(t) = \omega t + \varphi(t)$ which gives

$$\dot{\varphi} = \varepsilon^{1/2}\cos(\omega t + \varphi(t))\, F(t)\,. \tag{4}$$

We take for $F(t)$ a Gaussian process with zero mean and correlation function

$$\langle F(t)F(t')\rangle = \Delta(t-t') = \frac{c}{2\gamma}e^{-\gamma|t-t'|}\,. \tag{5}$$

This colored noise can be realized adding to (1) a second equation

$$\dot{F}(t) = -\gamma F(t) + c^{1/2}\xi(t)\,, \tag{6}$$

where $\xi(t)$ is a δ–correlated white noise of zero mean and γ and c are positive constants, since then the stationary correlation of $F(t)$ is (5). When we consider the case of white noise in (1) we have to replace $\Delta(t-t')$ by $\delta(t-t')$ which corresponds to put $c = \gamma^2$ and to take the limit $\gamma \to +\infty$.

We set up now our perturbation scheme. Let $G[\varphi]$ be a periodic functional of $\varphi(t)$, $t \in [t_0, T]$, i.e. $G[\varphi(t) + 2\pi n(t)] = G[\varphi(t)]$, $n(t) \in Z$, and let $\varphi_F(t)$ be the solution of (1) for a given fixed function $F(t)$ and initial condition $\varphi(t_0) = \varphi_0$. Then we have the following identity [3]

$$G[\varphi_F] = \int_{\gamma_1(0)} \mathcal{D}\varphi \prod_{t\in[t_0,T]} \delta(\dot{\varphi} - \varepsilon^{1/2}\cos(\omega t + \varphi(t))\, F(t))\; G[\varphi]\; \delta(\varphi(t_0) - \varphi_0)\,, \tag{7}$$

where $\gamma_1(0)$ stands for prepoint discretization in the functional integral in (7) and this is the reason why no Jacobian appears in (7) [5]. The integration $\mathcal{D}\varphi = \prod_t d\varphi(t)$ is for $-\infty < \varphi(t) < \infty$. Using the functional δ–function $\prod_t \delta(K(t)) = \int \mathcal{D}p\; exp[i\int dt\; p(t)\, K(t)]$ in (7) we obtain (we omit $\gamma_1(0)$ from now on since we shall always work in the prepoint discretization)

$$G[\varphi_F] = \int_{\varphi(t_0)=\varphi_0} \mathcal{D}\varphi\, \mathcal{D}p \exp\left[i\int_{t_0}^{T} dt\;\; p[\dot{\varphi} - \varepsilon^{1/2}\cos(\omega t + \varphi(t))\, F(t)]\right] G[\varphi]\,. \tag{8}$$

The average $\langle G[\varphi_F]\rangle$ of (8) over the different realizations of $F(t)$ is now realized using the probability

$$\mathcal{P}[F]\,\mathcal{D}F \approx \exp\left[-\frac{1}{2}\int_{t_0}^{T} dt \int_{t_0}^{T} dt'\; F(t)\,\Delta^{-1}(t-t')\,F(t')\right] \prod_{t\in[t_0,T]} dF(t)\,, \tag{9}$$

where Δ^{-1} is defined by $\int dt' \Delta(t_1 - t')\Delta^{-1}(t' - t_2) = \delta(t_1 - t_2)$. From (8) and (9) we obtain after doing the Gaussian integral over $F(t)$ the result

$$\begin{aligned} \langle G[\varphi_F]\rangle &= \int_{\varphi(t_0)=\varphi_0} \mathcal{D}\varphi\,\mathcal{D}p\;G[\varphi]\;\exp\Bigg[i\int_{t_0}^{T} dt\;p\,\dot\varphi \\ &\quad - \frac{\varepsilon}{2}\int_{t_0}^{T} dt \int_{t_0}^{T} dt'\;p(t)\,\cos(\omega t + \varphi(t))\,\Delta(t-t')\,p(t')\,\cos(\omega t' + \varphi(t'))\Bigg] . \end{aligned} \tag{10}$$

The generating functional is defined by (putting $g(t) \equiv \cos[\omega t + \varphi(t)]$)

$$\begin{aligned} \mathcal{Z}[j, j^*] = \int_{\varphi(t_0)=\varphi_0} \mathcal{D}\varphi\,\mathcal{D}p\,\exp\Bigg[i\int_{t_0}^{T} dt\,p\dot\varphi &- \frac{\varepsilon}{2}\int_{t_0}^{T} dt\,dt'\;p(t)g(t)\Delta(t-t')p(t')g(t') \\ &+ i\int_{t_0}^{T} dt\,[j(t)\varphi(t) + j^*(t)p(t)]\Bigg] , \end{aligned} \tag{11}$$

and this functional only makes sense for sources $j(t)$ of the form

$$j(t) = \sum_i n_i\,\delta(t - t_i)\,, \qquad n_i \in Z\,, \tag{12}$$

since the functional $G[\varphi]$ in (10), which here is $\exp[i\int dt j(t)\varphi(t)]$, must be periodic. Correlation functions are calculated using (10) for different choices of $G[\varphi]$. We are interested in

$$\langle\cos\theta(t_1)\cos\theta(t_2)\rangle = \frac{1}{2}Re\left[\langle e^{i[\theta(t_1)+\theta(t_2)]}\rangle + \langle e^{i[\theta(t_1)-\theta(t_2)]}\rangle\right] , \tag{13}$$

and one easily checks

$$\langle e^{i[\theta(t_1)\pm\theta(t_2)]}\rangle = e^{i\omega(t_1\pm t_2)}\,\mathcal{Z}[j(\bullet) = \delta(\bullet - t_1) \pm \delta(\bullet - t_2), j^* = 0]\,. \tag{14}$$

The perturbation expansion for $\mathcal{Z}[j, j^*]$ is set up writing $g(t)g(t')$ in the form

$$\frac{1}{2}\left[\cos(\omega(t+t') + \varphi(t) + \varphi(t')) + \cos(\omega(t-t') + \varphi(t) - \varphi(t')) - 2\alpha\right] + \alpha\,, \tag{15}$$

where α is a parameter which will be fixed in the expansion by a suitable condition as we shall explain later. We remark that in the white noise case $\Delta(t) \to \delta(t)$ one has

$$g(t)g(t')\Delta(t-t') \longrightarrow \frac{1}{2}\cos[2(\omega t + \varphi(t))] + \frac{1}{2} \tag{16}$$

and we take there $\alpha = 1/2$. We calculate $\mathcal{Z}[j, j^*]$ given by (11) making an expansion in $K(t, t') \equiv g(t)g(t') - \alpha$ (the square bracket in (15)). One has

$$\begin{aligned} \mathcal{Z}[j, j^*] &= \sum_{n=0}^{\infty}\frac{1}{n!}\left(-\frac{\varepsilon}{2}\right)^n \prod_{j=1}^{n}\int_{t_0}^{T} dt_j\,dt_j'\;\Delta(t_j - t_j') \\ &\quad\times\int_{\varphi(t_0)=\varphi_0}\mathcal{D}\varphi\,\mathcal{D}p\;p(t_j)\,p(t_j')\,K(t_j, t_j')\;\exp\Bigg[i\int_{t_0}^{T} dt\;p\dot\varphi \\ &\quad - \frac{\alpha\varepsilon}{2}\int_{t_0}^{T} dt\,dt'\;p(t)\,\Delta(t-t')\,p(t') + i\int_{t_0}^{T} dt\;[j(t)\varphi(t) + j^*(t)p(t)]\Bigg] . \end{aligned} \tag{17}$$

We define the "free" generating functional

$$\mathcal{Z}_0[j,j^*] = \int_{\varphi(t_0)=\varphi_0} \mathcal{D}\varphi\,\mathcal{D}p\ \exp\left[i\int_{t_0}^{T} dt\ p\dot{\varphi} \ - \ \frac{\alpha\varepsilon}{2}\int_{t_0}^{T} dt\,dt'\ p(t)\Delta(t-t')p(t') \right.$$
$$\left. + \ i\int_{t_0}^{T} dt\,[j(t)\varphi(t) + j^*(t)p(t)]\right] ,(18)$$

which can be calculated exactly since it is a Gaussian functional integral. In terms of $\mathcal{Z}_0[j,j^*]$ all terms in (17) can be calculated (for a similar approach to treat colored noise we refer to [6] [7]). As an example we take the term with $n = 1$ in (17) and consider the contribution of the first term in $K(t,t')$ (see (15)) which we write

$$K^{(1)}(t,t') = \frac{1}{4}\{\exp[i(\omega(t+t') + \varphi(t) + \varphi(t'))] + c.c.\}\,, \tag{19}$$

The contribution to $\mathcal{Z}[j,j^*]$ of the first term in (19) (for $n = 1$ in (17)) is then simply

$$\left(-\frac{\varepsilon}{8}\right)\int_{t_0}^{T} dt_1\,dt_1'\quad \Delta(t_1-t_1')\ e^{i\omega(t_1+t_1')}\ \frac{1}{i}\frac{\delta}{\delta j^*(t_1)}\ \frac{1}{i}\frac{\delta}{\delta j^*(t_1')}$$
$$\times\ \mathcal{Z}_0[j(\bullet)\to j(\bullet) + \delta(\bullet - t_1) + \delta(\bullet - t_1'), j^*]\,, \tag{20}$$

and all other terms in (17) can be calculated in the same way. The contribution of (20) to $\langle\exp[i(\theta(\tau_1)\pm\theta(\tau_2))]\rangle$ is then obtained replacing there $j(\bullet) = \delta(\bullet-\tau_1)\pm\delta(\bullet-\tau_2)$, $j^* = 0$, according to (14). The Gaussian functional integral in (19) has the value [3] ($H(t)$ is the Heaviside step function $H(t) = 0, t < 0$, and $H(t) = 1, t > 0$)

$$\mathcal{Z}_0[j,j^*] = \exp\left[i\varphi_0\int_{t_0}^{T} dt\ j(t) \ - \ i\int_{t_0}^{T} dt\,dt'\ j(t)\,H(t-t')\,j^*(t') \right.$$
$$\left. - \ \frac{\alpha\varepsilon}{2}\int_{t_0}^{T} dt\,dt'\ j(t)\,D(t,t')\,j(t')\right]\,, \tag{21}$$

$$D(t,t') = \min(t,t') - t_0 - \Delta(t-t') + \frac{1}{2\gamma}(e^{-\gamma t} + e^{-\gamma t'})e^{\gamma t_0}\,. \tag{22}$$

We are interested in the stationary state which involves the limit $t_0 \to -\infty$. From (22) we see that we have in (21) a term

$$\exp\left\{\frac{\alpha\varepsilon}{2}\ t_0\left[\int_{t_0}^{T} dt\ j(t)\right]^2\right\}, \tag{23}$$

and consequently the limit $t_0 \to -\infty$ vanishes unless $\int dt\ j(t) = 0$. We see then that for the stationary state we have to consider in the perturbation expansion only terms with $j(t) = \sum_i n_i\delta(t-t_i)$, $\sum_i n_i = 0$, and in this case $\mathcal{Z}_0$ reduces to its stationary value $\mathcal{Z}_0^s$ given by

$$\mathcal{Z}_0^s[j,j^*] = \exp\left[-i\int_{-\infty}^{T} dt\,dt'\ j(t)\,H(t-t')\,j^*(t') - \frac{\alpha\varepsilon}{2}\int_{-\infty}^{T} dt\,dt'\ j(t)\,D^s(t,t')\,j(t')\right], \tag{24}$$

$$D^{s}(t,t') = \min(t,t') - \Delta(t-t') . \tag{25}$$

We use now this perturbation expansion to calculate the correlation function $C(t) = \langle \cos\theta(t) \cos\theta(0) \rangle$ in the stationary state and its Fourier transform $\Delta(k)$ defined by

$$\Delta(k) = \frac{2}{\pi} \int_0^\infty dt \; C(t) \; \cos(kt) . \tag{26}$$

In order to exhibit the relevant parameters in the problem we scale the time to $t' = \gamma t$, $\theta'(t') = \theta(t)$, $F'(t') = (\varepsilon^{1/2}\gamma^{-1})F(t)$. Then the basic equation (1) becomes

$$\frac{d\theta'(t')}{dt'} = \Omega + g(\theta'(t'))F'(t') , \quad \Omega = \frac{\omega}{\gamma} , \tag{27}$$

and

$$\langle F'(t_1')F'(t_2')\rangle = \Delta'(t_1' - t_2') = \frac{1}{2\Gamma^3} e^{-|t_1'-t_2'|} , \quad \Gamma = \frac{\gamma}{(\varepsilon c)^{1/3}} . \tag{28}$$

Inspection of the expansion of $C'(t') = \langle \cos\theta'(t') \cos\theta'(0) \rangle$ by the method outlined above shows that

$$C'(t') = \sum_{j=0}^{\infty} [C^{(j)}(\alpha,\Omega,\Gamma,t') \; \cos(j\Omega t') + S^{(j)}(\alpha,\Omega,\Gamma,t') \; \sin(j\Omega t')] , \tag{29}$$

where the functions $C^{(j)}$ and $S^{(j)}$ do not contain trigonometric functions. Due to this form the Fourier transform $\Delta'(k') = \gamma\Delta(k)$, $k = \gamma k'$, will have in an obvious notation the form $\Delta'(k') = \sum_j \Delta_j'(k')$ where Δ_j' will exhibit a peak at $k' = j\Omega$, $j \geq 1$. We are interested in the case $\Gamma^{-1} \ll 1$ and $\Omega = \mathcal{O}(1)$ which is realized for small intensity of the noise (ε small). The principal contribution to the main peak ($k' = \Omega$) will be

$$\Delta_1'(k' = \Omega) = \frac{1}{2\pi} \frac{1}{4(1+\Omega^2)} \Gamma^3 + \mathcal{O}(\Gamma^{-3}) , \tag{30}$$

and the principal contribution to the peak at $k' = 2\Omega$ (the most important new peak) is

$$\Delta_2'(k' = 2\Omega) = \frac{1}{16\pi} \frac{\Omega^2 - 1}{\Omega^2 + 1} + \mathcal{O}(\Gamma^{-3}) . \tag{31}$$

In this calculations we have fixed the parameter α imposing the condition that the principal contribution to the main peak at $k' = \Omega$ should be given by the contribution of the free generating functional $\mathcal{Z}_0^s$. This fixes $\alpha = [2(\Omega^2+1)]^{-1}$ which goes to $1/2$ as it should be in the white noise limit $\gamma \to +\infty$. The results (30) and (31) should be compared with the case $g(\theta) = \mathit{constant}$ in (1) which corresponds to translational invariance of (1) and then to the non generic case where rotational invariance is maintained in the normal form (2) as we have explained (isotropic noise). In this case the only peak in $\Delta'(k')$ is the main one at $k' = \Omega$ given by (30) and the value of $\Delta'(k' = 2\Omega)$ is $\mathcal{O}(\Gamma^{-3})$ corresponding to the tail of the main peak. We see then that when translational invariance is violated $\Delta'(k' = 2\Omega)$ becomes one order of magnitude bigger as shown by (31). This effect has been observed numerically by M. E. Brachet [8] wo has simulated (1) and also the normal form

(2) with a) $f(t; z, \bar{z}) = \bar{z}F(t)$, b) $f(t; z, \bar{z}) = zF(t)$. In case a) the peak at $k = 2\omega$ is clearly observed while in b), the isotropic situation, no peak is seen.

In the case of the white noise ($c = \gamma^2$, $\gamma \to +\infty$) in (1) the peak at $k = 2\omega$ disappears (the scaling we have been using is not appropriate in this limit) and the biggest new peak of $\Delta(k)$ is at $k = 3\omega$, but this effect turns out to be too small to be observed numerically [8]. Details of these calculations can be found in [9] and will be published elsewhere.

References

[1] P. Coullet, C. Elphick and E. Tirapegui, *Phys. Lett.* 111A, 277 (1985); C. Elphick, M. Jeanneret and E. Tirapegui, *J. Stat. Phys.* 48, 925 (1987).

[2] C. Nicolis and G. Nicolis, *Dynamics and Stability of Systems* 1, 249 (1986).

[3] F. Langouche, D. Roekaerts and E. Tirapegui, *Functional integration and Semiclasical expansions*, Reidel (1982).

[4] A. Spina and H. Vucetich, *Phys. Rev.* A36, 2914 (1987).

[5] E. Tirapegui, in *New Trends in Nonlinear Dynamics and pattern forming phenomena*, eds. P. Coullet and P. Huerre (Plenun Press, NATO Advanced Research Workshop, 1989).

[6] P. Colet, H. S. Wio and M. San Miguel, *Phys. Rev.* A39, 6094 (1989).

[7] H. S. Wio, P. Colet, M. San Miguel, L. Pesquera and M. A. Rodriguez, *Phys. Rev.* A40, 7312 (1989).

[8] M. E. Brachet, private communication.

[9] R. C. Buceta, Ph. D. thesis, Univ. Nacional de La Plata, La Plata, Argentina, 1990.

SOME PROPERTIES OF QUASI STATIONARY DISTRIBUTIONS IN THE BIRTH AND DEATH CHAINS: A DYNAMICAL APPROACH

P.A. FERRARI
Instituto de Matemática e Estatistica
Universidade de São Paulo, São Paulo, Brasil.

S. MARTINEZ
Departamento de Ingeniería Matemática
Facultad de Ciencias Físicas y Matemáticas
Universidad de Chile, Santiago, Chile.

P. PICCO
Centre de Physique Théorique, C.N.R.S.
Luminy, Marseille, France.

ABSTRACT. We study the existence of non-trivial quasi-stationary distributions for birth and death chains by using a dynamical approach. We also furnish an elementary proof of the solidarity property.

1. Introduction

Consider an irreducible discrete Markov chain $(X(n))$ on $S^* \cup \{0\}$ where 0 is the only absorbing state and S^* is the set of transient states. Let ν be a probability distribution. Denote by

$$\nu^{(n)}(x) = I\!P_\nu(X(n) = x | X(n) \neq 0) \tag{1.1}$$

the conditional probability that at time n the chain is at state x given that it has not been absorbed, starting with the initial distribution ν. A measure μ is called a Yaglom limit if for some probability measure ν we have: $\nu^{(n)}(x) \underset{n\to\infty}{\longrightarrow} \mu(x)$ for all $x \in S^*$.

Now assume that the transition probabilities $p(x,y) = I\!P(X(n+1) = y | X(n) = x)$ verify the following hypothesis:

$$p(0,0) = 1$$

$$P^* = (p(x,y) : x, y \in S^*) \text{ is irreducible}$$

$$\forall x \in S \text{ the set } \{y \in S : p(y,x) > 0\} \text{ is finite and non-empty}$$

Then it is easy to show that Yaglom limits μ verify the set of equations

E. Tirapegui and W. Zeller (eds.), Instabilities and Nonequilibrium Structures III, 177–187.

$$\forall x \in S^*, \quad \mu(x) = \sum_{y \in S^*} \mu(y)(p(y,x) + p(y,0)\mu(x)) \tag{1.2}$$

or equivalently the row vector $\mu = (\mu(x) : x \in S^*)$ satisfies

$$\mu P^* = \gamma(\mu)\mu \text{ with } \gamma(\mu) = 1 - \sum_{x \in S^*} \mu(x)p(x,0) \tag{1.3}$$

In general a quasi-stationary distribution (q.s.d.) is a measure μ which verifies (1.3). If μ is also a probability measure we call it a normalized quasi-stationary distribution (n.q.s.d.). Obviously the trivial measure $\mu \equiv 0$ is a q.s.d. It is easy to show that the irreducibility condition we have imposed on the Markov chain implies that for any non-trivial q.s.d., $\mu(x) > 0$ for all $x \in S^*$.

Some of the interesting problems of q.s.d. are concerned with the search for

- necessary and/or sufficient conditions on the transition matrices for the existence of non-trivial q.s.d.,
- domains of attractions of q.s.d.,
- evolution of $\delta_x^{(n)}$, δ_x being the Dirac distribution at point x.

For several kinds of Markov chains it has been proved that $\delta_x^{(n)}$ converges to a n.q.s.d. This was shown for branching process by Yaglom (1947), for finite state spaces by Darroch and Seneta (1965), for continuous time simple random walk on $\mathbb{N}$ by Seneta (1966) and for discrete time random walk on $\mathbb{N}$ by Seneta and Vere-Jones (1966).

For birth and death chains the existence of the limit of the sequence $\delta_x^{(n)}$ does not depend on x, and if the limit exists it is the same for all x. We provide in section 3 an elementary proof of this fact. Good (1968) gave a proof of this result based on some powerful results of Karlin and McGregor (1957); some technical details need additional explanations.

The problem of convergence of $\nu^{(n)}$ for ν other than Dirac distributions was initially considered by Seneta and Vere Jones (1966) for Markov chains with R-positive transition matrix.

For random walks it turns of that the Yaglom limit of $\delta_x^{(n)}$ is the minimal n.q.s.d. (this means $\gamma(\mu)$ is minimal). Then the study of the domains of attraction of non-minimal n.q.s.d. concerns the evolution $\nu^{(n)}$ for ν other than Dirac distributions. Recently we proved in [FMP] that the domains of attraction of non-minimal n.q.s.d. are non-trivial. More precisely we show that:

Theorem 1.1. Let μ, μ' be n.q.s.d. with $\gamma(\mu) > \gamma(\mu')$. Assume that ν satisfies:

$$\sup\{|\nu(x) - \mu(x)|\mu'(x)^{-1} : x \in S^*\} < \infty \text{ or } \nu = \eta\mu + (1-\eta)\mu' \text{ for } \eta \in (0,1]$$

then $\nu^{(n)} \xrightarrow[n\to\infty]{} \mu$. ■

Our main results deal with q.s.d. in birth and death chains. A first study concerning the description of the class of q.s.d.'s for birth and death process was made by Cavender (1978). Roughly, this class was characterized as an ordered one-parameter family and it was proved that any q.s.d. has total mass 0,1 or ∞.

2. Existence of Q.S.D. for Birth and Death Chains

2.1 GENERAL CONDITIONS FOR EXISTENCE

Consider a birth and death chain (X_n) on $\mathbb{N}$ with 0 as its unique absorbing state, so $p(0,0) = 1$. Denote $q_x = p(x, x-1)$ and $p_x = p(x, x+1)$, so $p(x,x) = 1 - p_x - q_x$ for all $x \in \mathbb{N}^*$.

For a sequence $\mu = (\mu(x) : x \in \mathbb{N}^*)$ the equations (1.2) take the form,

$$\forall y \in \mathbb{N}^* : (p_y + q_y)\mu(y) = q_{y+1}\mu(y+1) + p_{y-1}\mu(y-1) + q_1\mu(1)\mu(y) \tag{2.1}$$

If $\mu(1) > 0$ we get $\sum_{y=1}^{x} \mu(y) = 1 - \frac{1}{\mu(1)q_1}(q_{x+1}\mu(x+1) - p_x\mu(x))$ so a non-trivial q.s.d. is normalized iff $\mu(x) \xrightarrow[x\to\infty]{} 0$.

Now for $\gamma \neq p_1 + q_1$ define in a recursive way the following sequence $Z_\gamma = (Z_\gamma(x) : x \in \mathbb{N}^*)$,

$$Z_\gamma(1) = \gamma \tag{2.2}$$

$$\forall y \geq 2: \quad Z_\gamma(y) = f_{\gamma,y}(Z_\gamma(y-1)) \tag{2.3}$$

where

$$f_{\gamma,y}(z) = \gamma + p_y + q_y - p_1 - q_1 - \frac{p_{y-1}q_y}{z} \tag{2.4}$$

Associate to Z_γ the following vector $\mu_{(\gamma)} = (\mu_{(\gamma)}(x) : x \in \mathbb{N}^*)$

$$\mu_{(\gamma)}(1) = \frac{1}{q_1}(p_1 + q_1 - \gamma) \tag{2.5}$$

$$\forall x \geq 2: \quad \mu_{(\gamma)}(x) = \mu_{(\gamma)}(1) \prod_{y=1}^{x-1} \frac{Z_\gamma(y)}{q_{y+1}} \tag{2.6}$$

In [FMP] it was shown that a vector $\mu = (\mu(x) : x \in \mathbb{N}^*)$ with non-null terms verifies equations (2.1) iff there exists a $\gamma \neq p_1 + q_1$ such that $\mu = \mu_{(\gamma)}$.

In particular this last result implies that there exist non-trivial q.s.d. μ iff for some $\gamma < p_1 + q_1$ the sequence $Z_\gamma = (Z_\gamma(x) : x \in \mathbb{N}^*)$ is strictly positive. Then we search for conditions under which the orbit

$$Z_\gamma(y) = f_{\gamma,y} \circ f_{\gamma,y-1} \circ \cdots \circ f_{\gamma,2}(\gamma)$$

is strictly positive.

Assume for simplicity that $p_x + q_x = 1$ for all $x \in \mathbb{N}^*$ so the evolution functions $f_{\gamma,y}$ take the form,

$$f_{\gamma,y}(z) = \gamma - \frac{p_{y-1}q_y}{z} \tag{2.7}$$

Now make the following hypothesis: there exists a $\bar{q} \in (\frac{1}{2}, \frac{\sqrt{7}-1}{2})$ such that

$$\forall y \in \mathbb{N}^*, \quad \frac{1}{2} < \bar{q} - \frac{1}{2}\left(\bar{q} - \frac{1}{2}\right)^2 < q \le q_y \le q' < \bar{q} + \frac{1}{2}\left(\bar{q} - \frac{1}{2}\right)^2 < 1 \tag{2.8}$$

Denote $p = 1 - q$, $p' = 1 - q'$. Notice that if $\bar{q} = \frac{\sqrt{7}-1}{2}$ then $\bar{q} + \frac{1}{2}(\bar{q} - \frac{1}{2})^2 = 1$. The above condition (2.8) means that the birth and death chain is a perturbation of a random walk of parameter $\bar{q}$.

It can be shown that the hypothesis (2.8) implies the inequality

$$2\sqrt{pq'} < p' + q < 1$$

Call $g_\gamma(z) = \gamma - \frac{p'q}{z}$ and $h_\gamma(z) = \gamma - \frac{pq'}{z}$. It is easy to check that:

$$\forall y \in \mathbb{N}^*, \ z \ge 0: \quad h_\gamma(z) \le f_{\gamma,y}(z) \le g_\gamma(z) \tag{2.9}$$

Take $\gamma \in [2\sqrt{pq'}, 1)$, then $h_\gamma(z)$ has two fixed points (only one if $\gamma = 2\sqrt{pq'}$), a stable one $\xi = \frac{\gamma+\sqrt{\gamma^2-4pq'}}{2}$ and an unstable one $\eta = \frac{\gamma-\sqrt{\gamma^2-4pq'}}{2}$. Also $g_\gamma(z)$ has two fixed points, a stable one $\tilde{\xi} = \frac{\gamma+\sqrt{\gamma^2-4p'q}}{2}$ and an unstable one $\tilde{\eta} = \frac{\gamma-\sqrt{\gamma^2-4p'q}}{2}$.

Theorem 2.1. If condition (2.8) holds then there exist n.q.s.d. More precisely, if $\gamma \in [2\sqrt{pq'}, 1)$ then $\mu_{(\gamma)}$ is a non trivial q.s.d. and if $\gamma \in [2\sqrt{pq'}, p' + q]$ then $\mu_{(\gamma)}$ is a n.q.s.d.

Proof. Take $\gamma \in [2\sqrt{pq'}, 1)$. We have $Z_\gamma(1) = \gamma \ge \xi$. Therefore

$$Z_\gamma(y) = f_{\gamma,y} \circ \ldots \circ f_{\gamma,2}(Z_\gamma(1)) \ge h_\gamma^{(y-1)}(Z_\gamma(1)) \ge h_\gamma^{(y-1)}(\xi) = \xi > 0$$

Then $Z_\gamma(y) \geq \xi > 0$. Now $\gamma < 1$ implies $\mu_{(\gamma)}(1) > 0$ and expression (2.6) shows $\mu_{(\gamma)}(x) > 0$ for any $x \geq 2$, so $\mu_{(\gamma)}$ is a non trivial q.s.d.

Now let us prove that:

$$\forall y \in \mathbb{N}^*, \quad Z_\gamma(y) \leq \tilde{\xi} + (\gamma - \tilde{\xi}) \left(\frac{\tilde{\eta}}{\tilde{\xi}}\right)^{y-1} \tag{2.10}$$

Since $Z_\gamma(1) = \gamma$ the relation (2.10) holds for $y = 1$. Now we have $Z_\gamma(y) = f_{\gamma,y} \circ f_{\gamma,y-1} \circ \cdots \circ f_{\gamma,2}(\gamma) \leq g_\gamma^{(y-1)}(\gamma)$ where $g_\gamma^{(x)} = g_\gamma \circ \cdots \circ g_\gamma$ x times. Since $g_\gamma^{(y-1)}(\tilde{\xi}) = \tilde{\xi}$ we get from Taylor formula,

$$g_\gamma^{(y-1)}(\gamma) \leq \tilde{\xi} + (\gamma - \tilde{\xi}) \sup_{z \in [\tilde{\xi}, \gamma]} \left| \frac{\partial}{\partial t} g_\gamma^{(y-1)}(z) \right|$$

Now

$$\frac{\partial}{\partial z} g_\gamma^{(y-1)}(z) = \prod_{x=0}^{y-2} g'_\gamma(g_\gamma^{(x)}(z))$$

with $g_\gamma^{(0)}(z) = z$ and $g'_\gamma(z) = \frac{p'q}{z^2} = \frac{\tilde{\xi}\tilde{\eta}}{z^2}$.

Using the fact that g_γ is increasing and $\tilde{\xi}$ is a fixed point of g_γ we get easily that for all $0 \leq x \leq y-2$, and $z \in [\tilde{\xi}, \gamma]$, we have $g_\gamma^{(x)}(z) \geq \tilde{\xi}$. Therefore we get

$$\sup_{z \in [\tilde{\xi}, \gamma]} \left| \frac{\partial}{\partial t} g^{(y-1)}(z) \right| \leq \left(\frac{\tilde{\eta}}{\tilde{\xi}}\right)^{y-1}$$

Then property (2.10) is fulfilled.

Recall that $q_y \geq q$. Use the bound (2.10) to get from (2.6),

$$\mu_{(\gamma)}(x) \leq \mu_\gamma(1)(\prod_{y=1}^{x-1}(1 + \frac{\gamma - \tilde{\xi}}{\tilde{\xi}}(\frac{\tilde{\eta}}{\tilde{\xi}})^{y-1}))(\frac{\tilde{\xi}}{q})^{x-1}$$

Since $\sum_{y=1}^{\infty}(\frac{\tilde{\eta}}{\tilde{\xi}})^{y-1} < \infty$ we deduce that $C = \prod_{y=1}^{\infty}(1 + \frac{\gamma - \tilde{\xi}}{\tilde{\xi}}(\frac{\tilde{\eta}}{\tilde{\xi}})^{y-1}) < \infty$. So $\mu_{(\gamma)}(x) \leq C\mu_\gamma(1)(\frac{\tilde{\xi}}{q})^{x-1}$.

Now assume $\gamma \in [2\sqrt{pq'}, p' + q]$. Since $pq' > p'q$ we get:

$$\tilde{\xi} = \frac{1}{2}(\gamma + \sqrt{\gamma^2 - 4pq'}) \leq \frac{1}{2}((p'+q) + \sqrt{(p'+q)^2 - 4p'q}) < \frac{1}{2}((p'+q) + (q - p')) = q$$

Then $\frac{\tilde{\xi}}{q} < 1$ so $\mu_{(\gamma)}(x) \underset{x\to\infty}{\longrightarrow} 0$. Then $\mu_{(\gamma)}$ is a n.q.s.d. ■

2.2 LINEAR GROWTH CHAINS WITH IMMIGRATION

These processes are birth and death chains with

$$p_y = \frac{py+1}{(p+q)y+1}, \quad q_y = \frac{qy}{(p+q)y+a} \quad \text{for } y \in \mathbb{N}^*, \tag{2.11}$$

(so $p_y + q_y = 1$) and an absorbing barrier at 0, $p(0,0) = 1$.

We assume conditions

$$p > q \quad \text{and} \quad a < p+q \tag{2.12}$$

It can be shown that these inequalities imply that the sequence of functions $(f_{\gamma,y} : y \in \mathbb{N}^*)$ defined in (2.7), is increasing with y. The pointwise limit of this sequence, when $y \to \infty$, is $f_{\gamma,\infty}(z) = \gamma - \frac{pq}{(p+q)^2 z}$. Then we have:

$$f_{\gamma,2} \le \cdots \le f_{\gamma,y} \le f_{\gamma,y+1} \le \cdots \le f_{\gamma,\infty} \tag{2.13}$$

Observe that $f_{\gamma,2}$ plays the role of h_γ and $f_{\gamma,\infty}$ that of g_γ in (2.9).

Now, inequality $p_1 q_2 < 1$ is equivalent to

$$2(q-p)^2 + a(3p + a - 5q) > 0 \tag{2.14}$$

This condition is verified if q is big enough, for instance if $q > p + \frac{5}{4}a + \sqrt{a(p + \frac{17}{16}a)}$. We assume (2.14) holds.

Take $\gamma \in (2\sqrt{p_1 q_2}, 1)$ so $\xi = \frac{\gamma + \sqrt{\gamma^2 - 4p_1 q_2}}{2}$ belongs to the interval $(0, \gamma)$ and it is a fixed point of $f_{\gamma,2}$. Then $Z_\gamma(1) = \gamma$ and,

$$Z_\gamma(y) = f_{\gamma,y} \circ \cdots \circ f_{\gamma,2}(\gamma) > f_{\gamma,2}^{(y-1)}(\xi) = \xi > 0$$

Since $\gamma < 1$, from (2.5) and (2.6) we get $\mu_{(\gamma)}(y) > 0$ for any $y \in \mathbb{N}^*$. Hence $\mu_{(\gamma)}$ is a non trivial q.s.d.

Recall that $f_{\gamma,2} \le f_{\gamma,\infty}$ is equivalent to $\frac{pq}{(p+q)^2} < p_1 q_2$. Take $\gamma \in (\frac{2\sqrt{pq}}{(p+q)}, 2\sqrt{p_1 q_2})$. Then the point $\tilde{\eta} = \frac{\gamma - \sqrt{\gamma^2 - \frac{4pq}{(p+q)^2}}}{2}$ and $\tilde{\xi} = \frac{\gamma + \sqrt{\gamma^2 - \frac{4pq}{(p+q)^2}}}{}$ are respectively the unstable and the stable fixed points of $f_{\gamma,\infty}$. Replacing g_γ by $f_{\gamma,\infty}$ we get that condition (2.9) holds with $\tilde{\eta}$, $\tilde{\xi}$ the fixed points of $f_{\gamma,\infty}$. Then,

$$\mu_{(\gamma)}(x) \leq \mu_{(\gamma)}(1)\{\prod_{y=1}^{x-1}(1+\frac{\gamma-\tilde{\xi}}{\tilde{\xi}}(\frac{\tilde{\eta}}{\tilde{\xi}})^{y-1}\}\tilde{\xi}^{x-1}\prod_{y=1}^{x-1}\frac{1}{q_{y+1}} \tag{2.15}$$

Denote $C = \prod_{y=1}^{\infty}(1+\frac{\gamma-\tilde{\xi}}{\tilde{\xi}}(\frac{\tilde{\eta}}{\tilde{\xi}})^{y-1})$ which is finite.

We have $\prod_{y=1}^{x-1}\frac{1}{q_{y+1}} = (\frac{p+q}{q})^{x-1}\prod_{y=1}^{x-1}(1+\frac{a}{(p+q)(y+1)}) \leq (\frac{p+q}{q})^{x-1}\exp\{\frac{a}{p+q}\sum_{y=1}^{x-1}\frac{1}{y+1}\}$. Then $\prod_{y=1}^{x-1}\frac{1}{q_{y+1}} \leq (\frac{p+q}{q})^{x-1}(x-1)^{\frac{a}{(p+q)}}$. Hence

$$\mu_{(\gamma)}(x) \leq \mu_{(\gamma)}(1)C(\tilde{\xi}\frac{(p+q)}{q})^{x-1}(x-1)^{\frac{a}{p+q}} \tag{2.16}$$

It can be easily verified that our assumptions imply that $\tilde{\xi} < \frac{q}{p+q}$. Then $\mu_{(\gamma)}(x) \underset{x\to\infty}{\longrightarrow} 0$. Then, for $\gamma \in (2\frac{\sqrt{pq}}{p+q}, 2\sqrt{p_1 q_2})$ the q.s.d. $\mu_{(\gamma)}$ is normalized.

3. Solidarity Property for Birth and Death Chains

Concerning the convergence of point measures to some Yaglom limit, the deepest results have been established in [S2,SV-J] for random walks ($q_x = q, p_x = 1-q$) with continuous and discrete time. Here we shall show a solidarity process which asserts that it suffices to have the convergence for the probability measure concentrated at 1. Our proof is elementary, in fact it does not use any higher technique. We must point out that Good [G] has also shown this result but in his proof some technical steps have been overlooked.

Theorem 3.1. If $\delta_1^{(n)}$ converges to a q.s.d. μ then for any $x \in \mathbb{N}^*$, $\delta_x^{(n)}$ converges to μ.

Proof. For $x, n \in \mathbb{N}^*$ set:

$$\alpha_x(n) = \frac{\mathbb{P}_{x+1}(X(n-1) \neq 0)}{\mathbb{P}_x(X(n) \neq 0)}, \quad \beta_x(n) = \frac{\mathbb{P}_x(X(n-1) \neq 0)}{\mathbb{P}_x(X(n) \neq 0)},$$
$$\xi_x(n) = \frac{\mathbb{P}_{x-1}(X(n-1) \neq 0)}{\mathbb{P}_x(X(n) \neq 0)}$$

Observe that $\xi_1(n) = 0$ for all $n \in \mathbb{N}^*$, all other terms being > 0.

These quantities are related by the identity

$$\forall x \in \mathbb{N}^*, \quad \xi_{x+1}(n) = \beta_x(n)\beta_{x+1}(n)(\alpha_x(n))^{-1} \tag{3.1}$$

On the other hand from the equation

$$\begin{aligned} \mathbb{P}_x(X(n) \neq 0) = q_x \mathbb{P}_{x-1}(X(n-1) \neq 0) \\ + (1-p_x-q_x)\mathbb{P}_x(X(n-1) \neq 0) + p_x \mathbb{P}_{x+1}(X(n-1) \neq 0) \end{aligned} \tag{3.2}$$

we deduce that

$$\forall x, n \in \mathbb{N}^*, \quad q_x \xi_x(n) + (1-p_x-q_x)\beta_x(n) + p_x \alpha_x(n) = 1 \tag{3.3}$$

Also from definition we get

$$\beta_x(n) = (\mathbb{P}_x(X(n) \neq 0 | X(n-1) \neq 0))^{-1} = (1 - \delta_x^{(n-1)}(1)q_1)^{-1} \tag{3.4}$$

If the limit of a sequence $\eta(n)$ exits denote it by $\eta(\infty)$. So the hypothesis of the theorem is: $\forall z \in \mathbb{N}^*$, $\delta_1^{(\infty)}(z)$ exists.

Since $\delta_1^{(\infty)}(1)$ exists and belongs to [0,1] we deduce from (3.4) that $\beta_1(\infty)$ exists and belongs to $[1, \frac{1}{1-q_1}]$. From $\xi_1(n) = 0$ and (3.3) we get that $\alpha_1(\infty)$ exists and is bigger or equal than $\frac{1}{p_1}(1 - \frac{(1-p_1-q_1)}{1-q_1}) = \frac{1}{1-q_1}$.

Now let us show that,

$$\forall x \in \mathbb{N}^*, \quad \liminf_{n\to\infty} \alpha_x(n) > 0 \tag{3.5}$$

This holds for $x = 1$. Now from (3.2) evaluated at $x+2$ we deduce the inequality

$$\mathbb{P}_{x+1}(X(n-1) \neq 0) \leq q_{x+2}^{-1} \mathbb{P}_{x+2}(X(n) \neq 0)$$

On the other hand since $\mathbb{P}_y(X(n) \neq 0)$ increases with $y \in \mathbb{N}^*$ and decreases with $n \in \mathbb{N}^*$ we get the following relations

$$\mathbb{P}_x(X(n) \neq 0) \geq p_x \mathbb{P}_{x+1}(X(n-1) \neq 0)$$

$$\alpha_{x+1}(n) = \frac{\mathbb{P}_{x+2}(X(n-1) \neq 0)}{\mathbb{P}_{x+1}(X(n) \neq 0)} \geq \frac{\mathbb{P}_{x+2}(X(n) \neq 0)}{\mathbb{P}_{x+1}(X(n-1) \neq 0)}$$

Hence we obtain:

$$\begin{aligned} \alpha_x(n) = \frac{\mathbb{P}_{x+1}(X(n-1) \neq 0)}{\mathbb{P}_x(X(n) \neq 0)} &\leq (p_x q_{x+2})^{-1} \frac{\mathbb{P}_{x+2}(X(n) \neq 0)}{\mathbb{P}_{x+1}(X(n-1) \neq 0)} \\ &\leq (p_x q_{x+2})^{-1} \alpha_{x+1}(n) \end{aligned}$$

Then $\alpha_{x+1}(n) \geq p_x q_{x+2} \alpha_x(n)$. So $\liminf_{n\to\infty} \alpha_{x+1}(n) > 0$ and relation (3.5) holds.

Now let us prove by recurrence that:

$$\text{the limits } \alpha_x(\infty), \beta_x(\infty), \xi_x(\infty) \text{ and } \delta_x^{(\infty)}(z) \text{ for all } z \in I\!N^*, \text{ exist} \tag{3.6}$$

We show above that these limits exist for $x = 1$. Assuming that property (3.6) holds for $x \in \{1, ..., y\}$, we shall prove that it is also satisfied for $x = y + 1$. With this purpose in mind, condition on the first step of the chain to get,

$$\begin{aligned} I\!P_y(X(n) = z) = q_y I\!P_{y-1}(X(n-1) = z) + (1 - p_y - q_y) I\!P_y(X(n-1) = z) \\ + p_y I\!P_{y+1}(X(n-1) = z) \end{aligned} \tag{3.7}$$

Now from definitions of $\alpha_y(n)$, $\beta_y(n)$, $\xi_y(n)$ we have the following identities for $y \geq 2$:

$$\begin{aligned} I\!P_y(X(n) \neq 0) &= \alpha_y(n)(I\!P_{y+1}(X(n-1) \neq 0))^{-1} = \beta_y(n)(I\!P_y(X(n-1) \neq 0))^{-1} \\ &= \xi_y(n)(I\!P_y(X(n-1) \neq 0))^{-1} \end{aligned}$$

Develop $\delta_y^{(n)}(z) = I\!P_y(X(n) = z)(I\!P_y(X(n) \neq 0))^{-1}$ according to (3.7) and the last equalities to get:

$$\delta_y^{(n)}(z) = \delta_{y-1}^{(n-1)}(z) q_y \xi_y(n) + \delta_y^{(n-1)}(z)(1 - p_y - q_y)\beta_y(n) + \delta_{y+1}^{(n-1)}(z) p_y \alpha_y(n) \tag{3.8}$$

This last equality holds for any $y \geq 1$ (recall $\xi_1(n) = 0$).

Since $\delta^{(\infty)}(z), \xi_x(\infty), \beta_x(\infty), \alpha_x(\infty)$ exist for any $x \leq y$ and $z \in I\!N^*$, and, by (3.5), $\alpha_x(\infty) > 0$ we get that $\delta_{y+1}^{(\infty)}(z)$ exists for any $z \in I\!N^*$. On the other hand equality (3.4) implies that $\beta_{y+1}(\infty)$ exists. Then by (3.1) the limit $\xi_{y+1}(\infty)$ exists and equation (3.3) implies the existence of $\alpha_{y+1}(\infty)$.

From (3.5) and (3.4) we deduce $\alpha_x(\infty) > 0$ and $\beta_x(\infty) > 0$ for any $x \in I\!N^*$. So (3.1) implies $\xi_x(\infty) > 0$ for $x \geq 2$.

Then if $\delta_y^{(\infty)}(z) = 0$ for some $y, z \in I\!N^*$ we can deduce from equality (3.8) that $\delta_x^{(\infty)}(z) = 0$ for all $x \in I\!N^*$. So the q.s.d.'s which are the limits of $\delta_x^{(n)}$ are all trivial or normalized.

Assume that $\delta_1^{(n)}$ converges to a normalized q.s.d. μ. Let us prove by recurrence that $\delta_x^{(n)}$ converges to μ for all x.

Since the limits $\delta_y^{(\infty)}$ exists and $\delta_1^{(\infty)}(z) > 0$, when we evaluate (3.8) at $y = 1, n = \infty$ we get the following equation:

$$1 = (1 - p_1 - q_1)\beta_1(\infty) + \left(\frac{\delta_2^{(\infty)}(z)}{\delta_1^{(\infty)}(z)}\right) p_1 \alpha_1(\infty)$$

Comparing this equation with (3.3) evaluated at $x = 1$, $n = \infty$, and by taking into account that $\xi_1(\infty) = 0$ we deduce $\delta_2^{(\infty)}(z) = \delta_1^{(\infty)}(z)$ for any $z \in \mathbb{N}^*$.

Assume we have shown for any $y \in \{1, ..., y_0\}$ that: $\forall z \in \mathbb{N}^*$, $\delta_y^{(\infty)}(z) = \delta_1^{(\infty)}(z)$. Let us show that this last set of equalities also hold for $y_0 + 1$. Evaluate equation (3.8) at $y = y_0$, $n = \infty$ to get

$$1 = q_{y_0}\xi_{y_0}(\infty) + (1 - p_{y_0} - q_{y_0})\beta_{y_0}(\infty) + \left(\frac{\delta_{y_0+1}^{(\infty)}(z)}{\delta_{y_0}^{(\infty)}(z)}\right)p_{y_0}\alpha_{y_0}(\infty)$$

Comparing this equation with (3.3) evaluated at $x = y_0$, $n = \infty$ we deduce that $\delta_{y_0+1}^{(\infty)}(z) = \delta_{y_0}^{(\infty)}(z)$. Hence the recurrence follows and for any $x \in \mathbb{N}^*$, $\delta_x^{(n)}$ converges to $\mu = \delta_1^{(\infty)}$. ∎

Acknowledgments

We thank Antonio Galves and Isaac Meilijson for discussions. P.P. and S.M. acknowledge the very kind hospitality at Instituto de Matemática e Estatística, Universidade de São Paulo. The authors acknowledge the very kind hospitality of Istituto Matematico, Università di Roma Tor Vergata. This work was partially supported by Fundação de Amparo à Pesquisa do Estado de São Paulo. S.M. was partially supported by Fondo Nacional de Ciencias 0553-88/90 and Fundación Andes (becario Proyecto C-11050).

References

[C] J.A. Cavender (1978) Quasi-stationary distributions of birth and death processes. Adv. Appl. Prob., 10, 570-586.

[FMP] P. Ferrari, S. Martínez, P. Picco (1990) Existence of non trivial quasi stationary distributions in the birth and death chains. Submitted J. Appl. Prob.

[G] P. Good (1968) The limiting behaviour of transient birth and death processses conditioned on survival. J. Austral Math. Soc. 8, 716-722.

[KMcG1] S. Karlin and J. McGregor (1957) The differential equations of birth and death processes, and the Stieljes Moment Problem. Trans Amer. Math. Soc., 85, 489-546.

[KMcG2] S. Karlin and J. McGregor (1957) The classification of birth and death processes. Trans. Amer. Math. Soc., 86, 366-400.

[SV-J] E. Seneta and D. Vere-Jones (1966) On quasi-stationary distributions in discrete-time Markov chains with a denumerable infinity of states. J. Appl. Prob., 3, 403-434.

[S1] E. Seneta (1981) Non-negative matrices and Markov chains. Springer Verlag.

[S2] E. Seneta (1966) Quasi-stationary behaviour in the random walk with continuous time. Australian J. on Statistics, 8, 92-88.

[SW] W. Scott and H. Wall (1940) A convergence theorem for continued-fractions. Trans. Amer. Math. Soc., 47, 115-172.

[Y] A.M. Yaglom (1947) Certain limit theorems of the theory of branching stochastic processes (in russian) Dokl. Akad. Nank SSSR, 56, 795-798.

[29] W. ... and H. Weld (18??). A convergence theorem for continuous-time-Markov-processes. Amer. J. Statist., 7(2), 129-175.

[30] ... A. V. ... Iosi (19??). Continual limit theorems in the theory of branching processes. pp. ... Internal Dokl. Akad. Nauk SSSR, 66, 737-786.

PHOTON NOISE REDUCTION IN LIGHT SOURCES

M. Orszag and J. C. Retamal

Facultad de Física
Universidad Católica de Chile
Casilla 6177, Santiago 22
Chile

ABSTRACT. We review the theory of a maser (or a laser) in which a probability distribution for the atomic pump is introduced. Photon number noise reduction is possible for a regular pump statistics. We discuss the photon noise in a particular case, namely, the two photon micromaser.

1. INTRODUCTION

In the last decade, great part of the investigation in Quantum Optics has been oriented to the study of quantum noise reduction in light sources.

One of the most interesting physical picture in which quantum noise reduction takes place, is the "squeezing of the electromagnetic field"[1]. We speak of squeezing in the electromagnetic field when the fluctuations in one of the quadratures of the field has a value below of in a coherent state, or equivalently the vacuum level. At the present we know that squeezed states of the light can be generated in a variety of non linear processes, for example in four wave mixing, second harmonic generation, parametric amplification, optical bistability, etc[2].

Another physical mechanism that leads to quantum noise reduction, is the correlated spontaneous emission laser[3] (CEL). The analysis shows that fluctuations in the relative phase angle in a two mode laser can be reduced to zero. On the other hand, the noise in the absolute phase of a single mode laser can be reduced in a two photon CEL. In addition in such system, squeezing in the phase quadrature can take place[4].

Recently, quantum noise reduction has been predicted in a laser system in which the lifetime of the atoms is comparable to the measurement time. Such an effect is called the atomic memory effect[5].

A different noise reduction scheme in a laser system is possible. It can be proven that reduced atomic pump fluctuations, will result in a photon number noise reduction in the intracavity field[6].

By introducing a pump statistics for the atomic injection times that ranges

E. Tirapegui and W. Zeller (eds.), Instabilities and Nonequilibrium Structures III, 189–198.

between a poissonian and a regular distribution, we find that for regular pump the photon number noise is reduced below the poissonian limit[7]. The predictions in the master equation approach for a laser can be reproduced in the Langevin equation approach[8]. Here we present the model of a laser and a maser including the pump statistics. We discuss the noise in a particular case, namely the two photon micromaser.

2. Pump statistics and master equation

Let's assume that a dense flux of two level atoms with a regular distribution, arrives to an excitation region, so that each atom has a probability p of being exited from a ground state to the upper level (fig. 1). If we assume that R atoms are injected per unit time, then the number of atoms that cross the excitation region in a time Δt is $K = R\Delta t$, Δt is a time large as compared to the time interval between the arrival of two consecutive atoms. The probability that k atoms get excited during Δt is given by[7]

$$P(k,K) = \binom{K}{k} p^k (1-p)^{K-k}. \tag{1}$$

Now, let's assume that the atom j coming from the exitation region enters to the maser cavity (or laser) at the random time t_j and interacts with the cavity mode during a time τ. The change in the reduced density matrix for the radiation field is given by[12]

$$\rho(t_j + \tau) = M(\tau)\rho(t_j), \tag{2}$$

where $M(\tau)$ is an operator which depends on the particular model. In general, $M(\tau)$ is determined when we know the temporal evolution for the atom field system. Explicitly, equation (2) means

$$M\rho(t_j) = tr[U(\tau)\rho_{A-F}(t_j)U^{\dagger}(\tau)], \tag{3}$$

where the trace is over the atomic variables.

In a maser system the interaction time represents the transit time of the atom inside the cavity. If we assume that the velocity fluctuations vanish, τ can be considered as a fixed time. In a laser system the time τ must averaged with respect to the atomic lifetime distribution[12] so that equation (2) now reads

$$\bar{M}\rho(t_j) = \int_0^{\infty} d\tau \Gamma e^{-\Gamma\tau} M(\tau)\rho(t_j). \tag{4}$$

Now, assuming that k excited atoms enter the interaction region in the interval $[0,t]$, the reduced density matrix, at time t is given by

$$\rho^k(t) = M^k \rho(0). \tag{5}$$

(For a laser we replace M by $\bar{M}$).

Since the incoming atoms are distributed according to (1) we have, the density matrix at time t given by

$$\rho(t) = \sum_{k=0}^{K} P(k, K)\rho^{k}(t), \tag{6}$$

where now $K = Rt$.

If the time t is very large as compared with the average time between two atoms, $t >> r^{-1}$ equation (6) can be reduced to

$$\rho(t) = [1 + p(M-1)]^{K} \rho(0) \tag{7}$$

Taking the time derivative we obtain the generalized Master equations

$$\dot{\rho}(t) = \frac{r}{p} \ell n[1 + p(M-1)]\rho(t). \tag{8}$$

This corresponds to the temporal evolution equation for ρ due to the gain, but we must also consider the effects of the loss in our system.

As in ref. [12], we add the contribution due to the gain and loss (the loss part is obtained from the interaction between the system and the loss reservoir) and obtain

$$\dot{\rho}(t) = \frac{r}{p} \ell n[1 + p(M-1)]\rho(t) + L\rho, \tag{9}$$

where

$$L\rho(t) = \frac{\gamma}{2}(\bar{N}+1)[a\rho a^{\dagger} - a^{\dagger}a\rho + hc] + \frac{\gamma}{2}\bar{N}[a^{\dagger}\rho a - \rho a a^{\dagger} + hc]. \tag{10}$$

We can see from equation (9) that when $p \rightarrow 0$ and r remains finite (poissonian case)

$$\dot{\rho}(t) = r(M-1)\rho(t) + L\rho(t), \tag{11}$$

which corresponds to the usual Scully Lamb equation[12].

When $p \rightarrow 1$ equation (9) reduces to

$$\dot{\rho}(t) = r\ell n M\rho(t) + L\rho(t), \tag{12}$$

which corresponds to regular pumping.

In general equation (9) can be expanded in the form

$$\dot{\rho}(t) = [r(M-1) - \frac{rp}{2}(M-1)^{2} \ldots]\rho(t) + L\rho(t). \tag{13}$$

The relevant effects of the pump statistic comes from the second term in this expression[7].

3. Photon noise reduction in a two photon micromaser

The Hamiltonian for the two micromaser in the rotating wave approximation is given by:

$$H = \hbar\nu a^\dagger a + \epsilon_a\sigma_{aa} + \epsilon_b\sigma_{bb} + \epsilon_c\sigma_{cc} + \hbar g[a(a^\dagger_{ab} + \sigma^\dagger_{bc}) + h.c.], \tag{14}$$

$\sigma_{ii} =| i >< i |$ are the population operators for the three level atom (levels a, b, c, with a being the upper level), $\sigma_{ij} =| j >< i |$ are the polarization operators. For simplicity we assume equal coupling between the $a \rightarrow b$ and $b \rightarrow c$ transition. The frequency ν is resonant to one half of the $a \rightarrow c$ energy difference. The intermediate level b is detuned from the center of the $a - c$ transition by: $\hbar\Delta = \epsilon_b - (\epsilon_a + \epsilon_c)/2$.

In order to calculate the gain operator, we go to the interaction picture and calculate the temporal evolution operator for the system. According to equation (3) we obtain the following gain operator[11].

$$\begin{aligned} M(t_{int})\rho(t) =& (1 - aGa^\dagger)\rho(t)(1 - aG^\dagger a^\dagger) \\ &+ g^2\frac{\sin(\sqrt{\Lambda}t_{int})}{\sqrt{\Lambda}}a^\dagger\rho(t)a\frac{\sin(\sqrt{\Lambda}t_{int})}{\sqrt{\Lambda}} \\ &+ a^\dagger Ga^\dagger\rho(t)aG^\dagger a, \end{aligned} \tag{15}$$

where

$$G = (a^\dagger a + aa^\dagger)^{-1}\left[1 - e^{-\frac{i\Delta t}{2}}\left(cos\sqrt{\Lambda}t + \frac{i\Delta}{2\sqrt{\Lambda}}sen\sqrt{\Lambda}t\right)\right],$$

$$\Lambda = g^2(a^\dagger a + aa^\dagger) + \Delta^2/4.$$

The second term in equation (15) can be neglected in the approximation $(g/\Delta)^2 << 1$, in agreement to the experimental situation in ref. [9]. Therefore we can finally write.

$$\begin{aligned} (M-1)\rho(t) =& - aGa^\dagger\rho(t) - \rho(t)aG^\dagger a^\dagger \\ &+ aGa^\dagger\rho(t)aG^\dagger a^\dagger + a^\dagger Ga^\dagger\rho(t)aG^\dagger a. \end{aligned} \tag{16}$$

By introducing this expression in the equation (13), we obtain for the diagonal elements of the density matrix

$$\begin{aligned} \dot{\rho}_{NN} =& - r(1- | A_N |^2)\rho_{NN} + r | B_N |^2 \rho_{N-2,N-2} \\ &- \frac{rp}{2}(1- | A_N |^2)^2\rho_{NN} + \frac{rp}{2} | B_N |^2 (2- | A_N |^2 - | A_{N-2} |^2)\rho_{N-2,N-2} \\ &- \frac{rp}{2} | B_N |^2 | B_{N-2} |^2 \rho_{N-4-N-4} \\ &- \gamma N\rho_{NN} + \gamma(N+1)\rho_{N+1,N+1}, \end{aligned} \tag{17}$$

where

$$
\begin{aligned}
A_N &= 1 + \frac{N+1}{2N+3}(e^{i\Omega(N)t_{int}} - 1),\\
B_N &= \frac{\sqrt{(N+1)(N+2)}}{2N+3}(e^{i\Omega(N)t_{int}} - 1),\\
\Omega(N) &= (2N+3)g^2/\Delta,\\
| B_N |^2 &= 1 - | A_{N-2} |^2 .
\end{aligned}
$$

When $p = 0$, equation (17) it reduces to the known master equation found in ref. [10] at zero temperature and without atomic velocity fluctuations.

The equations for the first two moments $< N >$ and $v =< N^2 > - < N >^2$ of the photon statistics are readily found to be:

$$
\begin{aligned}
< \dot{N} > &= 2r < \alpha_N > - \gamma < N >\\
\dot{v} &= 4r < \Delta N \alpha_N > + 4r < \alpha_N > - 4rp < (1- | A_N |)^2 (1- | A_{N+2} |)^2 >\\
&\quad - 2\gamma v + \gamma < N >,
\end{aligned} \tag{18}
$$

where $\alpha_N = (1- | A_N |^2)\left[1 + \frac{p}{2}(| A_{N+2} |^2 - | A_N |^2)\right]$.
The steady state condition in photon number reads

$$
2r < \alpha_N > = \gamma < N > . \tag{19}
$$

Depending on the interaction time the system can place itself in more than one steady state, which is the usual situation in micromasers. The steady state points are only stable when the slope of the gain curve is smaller than the slope of the loss curve. This condition can be written as:

$$
\alpha'_s < 1, \tag{20}
$$

where $\alpha'_s = d\alpha/dn |_{n=n_s}$, $n = N/2N_{ex}$, $N_{ex} = r/\gamma$.

Combining equations (18) and (19), we can obtain an approximated expression for the steady state normalized photon number variance, expanding α_N up to first order about N_S.

$$
v_s(\phi,p) = \frac{3}{2(1-\alpha'_s(\phi,p))}\left[1 - \frac{4pN_{ex}}{3\langle N\rangle_s}\langle(1- | A_N |^2)(1- | A_{N+2} |^2)\rangle\right], \tag{21}
$$

where $v_s(\phi,p) = v(\phi,p)/ < N >_s$, and $\phi = 2N_{ex}g^2 t_{int}/\Delta$. The pump statistics corrections to the usual variance are apparent from equation (21). When p is increased, the subpoissonian character of the photon statistics is enhanced.
Define

$$R = \frac{v_s(\phi,p)}{v_s(\phi,0)} = \frac{1-\alpha'_s(\phi,0)}{1-\alpha'_s(\phi,p)}\left[1 - \frac{4pN_{ex}}{3\langle N\rangle_s}\langle(1-\mid A_N\mid^2)(1-\mid A_{N+2}\mid^2)\rangle\right], \quad (22)$$

as the ratio between the steady state normalized photon number fluctuations for any pump statistics and the Poissonian one. When $N_S >> 1$ equation (22) can be simplified to

$$R = \frac{1-\alpha'_s(\phi,0)}{1-\alpha'_s(\phi,p)}\left[1 - \frac{2}{3}p\frac{<N>_s}{2N_{ex}}\right]. \quad (23)$$

In what follows, we will discuss some results obtained by solving numerically equation (17).

In figure 2, we have plotted the normalized photon number variance (v_s) for $p = 0$ and $p = 1$. We observe the usual result for $p = 0$.[10]

For $p = 1$, the strongest reduction is observed between $\phi \approx 1.26\pi/2$ and $\phi \approx 1.44\pi/2$. In this region, we observe large fluctuations for the $p = 0$ case, because the photon statistics has two peaks. This can be seen from the figure 4, where we have plotted ρ_{nn} versus n for $\phi = 1.28\pi/2$.

In figure 3 we have plotted R versus ϕ. The dotted curve corresponds to equation (23). The minimum of this curve, at $\phi = 1.28\pi/2$ corresponds to a maximum photon number variance reduction of 95.6%. As one gets away from this region, the reduction becomes smaller. If we go above threshold, equation (23) can be approximately simplified to:

$$R = 1 - \frac{2}{3}pn_s. \quad (24)$$

For example, for $\phi \sim 1.5\pi/2$, where $n_s \simeq 0.825$, we have $R = 0.45$. From figure 3 at this ϕ in the dashed line $R = 0.428$. As we can see here as well as in the figure 3, there is a good agreement between the numerical results and equation (24).

There is a second subpoissonian region around $\phi \simeq 9.8\pi/2$ (Figure 2). This appears, because one of the two peaks in the photon statistics becomes exceedingly small, as we can see in figure 5.

To summarize, we found that a regular pumping statistics generates a strong photon number noise reduction in the two photon micromaser, for certain values of the reduced interaction time. This feature is mainly present when one of the peaks in the photon statistics clearly dominates over the others. On the other hand, the photon-number fluctuations increase for those values of ϕ for which two or more peaks of similar heights are observed.

Acknowledgements

The authors would like to thank Mr. Luis Roa for his assistance in preparing the figures. One of us (M. O.) would like to acknowledge FONDECYT (Project 363/88). (J. R.) acknowldeges a CONICYT Doctoral fellowship.

References

[1] For review of fundamental concepts see for example D. Walls, (1983) Nature 306, 141 and R. Loudon. P. L. Knight, (1987) Jour. of Mod. Opt. 34, 709.
[2] For experimental and theoretical review see H. J. Kimble and D. F. Walls (1987) J. Opt. Am. Soc. B. 4, 6.
[3] M. Scully, (1985) Phys. Rev. Lett. 55, 2802.
[4] For a review see M. Orszag, J. Bergou, W. Schleich, M. Scully, (1989) "Squeezed and nonclassical light", edited by P. Tombesi and E. R. Pike (Plenum Publishing Corp.).
[5] C. Benkert , G. Susmann, M. Scully, (1988) Phys. Rev. Lett. 11, 1014.
[6] Y. M. Golubev, I. V. Sokolov, (1984) Sov. Phys. JEPT 60, 234.
[7] J. Bergou, L. Davidovich, M. Orszag, C. Benkert, M. Hillery, M. Scully, (1989) Opt. Commnication 72, 82 (1989), Phys. Rev. A. 40, 5073.
[8] J. Bergou, L. Davidovich, M. Orszag, C. Benkert, M. Hillery, M. Scully, (1990) Phys. Rev. A 42, 4.
[9] M. Brune, J. M. Raimond, S. Haroche, (1987) Phys. Rev. A 35.
[10] L. Davidovich, J. M. Raimond, M. Brune, S. Haroche, (1987) Phys. Rev. A 36 3771.
[11] M. Orszag, J. C. Retamal (submitted for publication).
[12] M. Sargent, M. Scully, (1974) W. E. Lamb, Laser Physics (Addison Wesley Reading MA).

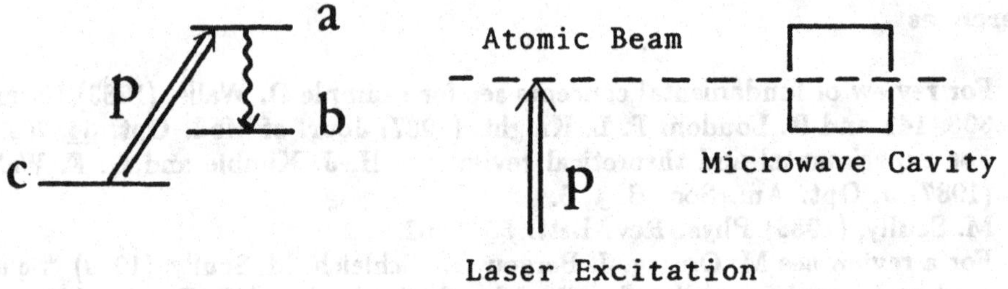

Figure 1.

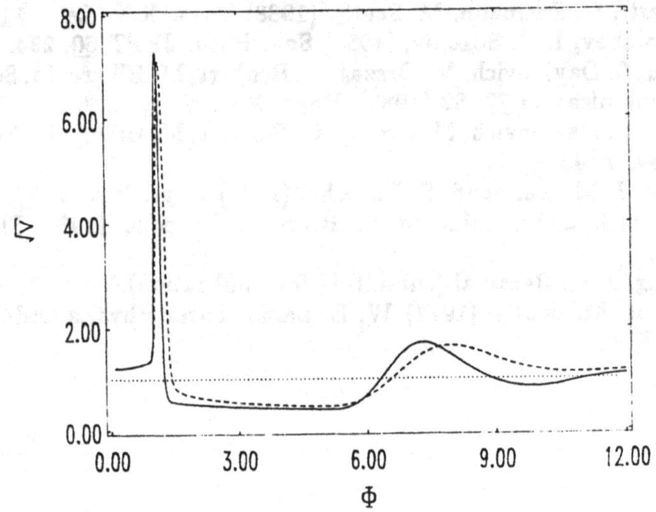

Figure 2. Root square of normalized photon number fluctuations vs the reduced interaction time ϕ (measured in units of $\pi/2$). The dashed line (---) corresponds to p = 0, the solid line corresponds to p = 1. $N_{ex}=30$.

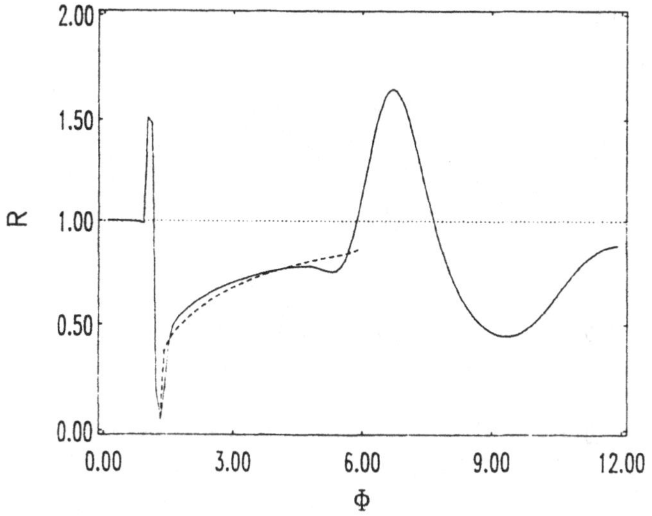

Figure 3. Ratio R vs ϕ (measured in units of $\pi/2$). The solid line corresponds to numerical solution of equation (10). The dashed line (---) corresponds to approximated equation (19) for $\phi\varepsilon[1.29,6.78]$. N_{ex} = 30.

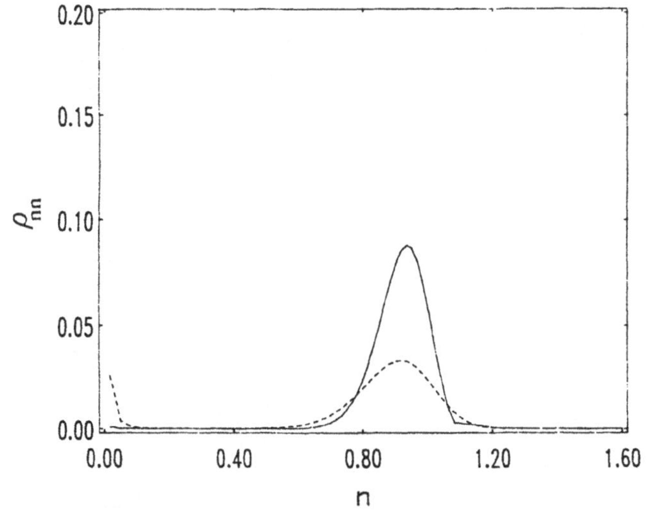

Figure 4. Photon statistics vs reduced photon number for $\phi\sim1.28\pi/2$. The dashed line (---) corresponds to p = 0 and the solid line corresponds to p = 1. N_{ex} = 30.

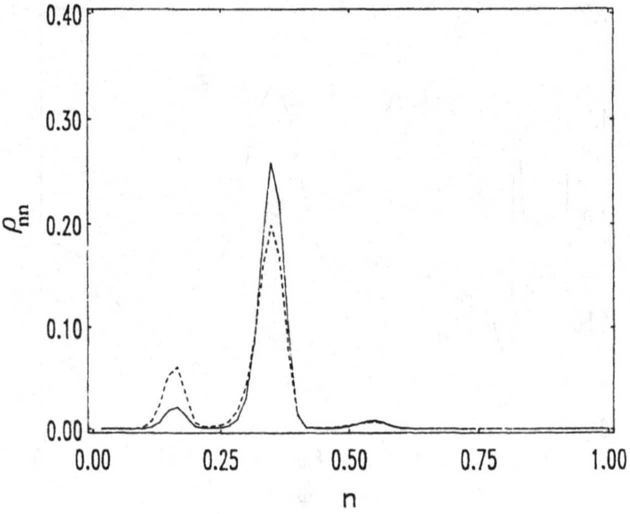

Figure 5. Photon statistics vs reduced photon number for $\phi \sim 9.8\pi/2$. The dashed line (---) corresponds to p = 0 and the solid line corresponds to p = 1. N_{ex} = 30.

PART III.

INSTABILITIES IN NONEQUILIBRIUM STRUCTURES

SPATIO-TEMPORAL PROPERTIES OF CENTRIFUGAL INSTABILITIES

Innocent MUTABAZI* and José Eduardo WESFREID
Laboratoire d'Hydrodynamique et Mécanique Physique,
ESPCI (URA CNRS n°857)
10, rue Vauquelin, F-75231Paris Cédex 05, France

ABSTRACT.

General properties of the centrifugal instabilities are presented with emphasis on spatio-temporal behavior observed in large aspect ratio systems. A tentative unification those properties common to Taylor and Dean vortex flows is proposed for the small gap approximation.

1. Introduction

Centrifugal instabilities occur in the flow with curved streamlines as a result of the unbalanced centrifugal force by the radial pressure gradient. Such flows are the flow in the gap between two rotating independently cylinders, the flow between the curved channel, the flow in the boundary layer over a concave wall.

Centrifugal instabilities produce longitudinal rolls aligned along the main direction of the base flow. Depending on the flow boundaries, one distinguishes three main types of centrifugal instabilities : the Taylor-Couette instability which appears in the flow between two rotating coaxial cylinders (Couette flow), the Dean instability in the flow in curved channel (Poiseuille flow) and the Taylor-Görtler instability in the flow in the boundary layer on a concave wall (Fig.1).

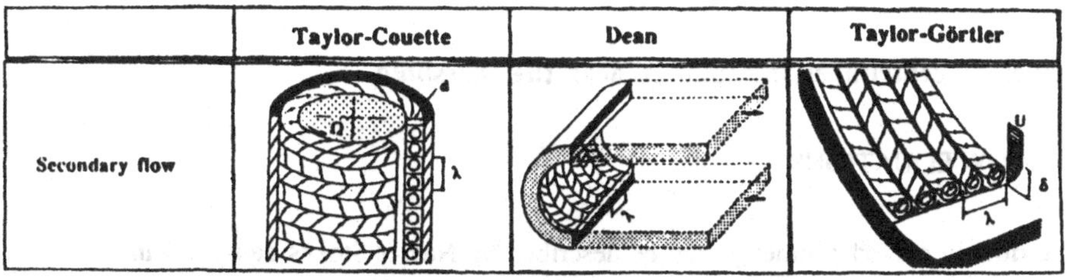
Taylor-Couette
Dean
Taylor-Görtler
Secondary flow
d
λ
U
δ

Fig.1 : Configurations with longitudinal roll pattern

The interest of investigation of centrifugal instabilities is twofold : from the fundamental view of the research, as their threshold has small Reynolds number (Re ~

* Present address : Departement of Physics, The Ohio State University, 174W 18th Avenue, Columbus, OH 43210, USA

E. Tirapegui and W. Zeller (eds.), Instabilities and Nonequilibrium Structures III, 201–216.

100), they present a large range of values in which they can be studied without coupling with other types of instabilities. From practical point of view, they play important role in heat and mass transfert, in aerodynamic systems and in turbomachinery, in pipe flows and in ducts used commonly in hydraulics.

Centrifugal instabilities have been well studied both theoretically and experimentally since the pionnering works of G.I.Taylor[1]. A complete description of these instabilities is given in [2,3]. Actually, many interesting results have been obtained in particular in the Taylor-Couette instability and almost all regimes flow have been discovered experimentally [4,5,6]. The Dean instability [7] has been recently investigated in systematic way because of its practical interest such as the flow in pipes, in ducts[8]. The Taylor-Görtler instability has been discovered by H.Görtler in 1940 [9], it belongs to the so-called *open systems*, the control parameter depends on the boundary layer thickness which grows continuously with the longitudinal coordinate. This variation of the control parameter with the longitudinal coordinate allows to reach different flow regimes in the different parts of the flow with a fixed upstream velocity. Even this instability has undoubtfull practical interest for aerodynamics problems, only few experimental and theoretical results are available [10] and they don't provide yet an accurate understanding of that instability. Different systems with practical interest have been conceived experimental by Wimmer [11] to study different centrifugal instabilities near and between bodies of revolution.

One interesting feature of hydrodynamic systems with centrifugal instabilites is that one encounters all temporal and spatial properties by appropriate choice of the control parameters which are changed more easily than in other systems. For example, variation of the rotation ratio μ of the cylindres gives transition from Taylor vortex flow to spiral vortex flow in the same experimental run.

2. Properties of centrifugal instabilities near the threshold

2.1. BASE FLOW EQUATIONS

The flow in curved geometry may be described by Navier-Stokes equations in cylindrical coordinates (r, θ, z) [12]. The stationary base flow is unidirectional (0,V(r), 0) and satisfies the following equations :

$$\frac{1}{\rho}\frac{\partial P}{\partial r}=\frac{V^2}{r} \qquad \left(\frac{d^2}{dr^2}+\frac{1}{r}\frac{d}{dr}-\frac{1}{r^2}\right)V=\frac{1}{\rho\nu}\frac{\partial P}{\partial \theta} \tag{1}$$

where ρ and ν represent the density and the kinematic viscosity of the working fluid. The first equation expresses the equilibrium between the radial pressure gradient and the centrifugal force. The azimuthal pressure gradient may be zero through the flow (Couette case) or constant (curved channel Poiseuille flow).The general solution of (1) is given by

$$V(r) = A\,r + \frac{B}{r} + C\,r\ln\frac{r}{R} \qquad \frac{1}{\rho\nu}P(r,\theta) = C\;\theta + \int_R^r \frac{V^2}{r'}\,dr' \tag{2}$$

with $C = (1/\rho\nu)\,\partial P/\partial\theta$ = const, the coefficients A, B are determined from the boundary conditions which depend on particular system under consideration. The model flow geometry in which such flow can be realized consists of two coaxial cylindrical surfaces (of radii R and R + d) with or without external pressure gradient.

Without loss of generality, the small gap approximation ($\delta = d/R \ll 1$) is used throughout the paper since it is known that it does only surestimate the threshold value of the instability. The base flow velocity is scaled by V_a, the perturbative radial and axial velocity components are scaled by ν/d while the azimuthal velocity component is scaled by $(\nu/d)\delta^{1/2}$. The characteristic velocity of the base flow V_a is the rotation velocity ΩR of the inner cylinder for Couette flow and the mean velocity across the channel $\langle V \rangle$ for the curved channel Poiseuille flow. The radial and axial coordinates are scaled by d : $r = R + d.x$, $z = z.d$ and the angular coordinate is scaled by $\theta = y\,\delta^{1/2}$. The control parameter of the flow is the Taylor number defined as

$$Ta = \frac{V_a d}{\nu}\left(\frac{d}{R}\right)^{1/2}$$

2.2. INVISCID STABILITY CRITERION

The stability condition of the curved streamline flow under centrifugally-driven perturbations is given by the Rayleigh stability criterion which states that the inviscid flow is unstable when the kinetic momentum of the fluid particle decreases outwards in the radial direction (Fig.1). This is expressed by the Rayleigh discriminant $\Phi(x)$ which measures the resulting specific force acting on fluid particle and is defined as

$$\Phi(x) = \delta\,\frac{dV^2}{dx} \tag{3}$$

The flow is potentially unstable if $\Phi(x) < 0$ and stable if $\Phi(x) > 0$. For nonmonotonic velocity profile, the flow may consist of both unstable and stable parts separated by a nodal surface given by $\Phi(x) = 0$. Application of this criterion to Couette circular flow, Poiseuille curved flow and to a combination of those two velocity profiles show that those flows contain at least one potentially unstable layer and one stable part of the fluid. When the instability sets in the unstable layers, it may penetrate by advection in the stable layer and such a mechanism is called *penetrative instability* [13,14]. This is the case for centrifugal force driven instabilities where the rolls fill the whole gap even if there is one unstable part of the flow.

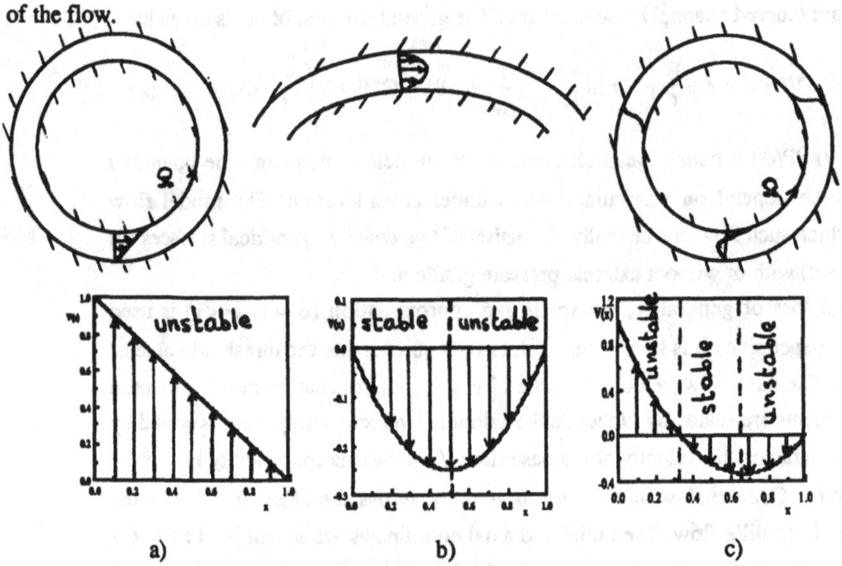

Figure 2: Base flow velocity profile and Rayleigh stability criterion : a)Couette circular flow, b)Poiseuille curved flow, c)Combination of Couette and Poiseuille velocity profiles in Taylor-Dean system

In centrifugally driven instability, viscosity plays a stabilizing effect by determining the threshold of instability and the higher cutoff in the wavenumber spectrum [15].

2.3. STABILITY THEORY FOR CENTRIFUGAL INSTABILITIES

The general equations describing the centrifugal instability in the small gap approximation admit solutions in the form of expansion of the Fourier-Laplace modes $\exp\{st + i(qz+py)\}$, where q and p are the axial and azimuthal wavenumbers respectively and are real quantities because of the boundness of the perturbations at infinity. Available experimental results have shown that two cases are possible : axisymmetric perturbations ($s = 0$, $p = 0$: principle of exchange of stabilities) giving rise to stationary roll pattern (e.g : Taylor vortex flow) and non axisymmetric perturbations ($s = i\omega$, $p \neq 0$) which give traveling roll pattern (e.g : spiral vortex flow).

2.3.1. Axisymmetric stationary states

Axisymmetric rolls pattern is given by the velocity field

$$v'(x,z) = \{u(x)\cos qz,\ v(x)\cos qz\ ,\ w(x)\sin qz\} \tag{4}$$

where the radial and azimuthal velocity components satisfy the following differential equations system obtained after substitution of (4) into linearized Navier-Stokes equations :

$$A\Psi = \mathrm{Ta}\, B\, \Psi \tag{5}$$

$$A = \begin{bmatrix} (D^2-q^2)^2 & 0 \\ 0 & D^2-q^2 \end{bmatrix} \quad B = \begin{bmatrix} 0 & 2q^2V \\ DV & 0 \end{bmatrix} \quad \Psi(x) = \begin{bmatrix} u(x) \\ v(x) \end{bmatrix}, \quad D = \frac{d}{dx} \tag{6}$$

The axial velocity component is given by $w = -Du/q$ from the continuity condition .

The boundary conditions are given by :

$$u = Du = v = 0 \text{ at } x = 0 \text{ and } x = 1 \tag{7}$$

The critical state (q_c, Ta_c) is obtained from the condition $<\Psi^a|L|\Psi> = 0$. Here Ψ^a is the solution of the adjoint problem associated with (5), the operator L is given by

$$L = \begin{bmatrix} -2(D^2-q_c^2) & -2Ta_cV \\ 0 & -1 \end{bmatrix} \tag{8}$$

and where the scalar product $<f,g>$ is defined as follows $<f,g> = \int_0^1 f\, g\, dx$. Numerical calculations have been done for Taylor-Couette, Dean and Taylor-Dean systems, the results of critical parameters are given in the Table 1.

Theoretical predictions agree well with experimental results in extended systems both for Taylor-Couette, Dean and Taylor-Dean flows. Moreover, it has been found that the radial u and spanwise (axial) w velocities have amplitudes an order of magnitude smaller than the streamwise v velocity.

System flow	q_c	Ta_c
Taylor-Couette(μ=0)	3.12	41.2
Dean	3.95	35.92
Taylor-Dean (μ=-2)	3.33	40.3

Table 1 : Critical parameters for axisymmetric stationary modes

Near the onset of the instability ($\varepsilon \ll 1$) and for small $q-q_c$, the envelope of roll pattern may be described by the Ginzburg-Landau equation

$$\tau_0 \frac{\partial A}{\partial t} = (\varepsilon + \xi_0{}^2 \frac{\partial^2}{\partial z^2} - g\,|A|^2)\,A \qquad (9)$$

where $\varepsilon = (Ta - Ta_c)/Ta_c$, τ_0 is the characteristic time of the perturbation, ξ_0 is their coherence length and g is the Landau constant which determines the nonlinear saturation of the amplitude, its value depends on the normalization of the linear eigenvectors. For the Taylor vortex flow, those constants have been calculated by different authors[16-20] and have been verified experimentally by Pfister and Rehberg [21] and by Ahlers et *al* [22]. Both experiment and theory show that Taylor-Couette instability in extended geometry is supercritical bifurcation i.e. $g > 0$, and in that case imperfections due to Eckmann rolls play less important role in extended Taylor-Couette system.

For Dean instability, nonlinear analysis has been performed independently by Daudpota et *al* [23] and Finlay et *al* [24] who showed that Dean instability is supercritical, but no experimental results have been performed to verify the critical nature of that instability. Recent experiments by Ligrani & Niver [25] and by Matsson & Alfredsson [26] have indicated difficulties to approach the threshold value, more sophisticated measurement methods need to be developed to have an accurate value of the threshold.

System flow	τ_0	ξ_0	g
Taylor-Couette(μ=0)	0.038	0.133	0.171
Dean	0.023	0.207	2.903
Taylor-Dean(μ=-2)	0.03	0.020	0.018

Table 2 : Coefficients of the amplitude equation for centrifugal instabilities

For Taylor-Dean system, the constants of the Ginzburg-Landau equation have been calculated for different values of the rotation ration μ, and it has been shown that the Landau constant vanishes for $\mu = -1.18$ [27]. In the phase diagram (μ,Ta_c), this point separates the supercritical branch states and the subcritical branch states and it is called *tricritical point*. Experimentally this point is very close to the stationary-stationary codimension 2 point [28].

2.3.2. Nonaxisymmetric rolls patterns

Centrifugal instabilities may manifest in the form of nonaxisymmetric rolls pattern traveling in both axial and longitudinal directions, this is the case for the spiral vortex flow observed in the Couette flow between counterrotating cylinders and for the traveling inclined rolls pattern observed in the Taylor-Dean system for some range of the rotation ratio μ [28, 29]. In that case the velocity field is given by

$$v(x,y,z,t) = (u(x),\ v(x),\ w(x))\ \exp\{i(\omega t+qz+py)\} + c.c \tag{10}$$

where u(x),v(x) satisfy the following matrix equation :

$$[\mathbf{A} - Ta\,\mathbf{B} - i\,(\omega + p\,Ta\,V)\,\mathbf{K} + i\,p\,Ta\,D^2V\,\mathbf{E}\,]\,\Psi = 0 \tag{11}$$

with the operators **A**, **B** defined above and

$$\mathbf{E} = \begin{bmatrix} 1 & 0 \\ 0 & 0 \end{bmatrix} \qquad \mathbf{K} = \begin{bmatrix} D^2-q^2 & 0 \\ 0 & 1 \end{bmatrix} \tag{12}$$

This equation has been solved for the Couette flow by Krueger et *al* [30] and by Langford et *al* [31], it has been shown that for $\mu < -0.73$, non axisymmetric oscillating states where critical, what confirmed experimental Coles results [5]. In the case of the Taylor-Couette system, the azimuthal wavenumber p is an integer number because of the periodicity in θ in the azimuthal direction.The velocity field (10) may correspond to a traveling wave in both azimuthal direction with phase angular velocity ω/p and in axial direction with a phase velocity ω/q. However the linear stability does not distinguish between the traveling and standing waves, but experiments show that only the traveling roll pattern are realized at threshold (spiral vortex flow in Couette geometry and traveling inclined rolls in Taylor-Dean system), no standing wave has been observed at threshold without external modulation.

For Dean problem, Gibson and Cook [32] have shown that nonaxisymmetric perturbations are critical in the very small approximation limit $\delta < 2.179\ 10^{-5}$, but this situation is almost unrealizable in experimental situation.

For Taylor-Dean system, it has been shown that nonaxisymmetric modes become critical for $\mu \in [0.26, 0.62]$ and result in the competition between two destabilizing mechanisms. Non axisymmetric traveling rolls have been observed for an extended range of μ including this interval. This is due to the existence of recirculation rolls in the base flow close to the free surfaces, effects of which have not been considered in numerical calculations.

The origin of the critical time-dependent modes is not very clarified : in Couette system with counter-rotating cylinders, the base flow has a zero velocity surface inside the gap and one could relate these modes to the result of interaction between the destabilization of the potentially unstable layer and the penetrative mechanism in the potentially stable layer of the fluid. In Taylor-Dean system, they appear in the range of μ for which the flow has two potentially unstable layers of the flow and we relate them to the result of the competition of the destabilizing mechanisms. The effect of the recirculation rolls is apparently to introduce the azimuthal variation of the roll amplitude.

System flow	(q_c, p)	Ta_c	f
Taylor-Couette(μ =-0.75)	(3.47, 1)	58.6	0.03
Taylor-Dean (μ = 0.5)	(5.0 , 0.6)	141.0	18.68

Table 3 : Critical parameters for nonaxisymmetric modes

Tentative description of the envelope equation near critical points without spatial variation has been developed for spiral vortex flow by DiPrima & Grannick [33] and by Demay & Iooss [19]. For the Taylor-Dean inclined roll pattern, the problem has not yet been investigated. The envelope equation is the complex Ginzburg-Landau equation in moving frame of roll pattern :

$$\tau_0 \frac{dA}{dt} = \varepsilon\,(1+i\,c_0)\,A - g(1+i\,c_2)\,|A|^2 A \qquad (13)$$

The constants τ_0, ε, ξ_0 have the same meaning as for the axisymmetric modes.The coefficients c_i describe the oscillatory properties of the spiral vortex flow : c_0 gives the variation of the frequency with the variation of the control parameter, and c_2 is a correction due to the nonlinear amplitude dependence of the frequency. From the results of Demay & Iooss, one has the following values of the constants c_i for spiral vortex flow in the Couette geometry : $c_0 = 0.033$, $c_2 = 2.241$ for $\mu = -0.75$, $\delta = 0.0526$. They found that the bifurcation nature depends also on the gap size δ, for $\delta = 0.0526$, the bifurcation to spiral vortex flow is supercritical ($g > 0$) for $\mu < -0.8$ and subcritical ($g > 0$) for $-0.73 < \mu < -0.8$. But for $\delta = 0.33$, the transition to spiral vortex flow occurs supercritically for $\mu < -0.6$.

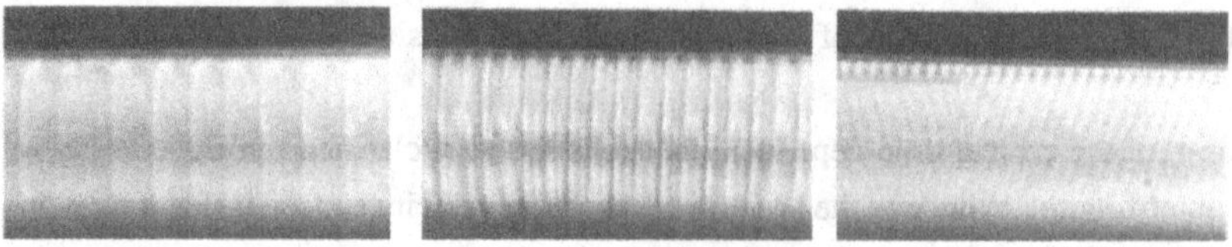

a) b) c)

Figure 3 : Taylor, Dean and traveling roll pattern in Taylor-Dean system : a) $\mu = -0.39$, Ta = 80.1, b) $\mu = 0.85$, Ta = 80.8, and c) $\mu = 0$, Ta = 98.3

3. Higher instabilities of roll pattern

3.1. ECKHAUS AND HALVING INSTABILITY

The neutral stability curve gives a large band of wavenumbers which can be selected by the pattern, however the wavenumber selected band is made smaller by the so called *Eckhaus instability*. It consists in local contraction and dilation of the roll pattern along the axis of the flow pattern. This phenomenon may be described by a weakly nonlinear theory by introducing the phase variable $\phi(t,x)$ described by the Pomeau-Manneville equation [34,35] :

$$\frac{\partial\phi}{\partial t} = D_{//}\frac{\partial^2\phi}{\partial z^2} + D_{\perp}\frac{\partial^2\phi}{\partial y^2} + \ldots \tag{14}$$

where ϕ is the phase allowed to vary along the axial direction and in the azimuthal direction. The Eckhaus boundary is given by the condition $D_{//} = 0$. The Eckhaus instability takes place when $D_{//} < 0$ while $D_{\perp} < 0$ corresponds to a zig-zag(ondulating) instability. For Taylor-Couette system, the coefficient $D_{\perp}$ have been computed by Tabeling[36] who showed that $D_{\perp} > 0$, meaning that the Taylor vortex flow was stable to zigzag instability. The experimental verification of the Eckhaus limit has been done by Burkhalter & Koschmieder[37,38] and by Ahlers and his coworkers[22,39,40]. When the Eckhaus stability boundary is crossed, the number of vortices in finite-length system usually changes by two, that is a pair of vortices is gained or lost.

Consequent theoretical investigation of the Eckhaus instability has been done recently by Riecke & Paap [41,42] in the light of experimental results by Ahlers et *al* for the Taylor vortex flow. The main result by Riecke & Paap is to have shown that the Eckhaus boundary for axisymmetric perturbations depends strongly on radii ratio η : for small gap approximation $\eta = 0.892$, transition to wavy vortices occurs for $\varepsilon = 0.12$ and this is in a good agreement with 3rd order amplitude expansion results. Lowering η increases the range of e for stable Taylor vortex flow and the locus of the transition to wavy vortex flow becomes strongly dependent of wavenumber. In addition Riecke & Paap showed that for $q < q_c$, the Eckhaus boundary is modified by the bifurcation to π-periodic solution $q/2$ which interact nonlinearly with the q solution (2π-periodic).This phenomenon is known as *halving or doubling instability*.

For the Dean vortex flow, Finlay et *al* [24] have shown that for fixed control parameter (Re), bifurcation to 2 pairs of vortices can occur as the wavenumber is decreased, when the control parameter is increased, similar bifurcations can occur and the secondary pair grows in size and strength as Re increases. To the best of our knowledge, there is no reliable experimental verification of Eckhaus instability for Dean vortex flow.

3.2. WAVY MODES ON LONGITUDINAL VORTICES

The secondary instability of the stationary vortex flow occurs via a wavy instability of the longitudinal vortices with a periodic time-dependence of frequency f_1 and a second wavenumber m in the azimuthal direction. The azimuthal wavenumber depends on the boundary conditions of the flow in the azimuthal direction : for Taylor-Couette, it is an integer number because the azimuthal extension is 2π-periodic while for the Dean vortex flow or for the Taylor-Dean system, it is a real number because the azimuthal extension in those 2 cases is $< 2\pi$. The wavy instablity has been observed in the Taylor-vortex flow and has been widely investigated[40,43,44]. Its threshold depends sensitively on the radii ratio, in the small gap approximation, it is close to $\varepsilon = 0.12$, while for $\eta = 0.5$, it is about $\varepsilon > 1$. The azimuthal wavenumber m depends on the radii ratio η and of the rotation ratio μ. The threshold of the wavy vortices depends also on the aspect ratio Γ but in non defined well manner.

For Dean rolls, wavy-like modes have been reported by Ligrani & Niver [25] and recently by Mutabazi on Dean rolls observed in the Taylor-Dean system [27].

The wavy instability might be a common property of longitudinal vortices and it may be described in the small gap approximation by the following field [45] :

$$v(t,x,y,z) = A(t)v(x) \cos qz + B(t)\, v'(x)\, e^{i py} \sin qz \qquad (15)$$

where the amplitudes A(t) of the stationary axisymmetric vortex flow and B(t) of the wavy vortex flow, satisfy the following nonlinear equations at the lowest leading order :

$$\frac{dA}{dt} = \alpha\, \varepsilon\, A - g_1 |A|^2 A + g_{12} |B|^2 A \qquad (16\text{-a})$$

$$\frac{dB}{dt} = \beta\, \varepsilon\, B - g_2 |B|^2 B + g_{21} |A|^2 B \qquad (16\text{-b})$$

The coupling coefficients g_{ij} describe the interaction between stationary vortex flow and wavy vortex flow. The wavenumber p is integer number only for the Taylor-Couette system but it might take any real value for Dean vortices.

To complete this description of wavy vortices, Brand and Cross [43] have represented the velocity field using the phase variables ψ and Φ in the following form :

$$v = |A|\, e^{i\psi} + i|B| e^{i\psi}[e^{i(qy+\Phi)} + c.c] \qquad (17)$$

where the phases ψ and Φ give the position of the vortices along the axial and the azimuthal direction respectively, and they are described by two coupled linear equations :

$$\frac{\partial\psi}{\partial t} = D_1 \frac{\partial^2\psi}{\partial z^2} + C_1(q_y)\frac{\partial\Phi}{\partial z}, \qquad \frac{\partial\Phi}{\partial t} = D_2 \frac{\partial^2\Phi}{\partial z^2} + C_2(q_y)\frac{\partial\psi}{\partial z} \tag{18}$$

with q_y representing the azimuthal wavenumber of the wavy vortex state, $\partial\psi/\partial z$ gives the variation of the axial wavenumber and $\partial\Phi/dt$ gives the variation of frequency of the wavy roll pattern.

3.3. SECOND BIFURCATION OF THE TIME-DEPENDENT ROLL PATTERN

The main feature of time-dependent roll pattern observed in curved geometries is that the rolls do not possess the axial symmetry, this leading to a pattern traveling in both axial and azimuthal directions with different speeds prescribed by the frequency of rolls and the wavenumbers in both directions. The secondary mode of those non axisymmetric roll pattern is characterized by a second frequency and eventually by a second wavelength. The known cases are the interpenetrating spiral in counterrotating cylinders Couette system and the inclined rolls in the Taylor-Dean system.

The recently observed spatio-temporal modulation in Taylor-Dean system has allowed a better understanding of the secondary instability from time-dependent structure at the onset [46]. A second frequency and a second wavelength have been evidenced in the system and they become more dominant in the transition process going over the primary instability characteristics.

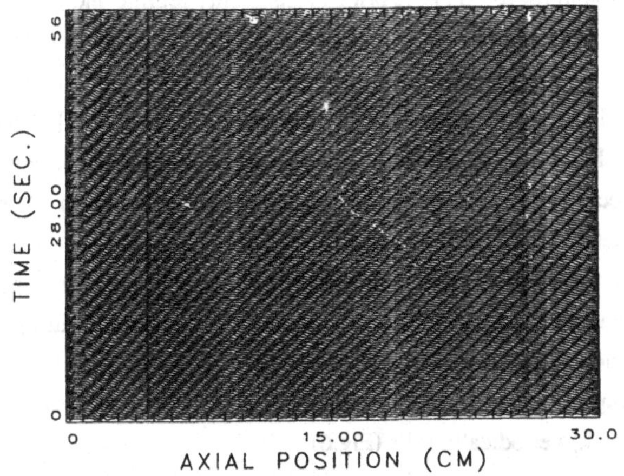

Figure 4-a : Spatio-temporal diagram for traveling roll nonmodulated pattern (μ = -0.1,Ta = 93.2)

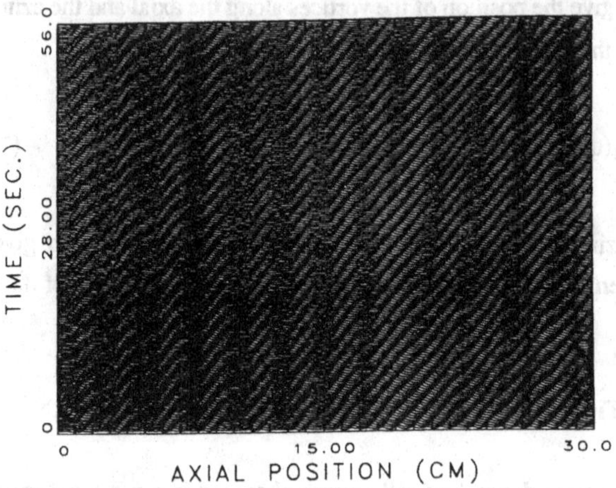

Figure 4-b : Spatio-temporal modulation of traveling inclined rolls in Taylor-Dean system (μ = -0.1, Ta = 96.1)

3.4. DEFECTS IN TIME-DEPENDENT ROLL PATTERN

In extended systems (with large aspect ratio), all time-dependent roll patterns (secondary instability of axisymmetric modes or time-dependent roll pattern as first instability) exhibit a common property of defects [47]. Those defects are of two kinds : the first type of defects occur randomly as succession of annihilation (collision) and regeneration of pairs of rolls. Those defects are local in time and in space (a point in the space-time plot). We call such type *spatio-temporal defects*. These defects have been observed in Taylor-Dean system for $\mu = 0$ in the primary instability regime and look like a result of the phase instability of a time-dependent roll pattern. They don't persist after the second instability has set in. These defects have been detected in many experiments using cross section of vortices and are named sometimes as splitting and merging [25]. Another important feature of these phenomena is that they are not periodic neither in space nor in time. The same type of defects has been observed in Rayleigh-Bénard convection when the oscillation of Bénard rolls set in as a secondary instability mode [48].

The second type is linear defects which separates two traveling in opposite directions, they are sometimes called dislocations, one distinguishes source emitting rolls traveling in opposite directions away from the defect and the sink which absorbs all roll traveling towards it. In the space-time plot, those defects look like a standing wave which emits (source) or absorbs (sink) periodically rolls (Fig.6).

The nature of dislocations is very sensitive to initial conditions and their number increases as the control parameter increases, however their role in the transition to chaos is not yet well elucidated. A tentative description of dislocation motion and spatio-temporal defects has been proposed bv Léga in numerical simulation of a pair of coupled complex Ginzburg-Landau equations [49].

The general feature of all defects is that the number of generating defects (source) is different of that of absorbing defects (sink or collision), this means that the appearance of defects in time-dependent roll pattern breaks the time inversion symmetry [48].

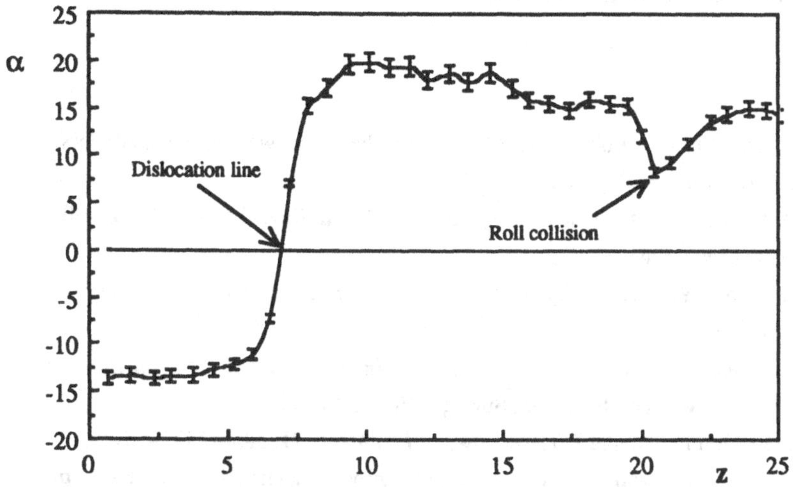

Figure 5 : Dislocation and collision deffect in traveling inclined roll pattern in Taylor-Dean system($\mu = 0$, $Ta = 101.9$), α is the inclination angle and z is the axial position in units of the gap size.

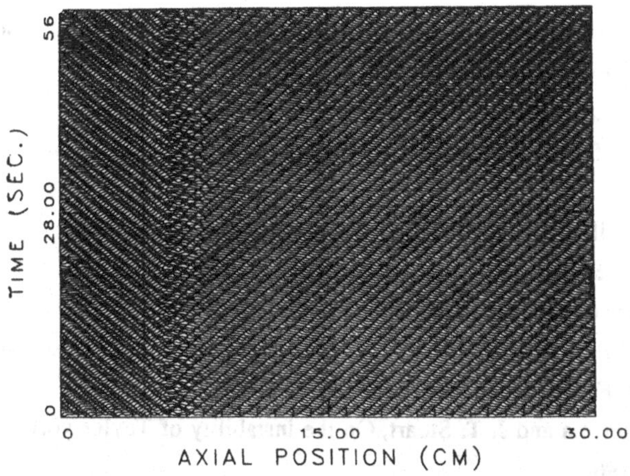

Figure 6 : Space-time diagram of a traveling roll pattern with a source emitting two traveling in opposite directions rolls ($\mu = 0.143$, $Ta = 130.3$)

4. Conclusion

Spatio-temporal properties of centrifugally driven instabilities have been collected in such a way to have a unified model of longitudinal vortices. Experimental data have been placed in a confirmed theoretical background giving a consistent way to understand the first stages of the transition to chaos in curved flows.

References

[1] G.I.Taylor, Stability of a viscous liquid contained between two rotating cylinders, *Phil.Trans. R. Soc. Lond.* **A 223**, 289 (1923)

[2] S. Chandrasekhar, *Hydrodynamic and Hydromagnetic Stability*, chapters VII & VIII, Oxford University Press, 1961

[3] P.G. Drazin and W.H. Reid, *Hydrodynamic Stability*, chapter III, Cambridge University Press, 1981

[4] H.L. Swinney and J.P. Gollub(ed.), *Hydrodynamic Instabilities and the Transition to Turbulence*, Topics in Applied Physics **45**, Springer-Verlag 45 (1981)

[5] A.Coles, Transition in circular Couette flow, *J.Fluid Mech.* **21**, 385-425 (1965)

[6] C.D.Andereck, S.S.Liu, and H.L.Swinney, Flow regimes in a circular Couette system with independently rotating cylinders, *J.Fluid Mech.* **164**, 155 (1986)

[7] W. R. Dean, Fluid motion in curved channel, *Proc. Roy. Soc. Lond. A* **121**, 402-420 (1928)

[8] S.A.Berger, L.Talbot and L.S. Yao, Flow in curved pipes, *Ann. Rev. Fluid Mech* **15**, 461 (1983)

[9] H.Görtler, On the three-dimensional instability of laminar boundary layers on concave walls, *NACA Technical Memorandum* **1375** (1954)

[10] H.Peerhossaini, *Thèse de doctorat es sciences*, Université de Paris 6 (1987)

[11] M.Wimmer, Viscous flows and instabilities near rotating bodies, *Prog. Aerospace Sci.***25**, 43(1988)

[12] L.D.Landau et E.M. Lifshitz, *Mécanique des Fluides*, éd. Mir, 1989

[13] G.Veronis, Penetrative convection, *Astrophys. J.* **137**, 641 (1963)

[14] L. Rintel, Penetratrive convective instabilities, *Phys. Fluids* **10**, 848 (1971)

[15] B.J. Bayly, Three dimensional centrifugal-type instabilities in inviscid two-dimensional flows, *Phys. Fluids* **31**, 56-64 (1988)

[16] A. Davey, R. C. DiPrima and J. T. Stuart, On the instability of Taylor vortices, *J. Fluid Mech.* **31**, 17-52 (1968)

[17] H.Yahata, Slowly-varying amplitude of the Taylor vortices near the instability Point I & II, *Prog. Theor. Phys.* **57**, 347 & 1490 (1977)

[18] R.C. DiPrima, W.Eckhaus and L.A. Segel, Nonlinear wavenumber interaction in near-critical two-dimensional flows, *J. Fluid Mech.* **49**, 705 (1971)

[19] Y.Demay et G.Iooss, Calcul des solutions bifurquées pour le problème de Couette-Taylor avec les deux cylindres en rotation, *J.Méc. Théo. Appl., numéro spécial* , 193-216 (1984)

[20] R.Graham andJ.A. Domaradzki, Local amplitude equation of Taylor vortices and its boundary condition, *Phys.Rev.*A **26** , 1572 (1982)

[21] G.Pfister and I.Rehberg, Space-dependent order parameter in circular Couette flow transitions, *Phys.Lett.* A **83**, 19 (1981)

[22] G. Ahlers, D.S.Cannel, M.A. Dominguez-Lerma and R. Heinrichs, Wavenumber selection and Eckhaus instability in Couette-Taylor flow, *Physica* **D 23**, 202 (1986)

[23] Q.I.Daudpota, P.Hall and T.Zang, On the nonlinear interaction of Görtler vortices and Tollmien-Schlichting waves in curved channel flows at finite Reynolds numbers, *J.Fluid Mech* **193**, 569(1988)

[24] W. H. Finlay, J. B. Keller and J. H. Ferziger, Instability and transition in curved channel flow, *J. Fluid Mech.* **194**, 417-456 (1988)

[25] P.M. Ligrani and R.D. Niver, Flow visualization of Dean vortices in curved channel with 40 to 1 aspect ratio, *Phys.Fluids*, **31**, 3605-3617 (1988)

[26] O.J.E. Matsson and P.H. Alfredsson, *J. Fluid Mech* **210**, 537(1990)

[27] I. Mutabazi, *Thèse de doctorat en Physique*, Université Paris 7 (1990)

[28] I. Mutabazi, J.J. Hegseth, C.D.Andereck and J.E.Wesfreid, Pattern formation in the flow between two horizontal coaxial cylinders with a partially filled gap, *Phys. Rev.*A **38**, 4752 (1988)

[29] I.Mutabazi , C.Normand, H.Peerhossaini and J.E.Wesfreid, Oscillatory modes in the flow between two horizontal corotating cylinders with a partially filled gap, *Phys.Rev.*A **39**, 763 (1989)

[30] E.R. Krueger, A. Gross and R.C. DiPrima, *J. Fluid Mech.* **24**, 521-538 (1966)

[31] W.F.Langford, R.Tagg, E.J. Kostelich, H.L.Swinney and M.Golubitsky, Primary instabilities and bicriticality in flow between counter-rotating cylinders, *Phys. Fluids* **31**, 776 (1988)

[32] R.D. Gibson and A.E. Cook, The stability of curved channel flow, *Q.Jl Mech. Appl. Math.* **27**(2), 149-160 (1974)

[33] R.C. DiPrima and R.N. Grannick, A nonlinear investigation of the stability of flow between Counter rotating cylinders, in *Instabilities in Continous Media*, ed.by Leipholz (1971)

[34] Y. Pomeau and P.Manneville, Stability and fluctuations of a spatially periodic convective flow, *J. de Physique Lett.***40**, L809-L612(1979)

[35] J.E.Wesfreid and V.Croquette, Forced phase diffusion in Rayleigh-Bénard convection, *Phys.Rev.Lett* **45**, 634 (1980)

[36] P.Tabeling, Dynamics of the phase variable in the Taylor vortex system, *J. de Physique Lett.* **44**, L665-L672 (1983)

[37] J.E. Burkhalter & E.L. Koschmieder, Steady supercritical Taylor vortices after sudden starts, *Phys.Fluids* **17**, 1929-1935, 1974,

[38] E.L. Koschmieder, Stability of supercritical Bénard convection and Taylor vortex flow, *Adv.Chem.Phys.* **32**, 109-133, 1975

[39] M.A. Dominguez-Lerma, D.S.Cannell and G. Ahlers, Eckhaus boundary and wavenumber selection in rotating Couette-Taylor flow, *Phys.Rev. A* **34**, 4956 (1986)

[40] G.Ahlers,D.S Cannell and M.A. Dominguez-Lerma, Possible mechanism for transitions in wavy Taylor-vortex flow, *Phys.Rev. A* **27**, 1225-1227 (1983)

[41] H.Riecke and H.G. Paap, Stability and wave-vector restriction of axisymmetric Taylor-vortex flow, *Phys. Rev.* **A 33**, (1986)

[42] H.G.Paap and H.Riecke, Wavenumber restriction and mode interactionin Taylor vortex flow : Appearance of a short-wavelength instability, *Phys. Rev.*A **41**, 1943 (1990)

[43] H.Brand and M.C.Cross, Phase Dynamics for the wavy vortex state of the Taylor instability, *Phys.Rev.*A **27**, 1237 (1983)

[44] P.Chossat, Y.Demay and G.Iooss, Primary and secondary bifurcation in the Couette-Taylor problem, *Japan J. Appl. Math* **2**, 37(1985)

[45] A. Davey, The growth of Taylor vortices in flow between rotating cylinders, *J. Fluid Mech.* **14**, 336-368 (1962)

[46] I.Mutabazi, J.J.Hegseth, C.D.Andereck and J.E.Wesfreid, Spatio-temporal pattern modulations in the Taylor-Dean system, *Phys. Rev. Lett.* **64**, 1729 (1990)

[47] J.Prost, E.Dubois-Violette and E.Guazzelli, Smectics : a model for dynamical systems, in *Cellular Structures in instabilities* ed. by J.E.Wesfreid and S.Zaleski, *Springer Lectures Notes in Physics* (1984)

[48] A.Pocheau, Phase dynamics attractors in an extended cylindrical convective layer, *J. Physique* **50**, 2059 (1988)

[49] J. Lega, *Thèse de doctorat en Physique*, Université de Nice (1989)

SHAPE OF STATIONARY AND TRAVELLING CELLS IN THE PRINTER'S INSTABILITY

M. RABAUD and V. HAKIM,
Laboratoire de Physique Statistique de l'Ecole Normale Supérieure
24 rue Lhomond, 75231 Paris Cedex 05, France.

ABSTRACT. We present shapes of deep experimental cells in viscous directional growth. For parity symmetric cells, the experimental shapes are compared with numerically computed shapes of stable Saffman-Taylor finger with surface tension. The cells that break the parity symmetry are compared with new analytical solutions of the Saffman-Taylor problem.

1. Introduction

Among the pattern forming system, interfacial instabilities are intensively studied. Classical archetypes are the free growth of a crystalline dendrite and the moving interface between two viscous fluids and their respective instabilities have been described by Mullins and Sekerka [1] and Saffman and Taylor [2,3]. Another classical experimental situation for crystal growth is the directional solidification one [4], where an externally imposed temperature gradient stabilizes the front. We have recently introduced a geometry where a viscous interface between air and oil is stabilized by a scalar field and form a one dimensional array of a large number of identical cells [5-7]. In part 2 we rapidly describe this geometry. The equations of motion and their analogy with the equations of directional growth of a crystal are presented in part 3. In part 4 a new analytical family of solution for the Saffman-Taylor problem without surface tension is found. The experimental, numerical and analytical shapes are compared in part 5.

2. Experimental Set-Up and Mains Experimental Results

The cell is formed with two horizontal glass cylinders ($L = 400$ mm long) of different radii ($R_2 = 33 \pm 0.01$ mm and $R_1 = 50 \pm 0.05$ mm) one inside the other but off-centered. Then there is a small gap on one generatrix, at the bottom of the apparatus (Fig. 1). This gap, b_0, can be adjusted with a resolution of 5 µm between 100 µm and 1 mm. These two cylinders rotate independently with tangential velocities V_1 and V_2, with a resolution of 1 mm/s. A small amount of oil is introduced between the cylinders and fills the gap. The oil insures a good wetting of the glass cylinders. We used a silicon oil, Rhodorsyl 47V100, with dynamical viscosity $\mu = 96.5\ 10^{-3}$ Kg/m.s and surface tension T=20.9 10^{-3} N/m at 25°C.

E. Tirapegui and W. Zeller (eds.), Instabilities and Nonequilibrium Structures III, 217–223.

The two air-oil interfaces are observed from below, through the outer cylinder, by a CCD video camera and video pictures can be digitized on a Macintosh IIx computer. A photographic camera is also used.

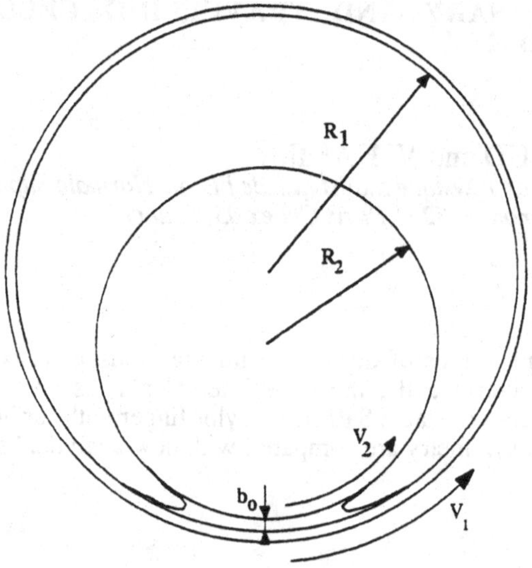

Figure 1. Sketch of a section of the experimental cell showing the two rotating cylinders coated by oil and the two menisci.

2.1. THRESHOLD

For low velocities of the cylinders we observe two linear and parallel dark lines. As the oil wet completely the plexiglass, there are two coating films of oil on the moving surfaces so the dark lines represent the limit of the domain where the gap is completely filled by oil. For larger velocities one of the two interfaces becomes wavy. The instability is supercritical and the front becomes a sinusoid of wavelength λ_c. A linear analysis [5], in the particular case where only one cylinder is rotating, give good predictions for the values of λ_c and for the critical velocity.

2.2. STATIONARY STATE

In case of one rotation only (one non rotating cylinder), the non linear evolution is simple because the cells are stationary. When the speed of rotation increases, the amplitude of the instability grows and the front departs from its sinusoidal shape, the wavelength λ of the pattern decreases at first rapidly then slowly. Well above threshold, the nonlinearity increases and the front becomes a series of parallel fingers separated by thin oil walls (Fig. 2). For all velocity changes, after a long transient (typically of the viscous diffusion time along the interface L^2/ν), the pattern adapts to take a optimum wavelength. Provided the structure has sufficient time to evolve, there is no measurable hysteresis in the wavelength versus velocity curve and the wave number dispersion is less than a few percents on an interface of more than 50 cells.

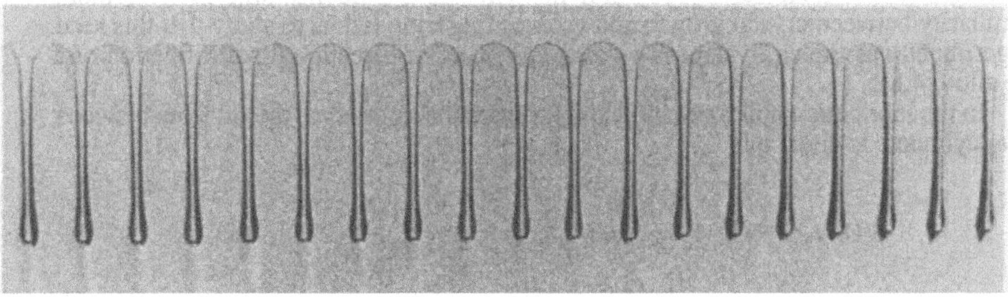

Figure 2. Photograph of a stable pattern of stationary cells.

2.3. TRAVELLING CELLS STATE

If the two cylinders rotate but in contra-rotation, the unstable interface is composed of patches of left or right propagating cells [6,7]. On Fig. 3 it is clear that the cells have broken their left-right symmetry and propagate in the direction of their bending. These domains are separated by source point that emit alternatively right moving and left moving cells, and sink point where there are alternatively absorbed. This propagation is associated with a slow large scale flow along the front : if one single source is present at the center, the fluid moves very slowly towards the extremities of the cylinders and accumulates there.

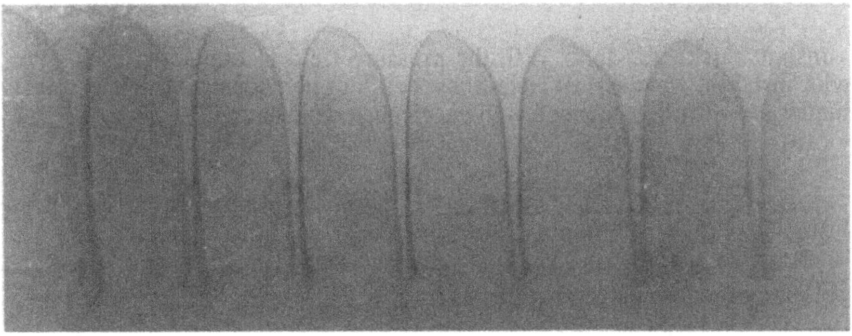

Figure 3. Photograph of a portion of a stable pattern of left propagating cells.

3. Equations of Motion

In a previous work [5] we showed, in the case of a rotating cylinder above an horizontal glass plate, that this cellular instability can be understood as a Saffman-Taylor instability : in the frame of reference of the boundaries a low viscosity fluid (e.g. air) is pushing a large viscosity one (e.g. oil). The thickness gradient of the cell add a stabilizing factor to the linear analysis and this will saturate the instability. Furthermore the similarity between crystal growth and viscous fingering led us to show that this kind of experiment was the equivalent in viscous fingering of directional solidification of a dilute alloy [4,8].

In the lubrication approximation and far from the interfaces, the oil flow between the two cylinders is given by

$$u(x,y,z) = \left(V_2 + \frac{V_1 - V_2}{b(x)} z\right)\hat{x} + \frac{z(z - b(x))}{2\mu}\vec{\nabla}P \qquad (1)$$

where 0x is the tangential direction, 0y is parallel to the axis of rotation of the cylinders and 0z is radial outward (transverse to the gap). The first term of Equ. 1 is a Couette flow due to tangential velocities of the cylinders V_1 and V_2, and the second one a Poiseuille flow due to a local horizontal pressure gradient.
The oil flux in the gap is then :

$$Q = \int_0^{b(x)} u dz = \frac{V_1+V_2}{2} b(x)\,\hat{x} - \frac{b^3(x)}{12\mu}\vec{\nabla}P \qquad (2)$$

The incompressibility of oil impose the flux conservation, div (Q) = 0, and gives the equation of the pressure field :

$$\nabla^2 P + \frac{3}{b}\frac{db}{dx}\frac{\partial P}{\partial x} = \frac{6\mu(V_1+V_2)}{b^3}\frac{db}{dx} \qquad (3)$$

For constant thickness, db/dx = 0, the pressure field is Laplacian as for the Saffman-Taylor instability in classical Hele-Shaw cell. For non constant thickness, this equation is similar to a diffusion equation with the characteristic length :

$$L(x) = \frac{b}{3\left(\frac{db}{dx}\right)} \qquad (4)$$

The pressure equation can be rewritten as :

$$\nabla^2 P + \frac{1}{L(x)}\frac{\partial P}{\partial x} = \frac{2\mu(V_1 + V_2)}{b^2 L(x)} \qquad (5)$$

This equation is similar (but with an opposite sign !) to the impurity concentration equation that govern the growth of a crystal [1] :

$$\nabla^2 C - \frac{V}{D}\frac{\partial C}{\partial x} = 0 \qquad (6)$$

The role of the diffusion length, given by the ratio of the diffusion coefficient D to the velocity propagation V, is here played by the characteristic length L(x) associated to the local gradient of the cell thickness. Here and there, the velocity of the interface is given by the gradient of a diffusing field and by a mass conservation law (conservation of oil flux or impurity flux). As in the case of the directional solidification, the large wavelengths are stabilized by an external parameter (here by the variation of cell thickness, there by a temperature gradient). In both cases, small wavelengths are stabilized by surface tension. The pressure field here is the equivalent of the impurity concentration field of directional growth.

The shape of the cellular interface in this kind of diffusive instability is linked to the value of a dimensionless number, the Péclet number which is the ratio of the wavelength of the pattern to the diffusion length. In our experiment, when the interface is formed by cells of wavelength λ, the Péclet number can be define as : $Pé = \lambda/L(x)$. At threshold the wavelength is large and Pé is about 5, but for larger velocity this number decreases (at high velocity, a typical value is 0.5).

4. New Analytical Solutions for Tilted Cells in the Saffman-Taylor Problem

In the absence of surface tension, Saffman and Taylor [2] have found a family of analytical solution for the stationary shape of a propagating air finger in a linear Hele-Shaw cell. The shape of a centered finger in the plane 0xy (where 0x is along the cell axis and the pressure gradient, and 0y transverse to it) is given by :

$$x = \frac{W(1-\alpha)}{\pi} \ln\left|\cos\left(\frac{\pi y}{\alpha W}\right)\right| \qquad (7)$$

W is the width of the channel, α the relative width of the finger ($0 < \alpha < 1$). In our case we found that one hypothesis of the computation can be relaxed. The air finger does not have to propagate along the 0x axis. Allowing the solution to be periodic in the 0y direction, continuity of finger shape can be restored, and a family of tilted solutions is found for any angle θ. After some algebra in the complex plane of the velocity potential, we found the parametric equations of the interface :

$$\begin{aligned} x &= \frac{W\cos^2\theta}{\pi}\left[u(1-\alpha)\,\mathrm{tg}\theta + (1-\alpha)\ln|\cos u|\right] \\ y &= \frac{W\cos^2\theta}{\pi}\left[u\left(\alpha + \mathrm{tg}^2\theta\right) + (1-\alpha)\,\mathrm{tg}\theta\,\ln|\cos u|\right] \pm k\pi \end{aligned} \qquad (8)$$

with $-\pi/2 < u < \pi/2$ and k an integer number. For $\theta = 0$ this set give back Equ. 7. α is the ratio of the width of the finger to the periodicity W of the pattern, θ is the angle of the direction of propagation of the fingers with the 0x axis.

So, without surface tension, the Saffman-Taylor problem admit a continuous family of solution of an array of tilted solutions as in our experiment for contra-rotating cylinders. A typical solution for $\theta = 10°$ and $\alpha = 0.82$ is presented on Fig. 5b. Taking into account surface tension effect, McLean and Saffman [9] have found numerically the profile of the non tilted Saffman-Taylor fingers. We believe that, by a similar procedure, tilted solutions with surface tension can be numerically obtained.

5. Experimental Results and Comparisons

There is a well-known analogy in the small Péclet limit between cells in directional growth and the Saffman-Taylor fingers [10,11]. As our experiment is a viscous instability similar to the directional growth, a close similarity of shape is not surprising. On Fig. 4 we superimpose a picture of experimental deep cell, obtained for a small Péclet number, with a numerical shape of the Saffman-Taylor fingers with surface tension having the same relative width α. This numerical shape have been obtained by M. Ben Amar using a code similar to the one described in Ref. 9. The agreement between the two is quite good. This agreement can probably be improved by using a numerical code that would take into account the effect of the Péclet number.

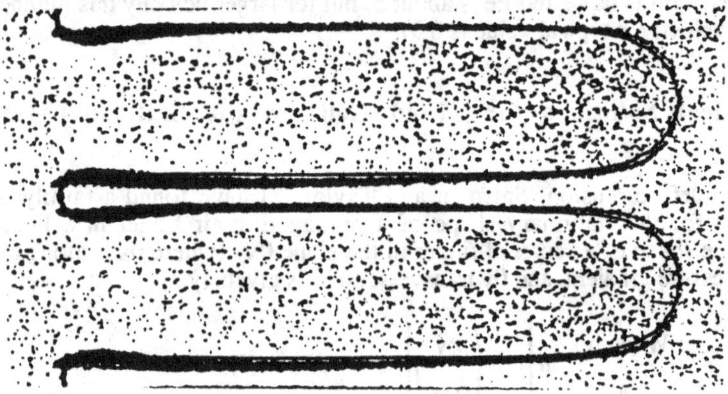

Figure 4. Superimposed on experimental cells (b_0 = 0.4 mm) a numerically calculated Saffman-Taylor profile corresponding to $\alpha = 0.82$.

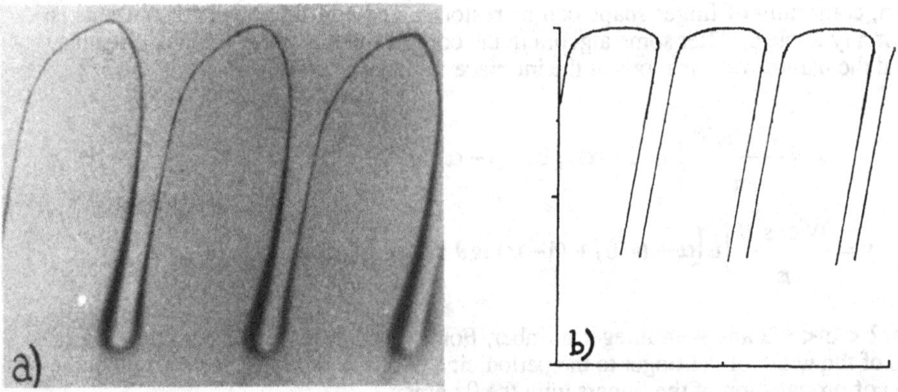

Figure 5. a) Photograph of three propagating cells. b) Analytically calculated Saffman-Taylor profiles (Equ. 8) corresponding to the same $\alpha = 0.82$ and $\theta = 10°$ values.

In case of propagating cells that break the parity symmetry (Fig. 3), we unfortunately do not have numerical shapes that takes surface tension into account. On Fig. 5 we compare an experimental tilted cell (a) with the shape given by Equ. 8 (b). As the experimental α value is large, surface tension effects are important and the analytical shape is a bad approximation (in the same way, Equ. 7 is a bad approximation of a symmetric finger for large value of α).

6. Conclusion

The deep cells of directional viscous fingering are well described by Saffman-Taylor finger profiles. This can be easily understood along the lines of Ref. 10 and 11. When the two cylinders are contrarotating, travelling cells appear. (The same analogy would lead us to expect that travelling cells with deep grooves could appear in directional solidification for some parameter range.) We have found new analytical travelling finger solutions in order to describe their shape. A better comparison with the experimental profiles should take into account surface tension and will require some numerical work. However the main puzzling experimental result remains unexplained : why do cells travel when the cylinders are in contrarotation ?

Acknowledgment : We want to thank M. Mashaal and M. Ben Amar for computing the shape of symmetric Saffman-Taylor fingers with surface tension, and Y. Couder, S. Michalland and H. Thomé for their participation to the experiment.

[1] Mullins W.W. and Sekerka R.F. (1964) J. of Appl. Phys. **35**, 444.
[2] Saffman P.G. and Taylor G.I. (1958) Proc. Roy. Soc. London, A **245**, 312.
[3] Bensimon D., Kadanoff L.P., Liang S., Shraiman B.I., and Tang C. (1986) Rev. of Modern Phys., **58**, 977.
[4] de Cheveigné S., Guthmann C., and Lebrun M.M. (1986) J. Phys. (Paris) **47**, 2095 and Bechhoefer J., Simon A., Libchaber A. and Oswald P. (1989) Phys. Rev. A **40**, 2042.
[5] Hakim V., Rabaud M., Thomé H. and Couder Y., to appear in the proceedings of Nato Advanced Research Workshop, (Cargèse 1988) "New Trends In Nonlinear Dynamics and Pattern Forming Phenomena : The Geometry of Nonequilibrium", editors P. Coullet and P. Huerre.
[6] Rabaud M., Michalland S. and Couder Y. (1990) Phys. Rev. Lett. **64**, 184.
[7] Couder Y., Michalland S., Rabaud M. and Thomé H., to appear in the proceedings of Nato Advanced Research Workshop, (Streitberg, September 1989) "Nonlinear Evolution of Spatio-Temporal Structures in Dissipative Continuous Systems", editors F.H. Busse and L. Kramer.
[8] Kurowski P. (April 1990) Thèse de l'Université Paris VII.
[9] McLean J.W. and Saffman P.G. (1981) J. Fluid Mech. **102**, 455.
[10] Dombre T. and Hakim V. (1987) Phys. Rev. A, **36**, 2811.
[11] Mashaal M., Ben Amar M. and Hakim V. (1990) Phys. Rev. A **41**, 4421.

SECONDARY INSTABILITY OF CELLULAR STRUCTURES IN DIRECTIONAL SOLIDIFICATION

Alain Karma

Physics Department, Northeastern University, Boston, MA 02115, USA

Pierre Pelcé

Laboratoire de Recherche en Combustion, Université de Provence - St. Jerome, 13397 Marseille CEDEX 13, France

ABSTRACT

We discuss the stability of large amplitude cellular structures observed in directional solidification. An oscillatory instability of cell tips is found above some critical growth rate in an effective interface model of the full free boundary problem which is derived in the small Peclet number limit. This instability may be responsible for the coherent type of sidebranching observed in the neighbourhood of the cell to dendrite transition.

1. INTRODUCTION

During a typical directional solidification experiment a thin sample containing a dilute binary liquid mixture is drawn at constant velocity U in the presence of a temperature gradient G [1]. The temperature gradient is fixed in the frame of the laboratory and, in steady-state, the solid-liquid interface is therefore constrained to grow at a rate which is equal to the drawing velocity of the sample. The position of the interface in the temperature gradient is fixed by the global conservation of impurity condition. That is, the interface will grow at a temperature T_o, such that the concentration of impurity in the solid phase at this temperature $C_S(T_o)$ is equal to the concentration of impurity in the sample C_∞, far away from the solid liquid interface.

When the velocity of the interface exceeds a threshold velocity U_1 the planar interface undergoes a morphological instability at finite wavenumber [2]. As the drawing velocity is increased further the interface is typically oberved to evolve into a spatially periodic cellular structures. At velocities a few times U_1 this structure is already highly nonlinear. The depth of the cells is several times the cell spacing, and the shape of the cells resembles the Saffman-Taylor fingers observed in the Hele-Shaw apparatus [3]. This resemblance in shape between solidification cells and viscous fingers is a consequence of the close mathematical resamblance of the equations of interface motion for the two problems (in the small Peclet number limit), the pressure field in the fingering problem playing a role analogous to the composition field in the solidification problem. This analogy was pointed out independently by Pelcé and Pumir [4] and Karma [5], and used by Dombre and Hakim [6] who provided a detailed analytical treatment of solidification fingers in the small Peclet number limit. The presence of a continuum of cell spacing predicted by Dombre and Hakim was confirmed numerically by Ben Amar and Moussallam [7].

E. Tirapegui and W. Zeller (eds.), Instabilities and Nonequilibrium Structures III, 225–231.

As the velocity of the interface is increased even further the steady-state array of deep cells is observed to evolve into a time dependent state above a second critical velocity U_2 [8]. It has not yet been determined experimentally if this second transition is really a sharp transition or if steady-state cells evolve smoothly into time dependent dendrites over a narrow range of velocity around U_2.

One aspect of the transition from cells to dendrites which seems at first surprising is that sidebranching events behind cell tips appear to be correlated on neighbouring cells. One of the experimental photograph of Eshelman, Seetharaman, and Trivedi [8] has been reproduced in Fig. 1. The correlation of sidebranches can be seen most clearly in the more anisotropic material (pivalic acid-ethanol system shown in Fig. 1a). The correlation extends over several cells and is marked by the appearance of equally spaced strips

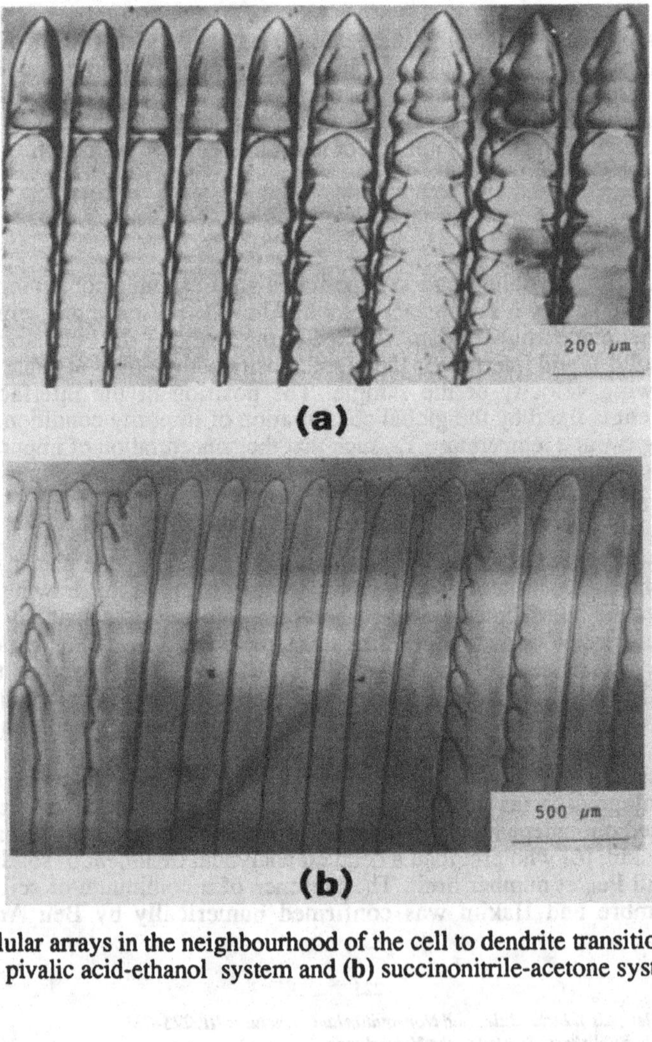

Fig. 1 Cellular arrays in the neighbourhood of the cell to dendrite transition [Ref. 8]: **(a)** pivalic acid-ethanol system and **(b)** succinonitrile-acetone system.

parallel to the solidification front. At this point a more detailed experimental analysis is needed to resolve the sharpness of the cell-dendrite transition and the presence of persistent oscillations. Nevertheless, there exists the possibility that in directional solidification cells evolve into dendrites as a result of linear oscillatory instability. This does not exclude the possibility that thermal or composition fluctuations, which are believed to be responsible for sidebranching in isolated dendrites, also play an important role in directional solidification. In particular, this mechanism should be dominant when the dendrite tips becomes less affected by the composition field of their neighbours at larger velocities.

2. EFFECTIVE INTERFACE MODEL

Motivated by the experimental results of Ref. 8, we have performed a linear stability analysis of steady-state cells in the small-Peclet-number limit, restricting our analysis to perturbations of the array of infinite wavelength [9]. The stability analysis of the highly nonlinear cellular structure is in general very difficult to treat analytically. However in the small Peclet number regime this analysis was made possible by first reducing the full two-dimensional free boundary problem to an essentially one dimensional problem where the cell tips act as point sources in a semi-infinite medium. The envelope of cell tips forms an "effective interface" on which new boundary conditions which depend on the underlying microstructure are applied.

To see how this reduction to a one dimensional problem is possible, consider perturbing slightly the position of the cell tips. Close to the tip of the cell the concentration field varies on the scale of the cell spacing Λ which is very small compared to the diffusive length $\mathbf{l} = \mathbf{2D/U}$ (here $\mathbf{D}$ is the coefficient of solute diffusivity in the liquid and the Peclet number $\mathbf{P} = \mathbf{U\Lambda/2D} \ll \mathbf{1}$). This perturbation will relax on a time scale $\Lambda^2/\mathbf{D}$ which is very small compared to the diffusive time scale $\mathbf{D/U^2}$ on which the essentially one-dimensional boundary-layer of composition ahead of the interface relaxes. Consequently, the tip region can be assumed to remain quasistationary on the slow diffusive time scale during the displacement of the cell tips.

Of course the tail region of the cells relaxes to a tip perturbation on a time scale which is comparable to the diffusive time scale. However we make the plausible assumption that the relaxation of the tail region will not affect the motion of the tip region. This is the essential assumption on which all of our analysis is based. When the tail region is ignored, the concentration flux released by the cells is governed solely by the position $\mathbf{z_0}$ of the cell tips (i.e. the tip composition) and the relative width λ of the cell (equal to the ratio of the width of solid occupied by the cell in the tip region to the cell spacing). The non-locality left in the free boundary problem is then contained in the dynamic response of the mono-dimensional boundary-layer of impurity ahead of the interface to the varying flux released by the cells. The equations governing the reduced free boundary problem for the effective interface are of the form [9]:

$$D\frac{\partial^2 u}{\partial z^2}+U\frac{\partial u}{\partial z}=\frac{\partial u}{\partial t} \qquad (1)$$

$$u_i = 1 - \frac{z_o}{l_T} \qquad (2)$$

$$-D\left[\frac{du}{dz}\right]_i = U\left[\frac{1}{2v}+\lambda\left(1-(1-K)\frac{z_o}{l_T}-\frac{1}{2v}\right)\right] \qquad (3)$$

$$\lambda = f(C) \quad ; \quad C=\frac{\Lambda^2 U}{D d_o}\left[1-(1-K)\frac{z_o}{l_T}-\frac{1}{2v}\right] \qquad (4)$$

where we have defined **u (z,t) = K (C(z,t) - C$_\infty$) / C$_\infty$ (1-K)** with **C(z,t)** the concentration of impurity in the liquid phase and **K** the partition coefficient. We have also defined the thermal length $\mathbf{l_T}$ **= m C$_\infty$ (1-K) / KG** where **m** is the absolute value of the slope of the liquidus, the capillary length $\mathbf{d_o}$, and the dimensionless velocity v which is the ratio of the thermal length to the diffusion length.

Eqn. 1 is the diffusion equation for the composition field in a frame moving at constant velocity **U**. Eqn. 2 is the boundary condition for the cell tip composition $\mathbf{u_i = u\ (z_o,t)}$. Note that the effect of surface tension on the tip composition are of order $\mathbf{d_o / \Lambda \sim P}$ and can be neglected. Eqn. 3 is the conservation of impurity condition for the effective interface which is simply the statement that the composition flux released by the cell has to be diffused away in the boundary-layer ahead of the interface. This flux depends on the tip composition and therefore on the tip position $\mathbf{z_o}$ but also on shape of the underlying microstructure which here is controlled by the relative cell width λ.

The dependence of λ on **U** and $\mathbf{z_o}$ is contained in eqn. 4. The function **f(C)** is determined by solving the two-dimensional steady-state problem in the cell tip region. As mentioned previously, in this region, the concentration field varies on the scale of the cell spacing Λ which is very small compared to the diffusive length **l = 2D/U**. As a result, the two-dimensional diffusion equation in the moving frame can be approximated by the Laplacian. The free-boundary problem in the tip region is then formally identical to the Saffman-Taylor free boundary problem which was solved by McLean and Saffman [10]. The function **f(C)** is a decreasing function of **C** which asymptotes 1/2 at large **C** and has been determined numerically by these authors. When a finite amount of surface tension anisotropy is present the function **f(C)** asymptotes to zero at large **C** [11].

3. OSCILLATORY INSTABILITY

The steady-state solutions of the reduced free-boundary problem (eqns. 1-4) are given by:

$$1 - \frac{z_o}{l_T} = \frac{K\lambda + (1-\lambda)/2v}{1-(1-K)\lambda} \qquad (5)$$

Since for a fixed velocity **U** the relative width λ is uniquely determined by the cell spacing Λ via the control parameter **C** and the function **f(C)** (eqn. 4), eqn. 5 represents a one-parameter of steady-state shapes with varying cell spacing and tip position.

A linear stability analysis of eqns. 1-4 for perturbations of the form:

$$\delta u = A \exp\left[\omega t + q(z - z_o)\right] \quad ; \quad \delta z_o = B \exp\left[\omega t\right] \qquad (6)$$

can be readily performed and yields a quadratic eigenvalue equation for the dimensionless amplification rate Ω (C,v) (only the branch of solution corresponding to Re(q) < 0 is physically allowed) which is given in Ref. 9. The physical origin of the oscillatory instability can be best understood by considering limiting cases of the eigenvalue equation for Ω (C,v). In the limit $\lambda = 1$ and **df(C)/dC = 0**, the growth rate is simply:

$$\Omega = -(1/v)(1 + 2K) \pm i\sqrt{2K/v} \qquad (7)$$

and the interface undergoes damped oscillations. The damping is due to the restabilizing effect of the temperature gradient on long-wavelength perturbations and the oscillations to the delayed response of the boundary-layer of composition ahead of the interface to changes in the tip composition. In the limit of vanishing temperature gradient or large velocity $1/v = 0$, the growth rate becomes:

$$\Omega = \frac{-C\, f(C)\, \dfrac{df(C)}{dC}}{\left[\dfrac{d}{dC} C\, f(C)\right]^2} \qquad (8)$$

which is positive since **f(C)** is a decreasing function of **C**. This limit corresponds to the growth of a crystal in a capillary tube analysed previously by Pelcé [12]. The destabilization comes from the fact that a perturbation which increases slightly the finger width λ will slow down the tip, because more solute is being rejected, and consequently will continue to develop since a slower tip yields an even larger finger width. This type of instability has been observed experimentally during a crystal growth in a capillary tube experiment [13].

In general, the experimental system operates at a point which lays in between the two limiting cases described by eqns 7 and 8, and in this range there exists a critical value v_c at which the oscillatory instability will first appear. To determine if this oscillatory mode is really responsible for the observed cell to dendrite transition and coherent sidebranching several questions remain to be answered. Firstly, we have only considered here perturbations of the cellular array of infinite wavelength. Long-wavelength modes, which are known in the eutectic system [14] to destabilize the array for small cell spacings, can interact with the oscillatory mode. For instance, it may happen that the oscillatory mode is only unstable for cell spacings which are unstable to long-wavelength modes. In this case, coherent sidebranching would only be observed as a transient dynamical state as the array evolves to larger cell spacings by elimination of cells. Secondly, our calculation is restricted to the small Peclet number limit. In the experiment of Ref. 8, this number is about 0.5, and finite Peclet number effects may affect significantly the stability boundary of the oscillatory mode.

Recently, Kessler and Levine [15] have performed numerically a stability analysis of large amplitude cellular structures in the symmetric model of directional solidification (which assumes equal diffusivity of impurities in the liquid and in the solid phases) and compared their results to our analytical predictions. They find no evidence for an oscillatory instability and claim, on the basis of their results, that the assumption in our analysis that the relaxation of the tail region does not affect the tip region must be incorrect. Although their numerical results for the symmetric model may indeed be correct, they can not be compared directly to our results for the physically more realistic one-sided model. Because of solid diffusion, the coupling to the tail region may be more important in the symmetric model and restabilize the oscillatory mode. In addition, these authors have performed their stability analysis for relatively large value of the dimensionless surface tension parameter $\mathbf{d_0}/\lambda$ (which is less than 0.001 in the experiment). The oscillatory mode may be stable for such large surface tension parameter but unstable in the experimentally relevant range.

An analysis similar to the one of Kessler and Levine remains to be performed for the one-sided model of directional solidification to resolve these issues. However, even in the symmetric model, for large enough velocity one would expect an oscillatory instability to occur. This is because, in this model also, the equations of interface motion reduce to the unstable limit of crystal growth in a capillary tube at infinitly large values of v.

REFERENCES

[1] J. S. Langer, Rev. Mod. Phys. **52**, 1(1980).

[2] W. W. Mullins and R. F. Sekerka, J. Appl. Phys. **35**, 444 (1964).

[3] P. G. Saffman and G. I. Taylor, Proc. Roy. Soc. **A245**, 312 (1958).

[4] P. Pelcé and A. Pumir, J. Cryst. Growth **73**, 337 (1985).

[5] A. Karma, Phys. Rev. **A34**, 4353 (1986).

[6] T. Dombre and V. Hakim, Phys. Rev. **A36**, 2811 (1987).

[7] M. Ben Amar and B. Moussallam, Phys. Rev. Lett. **60**, 317 (1988).

[8] M. A. Eshelman, V. Seetharaman, and R. Trivedi, Acta Metall. **36**, 1165 (1988).

[9] A. Karma and P. Pelcé, Phys. Rev. **A39**, 4162 (1989); Europhys. Lett. **9**, 713 (1989).

[10] J. W. McLean and P. G. Saffman, J. Fluid Mech., **102** 445 (1981).

[11] A. T. Dorsey and O. Martin, Phys. Rev. **A35**, 3989 (1987).

[12] P. Pelcé, Europhys. Lett. **7**, 453 (1988).

[13] J. Bechhoefer, H. Guido, and A. Libchaber, C. R. Acad. Sci. Paris **306**, Serie II, 619 (1988).

[14] J. S. Langer, Phys. Rev. Lett. **44**, 1023 (1980).

[15] D. Kessler and H. Levine (Preprint, 1989).

VELOCITY FIELD STRUCTURE AND SEMIQUANTITATIVE ANALYSIS OF TRACER DISPERSION IN A TAYLOR-VORTEX FLOW OF WIDE GAP

A. Barrantes- A. Calvo -M. Rosen*
and J. E. Wesfreid**

ABSTRACT. An experimental study of the velocity field structure of stationary Taylor Vortex Flow of Wide Gap is performed by visualization in a range of high Reynolds numbers. The tracer dispersion in the flow field shows a short wavelength modulation throughout the structure.
A semiquantitative study of the tracer dispersion is done, finding the existence of a regime of anomalous diffusion, which is compared with the theoretical model of Pomeau, Pumir and Young [1] for this particular flow field.

1- INTRODUCTION

We present here a global description of a Taylor Vortex Flow of Wide Gap. It was used a Taylor-Couette apparatus with the outer cylinder and end plates at rest.

It was used destilled water as the working fluid (kinematic viscosity $\nu = 0.01\ cm^2 / sec$). Its temperature was mainteined at 20.0 + 0.1 °C by water circulation through an external sleeve. The lenght of the fluid-filled was fixed by axially moving and non-rotating rings. The total lenght L is variable up to 320 mm. Most of the results were obtained with an aspect ratio $\Gamma = L / d = 9.91$.

The radius of the inner cylinder is R_1 = 28,25 mm and the gap between both cylinders is d = 28.25 mm , resulting a radius ratio $\eta = 1/2$.

For this geometry ten Taylor rolls were formed in a range of Reynolds number between Re = 480 and Re = 600 (the theoretical critical Reynolds number is 68,19 [2]). It was used as the Reynolds number $Re = \Omega d^2 / \nu$, where Ω (rad/sec) is the angular speed of the inner cylinder, and $\nu (cm^2/sec)$ is the kinematic viscocity. The observation of the fluid field was performed by visualization using a dilute solution of fluoresceine in water. The concenration gradients of the tracer allowed us to observe the separatrices between rolls, showing particullarly short wavelength modulation, with alternative thick and thin separatrices.

The experimental results were compared with the numerical simulations of streamlines obtained by Fassel and Booz[3].

A semiquantitative analysis of the tracer dispersion in the Taylor vortex flow was performed and comparedwith thetheoretical prediction of Pomeau, Pumir and Young about the anomalous diffusion process [1] in a linear array lattice of two dimensional stationary cells.

* Dto. de Fisica- FIUBA- Paseo Colon 850- (1063) Bs. As. - ARGENTINA
** ESPCI- LHMP- 10 Rue Vauquelin- 75231- Paris 05- FRANCE

E. Tirapegui and W. Zeller (eds.), Instabilities and Nonequilibrium Structures III, 233–238.

2- EXPERIMENTAL RESULTS

We studied the dispersion process in the flow pattern of ten cells.

A pulse of dye was injected in different points of a peripherical streamline, such as a separatrix or close to the boundary layer of the inner cylinder.

It is considered now the case where the inyection at time $t = 0$ is localized close to the inner cylinder.

The dispersion of the tracer is shown in figure 1 in a sequence of pictures which were taken at different times after the inyection.The main feature of the transport process can be seen: the tracer is carried from the inyected cell up to the end of the apparatus, spreading along the separatrices and streamlines close to the wall of both cylinders. By the same time, the tracer initially localized in the peripherical streamlines starts to contaminate the cells.

An enhacement of the coloured layer along the separatrices is produced, and the same happens with the coloured layer close to the cylinders.

Figure 2 shows the state of the system 60 minutes after the inyection. It is shown that there exist two types of separatrices. We have called them "thin" and "thick" separatrices. They correspond to the outflow streamlines (towards the outer cylinder) and to the inflow streamlines (towards the inner cylinder) respectively.

It can be seen the cells structure loose its symetry with respect to a vertical and horizontal axis through the cell centre. The centres of the streamlines pattern are displaced from the cell centre and are shifted towards the cell boundaries. The centre of the pattern additionally moves towards the outer cylinder.

We present here the results of five experiments.

In each one a sequence of 15 pictures was taken. The evolution of the concentration profile as a function of time was obtained using densitometry techniques.

Typical results are shown in figure 3, the maxima of the concentration are located on the separatrices. It can be seen the enhancement of the coloured layers around the separatrices.

The results of one of the experiments is shown in figure 4, where the half-width of two of the peaks (one belonging to a thin separatrix and the other to a thick one) was measured as a function of time and plotted in log-log scale.

In both graphs a different behaviour of the temporal evolution of the width of the coloured layer, ε , for the thin and thick separatrices is well observed for times bellow 700 seconds.

3- DISCUSSION:

The observation of the flow filled agrees with the numerical results of the streamlines pattern in a Taylor vortex flow of wide gap obtained by Fassel and Booz [2] , (figure 5). The outflow streamlines are more compressed than the inflow streamlines.

It was observed that the typical velocity in the outflow streamlines (thin separatrices) is faster than the typical velocity in the inflow ones (thick separatrices). Therefore, the transit time along the former, $\tau_{thin} = 3$ sec, is smaller than the transit time along the latter, $\tau_{thick} = 5$ sec.

During the transit time along each separatricex, the tracer invades the cell and reaches, by molecular diffusion a distance $\varepsilon_{thin} = (D_m \tau_{thin})^{1/2}$ from the thin separatrix and a distance $\varepsilon_{thick} = (D_m \tau_{thick})^{1/2}$ from the thick one.Thus the enhancement of the width of the coloured layers in both separatrices have initially different behaviours.

Now we compare our results with the theoretical prediction of the anomalous diffusion process of Pomeau et al [1] . They studied the contamination of a passive tracer in a linear array of two

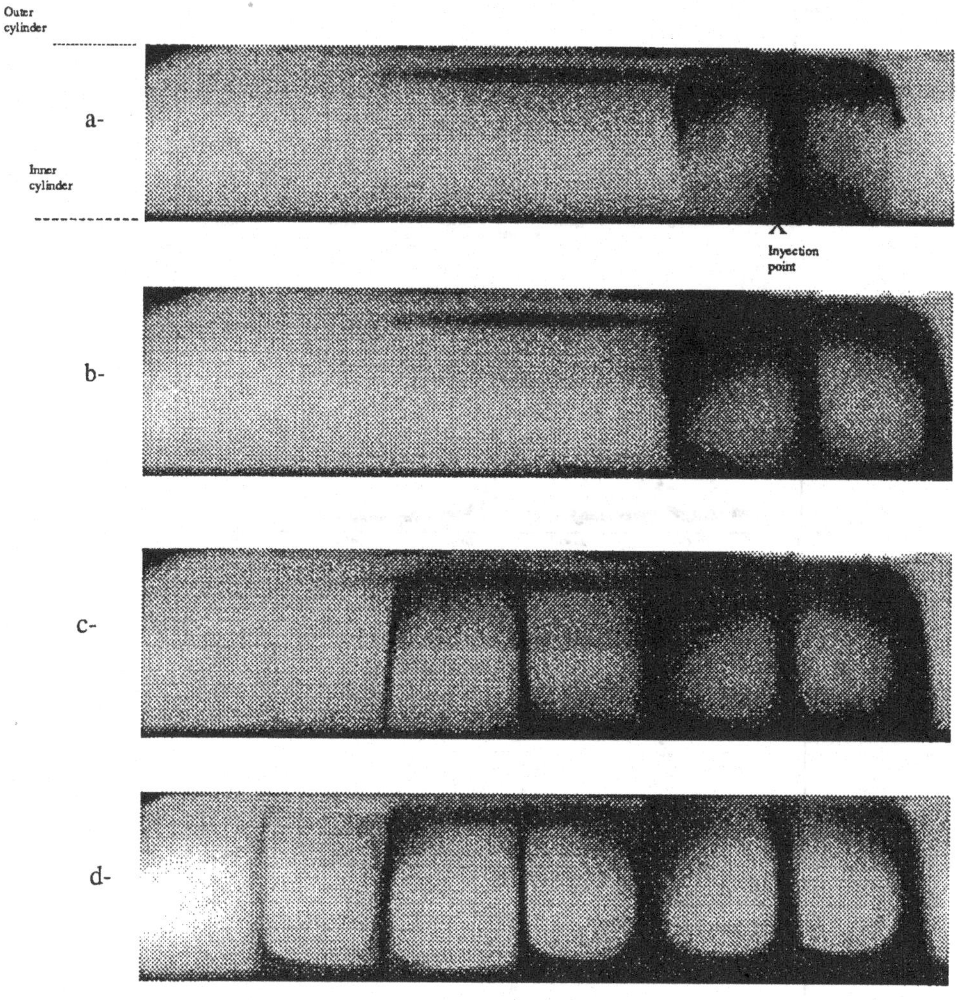

Figure 1: A sequence of four pictures at different times after the inyection of the tracer at a Re= 540. The inyection point is localized close to the inner cylinder. a- 32 sec; b- 73 sec; c- 123 sec and d- 240 sec.

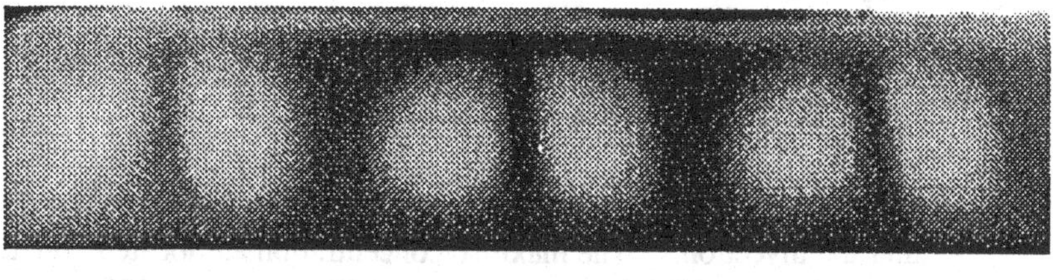

Figure 2: The state of the system 3600 seconds after the inyection.

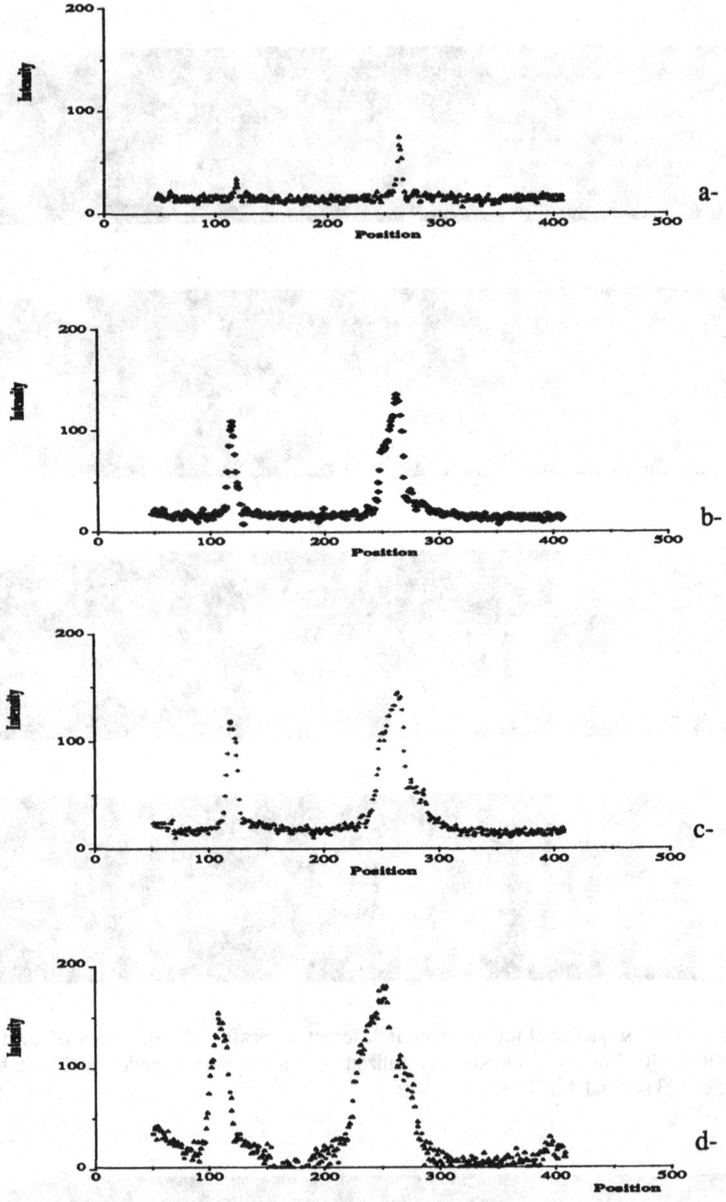

Figure3: The evolution of the concentration profile at different times after the inyection. The maxima concentration are located on the separatrices. a- 180 sec; b- 540 sec; c- 780 sec and d- 1800 sec.

a -

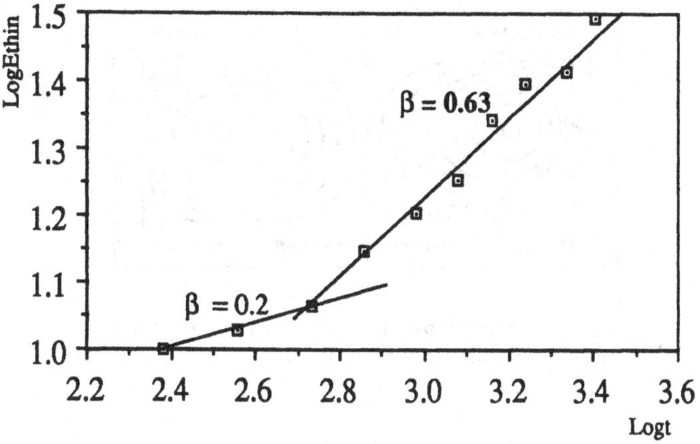

b -

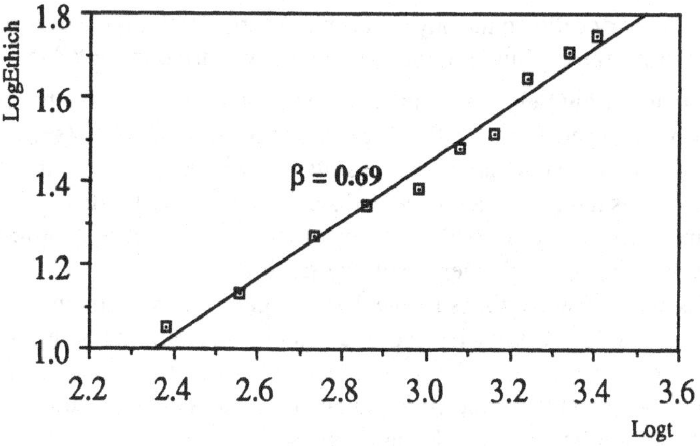

Figure 4: Evolution of the width of the coloured layer obtained by measuring the half-width of the peaks and plotted in log-log scale. a - Thin separatrix and b - Thick separatrix. (Re = 540 and the inyection point located in the separatrix between the 5th and 6th cell)

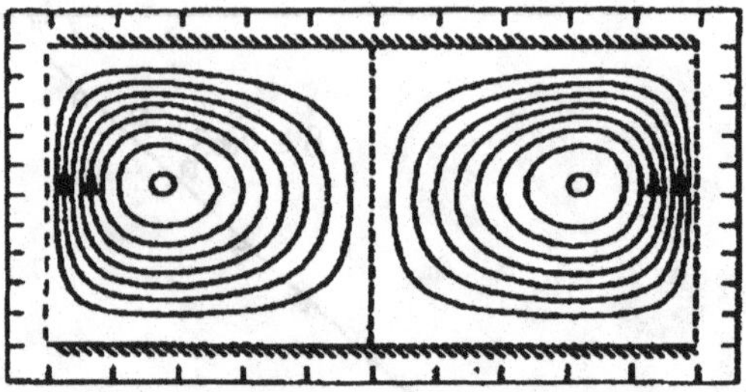

Figure 5: Streamline pattern of the Taylor Vortex Flow in the case of η = 1/2 for a Re = 150.

dimensional stationary cells, in the physical limit of large Peclet number. In the case of rigid boundary conditions the evolution of the thickness ε with time is $\varepsilon = t^{2/3}\,Pe^{-2/3}$

In our particular flow field the exponents in the law $\varepsilon = t^{\beta}$ for the thin and thick separatrices were found between 0.6 and 0.7 for times above 700 seconds (see figure 4).

Although the qualitative nature of our results, they agree with the theoreticall exponent obtained under the asumption of an idealized symetric velocity field.

But for times below 700 seconds, the exponent for the thin separatrices is : 0.2, meanwhile the exponent for the thick ones is still the same value.

This results shows that for times below 700 seconds the dispersion process within the cells is highly affected by the asymetry of the velocity field around both separatrices.

In futures experiments we will try to obtain the accurate exponents in the distinct laws of the evolution of the width of the coloured layers around both separatrices taking into account the presence of different local Peclet numbers.

REFERENCES:

[1] Y. Pomeau, A. Pumir and W. Young, submitted to *Phys. Fluids* (1988).

[2] A. Davey, *J. Fluid Mech.* 31 (1962) 17.

[3] H. Fasel and O. Booz, *J. Fluid Mech.* 138 (1988) 21.

LASERS AS A TEST BENCH FOR THEORIES OF NON-EQUILIBRIUM STRUCTURES

J. R. Tredicce, E. J. D'Angelo, C. Green, G. B. Mindlin, L. M. Narducci, G. L. Oppo*, and H. Solari
Department of Physics
Drexel University
Philadelphia, Pa. 19104
U.S.A.

I. Introduction

The appearance of intensity pulsations at the output of a laser was noted immediately after the first successful operation of the ruby laser. At first this phenomenon was thought to be the consequence of external noise, optical inhomogeneities in the active medium or mechanical instabilities; later it became apparent that, while the stability of the output intensity was surely conditioned by instrumental imperfections, self-pulsing could also be the consequence of the nonlinear interaction between the electromagnetic radiation and the active atoms [1]. In 1975 H. Haken [2a] showed that the single mode equations of the laser are isomorphic to the Lorenz equations [2b]. Because the latter at that time were already known to include chaotic oscillations among its possible solutions, Haken's proof endowed the laser with a complex temporal phenomenology of its own. In the early 1980's the first experimental observations of period doubling, bifurcations and chaos began to appear in the literature [3,4]. Maybe, the main reason for the delay was the fact that during two decades laser physicists were worried more on the design of new and more stable lasers than in the study of laser physics. Since 1982 a a considerably body of evidence has been uncovered leading to the observation of period doubling, quasi-periodicity, intermittency, crises, Shil'nikov instabilities and many other phenomena in optical systems [5,6].

The study of these strange behaviors of lasers has been aided considerably by powerful new techniques which have been developed during the last several years for the purpose of characterizing time dependent signals and the transitions from ordered to highly complex states [7]. The use of power spectrums, Poincaré sections, bifurcation diagrams, return maps, dimension and entropy measurements, relative rotation rates [8], templates [9], and other methods have proven very useful not only in understanding time dependent phenomena in laser physics but also their connections with other branches of science such as hydrodynamics, mechanics, biology and chemistry.

At this point a natural question to ask is: In which direction should the study of laser instabilities evolve? As we asses the current status of the models and the corresponding measurements we see the need for better and more quantitative comparisons between theory and experiments; we need on the one hand more realistic theoretical calculations and on the other better measurements of the control parameters in order to improve practical applications. There is, from a more basic point of view, a new and broader issue that needs to be explored which concerns the spatial behavior of optical

E. Tirapegui and W. Zeller (eds.), Instabilities and Nonequilibrium Structures III, 239–248.

systems and lasers in particular: are lasers able to display the type of complex spatial structures such as one finds under fully developed turbulence conditions in hydrodynamic systems? Turbulence involves increasing complexity not only in time but also in space; one could argue that, just as dynamical chaos results from the presence of at least a few competing frequencies in a nonlinear system, spatial complexity may be obtained when competing spatial structures play a relevant role in the evolution of a dynamical system.

In this paper we give evidence that lasers can undergo complex spatio-temporal evolutions and that these systems can provide almost ideal tests of bifurcation theories in areas that touch upon one of the oldest unsolved problems in physics: the origin of turbulence. We have known for a long time that an optical resonator with spherical mirrors can support different modal structures in a direction roughly perpendicular to the axis of propagation of the beam. Most previous studies of the laser have relied heavily on the fact that practical lasers can operate mainly with a simple Gaussian transverse intensity distribution with small if any interaction with other transverse modes [10]. There are situations, on the other hand, where transverse mode-mode competition can be enhanced to the point that the role of the nonlinear coupling among modes acquires an essential significance. When this type of mode-mode competition becomes important not only the theoretical description is considerably more involved, but also new and unexpected phenomena develop. We have shown [11] that the interaction among transverse modes may lead to cooperative frequency locking, a kind of synchronous oscillation of modes with different natural frequencies, and in the absence of locking to a gradual increase in the spatial complexity of the total intensity distribution [12]. Here we offer a summary of these recent studies from both a theoretical and experimental point of view. In section II we give an overview of the system used in our investigations and in section III we discuss the main experimental results using a large bore CO_2 laser. In section IV we outline some qualitative theoretical arguments connected with the phenomenon of spontaneous symmetry breaking. We discuss some relevant numerical results in section V and finally provide a few concluding comments in section VI.

II. Experimental set-up to study transverse effects

A schematic diagram of our experimental system is shown in Fig.1. Our laser is a CO_2 laser in a Fabry-Perot configuration with the active medium enclosed in a 22 mm inner diameter tube whose design was carefully studied to avoid instabilities in the electric discharge. The main power supply is stabilized to better than 0.1% over the useful range of currents (3 mA to 12mA) when the pressure in the tube is between 10 and 20 Torr with 10% CO_2, 14% N_2 and 76% He. The tube is closed by two ZnSe Brewster windows that can be rotated to compensate for possible asymmetries of the cavity and also to provide an extra output beam to be used in checking the operational rotational transition (all measurements are done with the P20 transition of the 10.6 μm band). The cavity is of the Fabry-Perot

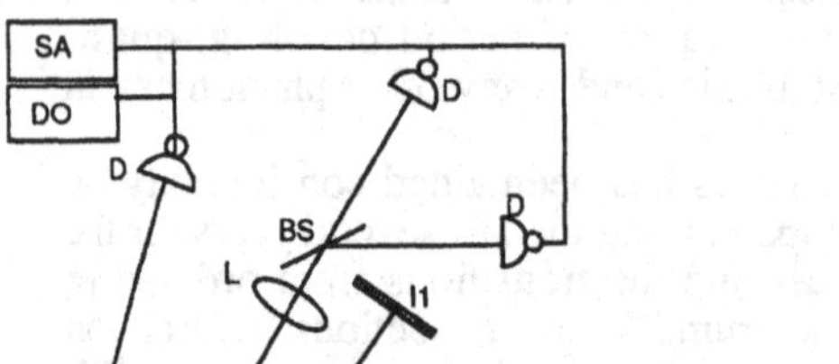

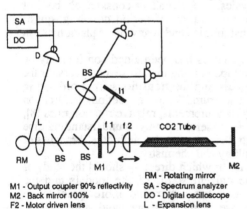

Fig.1: Schematic of the experimental setup

type formed by two plane mirrors at a distance of 1.7 m. The back mirror is gold coated and the output coupler is made of Ge.with a 90% reflectivity. All optical elements are two inches in diameter to avoid any truncation effects on the beam. Inside the cavity we plug two lenses; the first is mounted on a translation stage that is controled by a stepper motor. Its position can be controlled to within 0.001 inches. The second lens is fixed to the optical table. The entire optical system effectively acts as a cavity formed by two mirrors whose radii of curvature depends on the distance between the two lenses. In the telescopic configuration (distance between lenses equal to the sum of their focal lengths) the effective cavity is in a plane-plane configuration. For larger distances the cavity is stable until the effective radius of curvature of the mirrors becomes equal to half of their separation (concentric cavity). For distances smaller than the telescopic separation, the cavity is always unstable and the laser does not operate.

The variable distance between the two lenses allows us to control adiabatically the radius of curvature of the effective cavity. This in return has two effects:

1.it allows a change of the frequency difference among the resonant frequencies of the cavity.

2.it allows a change of the Fresnel number which is usually defined as the ratio between the aperture of the system and the beam waist for the Gaussian mode (lowest order mode) of the resonator.

The cavity resonance frequencies are given by the equation [12]:

$$\omega_{npm} = \frac{\pi c n}{2L} + \frac{2c}{L} \left(2p + |m| + 1 \right) \tan^{-1}\left(\frac{1}{\sqrt{2R_0/L - 1}}\right)$$

where R_0 is the radius of curvature of the mirrors and c is the speed of light. The index n defines the order of the longitudinal resonance associated with the distance L between the mirrors; the indeces p and m define the spatial structure of each mode in such a way that p is the number of rings (or maxima) of the intensity in the radial direction, while 2m is the number of local maxima in a full circle at a given radial position. For a given n there exists a full set of integers p's and m's from 0 to ∞, but higher values of p and m implies that higher gain is needed at large values of the radius. Thus the aperture of the system provides a natural way to select the maximum values of p and m for the chosen geometry.

If we keep in mind that the field is amplified by the atoms over a limited bandwidth (usually called $\gamma_\perp$), only those modes, whose frequency lies inside this frequency domain will exhibit gain and experience laser action. Therefore controlling the separation in frequency between transverse modes allows the selection of a given spatial pattern. However, due to the presence of a infinite number of n values, the selectivity will not be effective if it is not accompanied by a finite cross section of the tube. For tubes that have a small cross section , (which also implies a small Fresnel number) the selection is effective and the dynamical behavior of the system is described by only a few spatial modes. In this case, a large separation between low order transverse modes gives the best selection for a given pattern, while the output intensity distribution is sensitive to the total length of the resonator because this determines which mode has the largest gain. For large diameter tubes, instead, the resonant frequency does not play a critical role anylonger. However, the effective radius of curvature of the mirrors still controlles the size of the beam waist and the Fresnel number (FN). For small FN corresponding to a near plane-plane cavity, only the lowest order modes (p=m=0) can satisfy the threshold condition even if higher order modes are still resonant. When the FN is increased, more complex spatial structures are allowed. In our system the maximum value of the FN can be estimated at 42. It is worthwhile to observe that commercial lasers usually work in the range of FN from 0.5 to 3.

To analyze the output intensity of the laser, we use a beam spliter (10% reflectivity) to send the output to an infrared imaging plate. The plate displays the time average intensity in a transverse plane with respect to the direction of propagation. The laser radiation is also sent to a rotating mirror and from there to a HgCdTe detector. This is a photovoltaic detector with a 100 MHz bandwidth and a 1 mm^2 sensitive area. With this arrangement we monitor the intensity across the pattern for a given cross section of the beam. Two other HgCdTe detectors are used to detect the intensity as a function of time at two different spatial points. We use a lens to expand the beam before it reaches the detectors. This makes the beam broader for better spatial resolution and lowers the local intensity at the detector. The output signals are recorded with a Lecroy digital oscilloscope (125 MHz) and stored in a PDP11 computer for further processing.

III. Experimental results from a large bore diameter laser

In a near concentric cavity, the pattern (Fig. 2a) is radially symmetric and almost Gaussian in shape (Fig. 3a) while the intensity is constant in time at every point on the pattern (Fig 3b). Decreasing the distance between lenses (increasing FN), produces a spontaneous symmetry breaking and the coexistence in parameter space of two clearly distinguishable patterns (Fig. 2b and 2c) displaying two or four intensity maxima. The signal recorded from the HgCdTe detectors shows also that the intensity oscillates in time and that the oscillations are out of phase between two consecutive peaks in the angular direction (Fig. 4a). A further decrease in the distance between lenses increases the spatial and temporal complexity by adding a ring connecting the maxima (Fig.2d,g) and/or

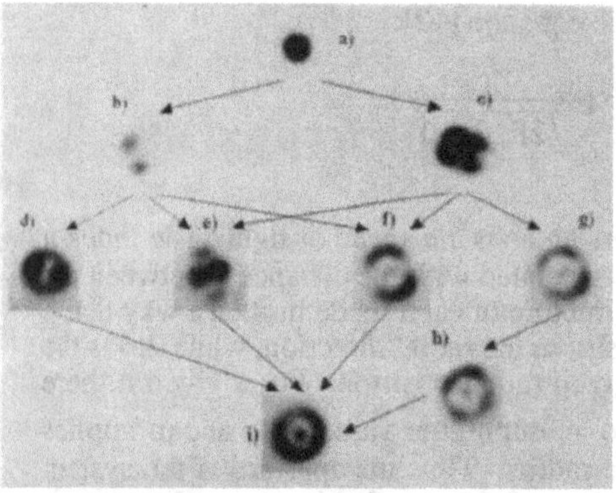

Fig 2: Spatial patterns of the average laser intensity as observed in the thermal plate

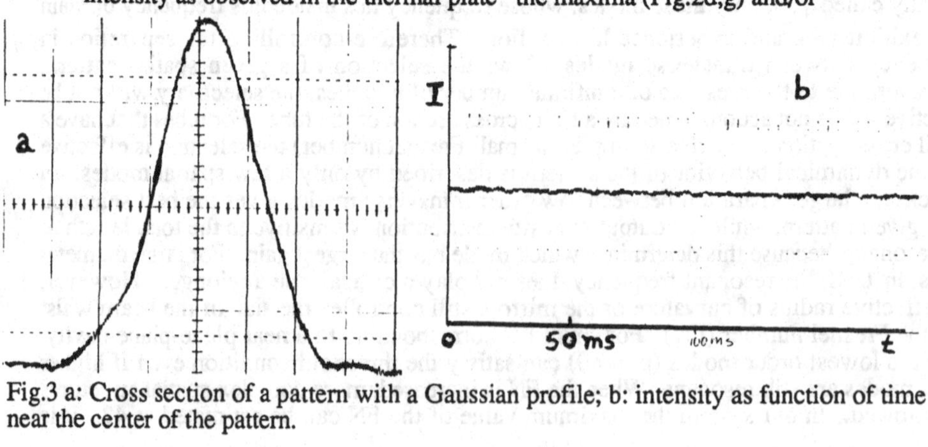

Fig.3 a: Cross section of a pattern with a Gaussian profile; b: intensity as function of time near the center of the pattern.

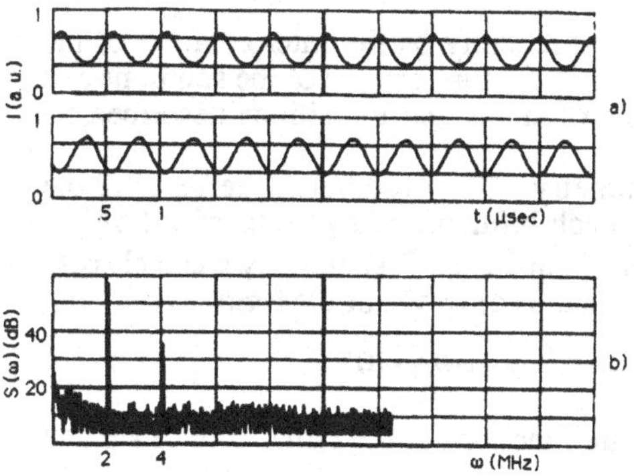

Fig.4 a: Intensity as a function of time and b: local power spectrum.for the pattern of Fig.2 b.

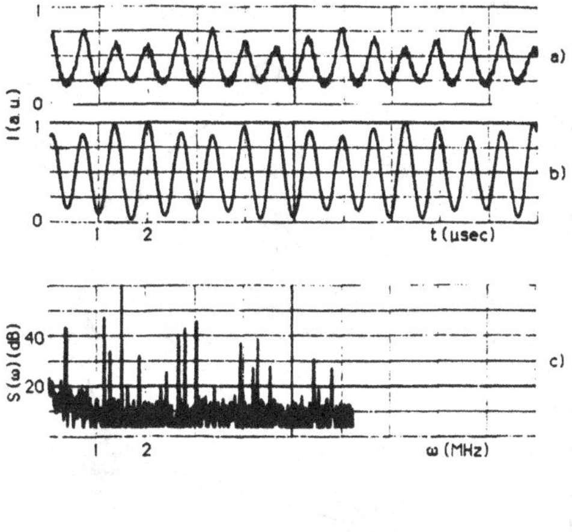

Fig.5 Idem as Fig.4 for pattern of Fig.2 e.

the existence of two frequencies in the time signal which becomes quasiperiodic (Fig. 5). The observation of the signal after the beam is reflected by the rotating mirror allows us to conclude that the pattern is characterized by the existence of a traveling wave (Fig. 6a-b). Following this state, the next pattern is a radially symmetric on the average but the time behavior at different spatial points shows periodic, quasiperiodic or even irregular behavior, indicating an increase in complexity (Fig. 2i). If we decrease further the distance between lenses, the apparent symmetry of pattern 2 is broken with the appearance of 6 or 8 maxima and a larger number of rings. The temporal behavior still remains aperiodic except for selected points of the pattern.

On approaching the confocal cavity configuration we see an expansion of the output pattern with the appearance of as many as 22 maxima. Irregular spatial patterns and spirals are more likely to occur as a result of increasing the FN. These states are reminiscent of a turbulent state in hydrodynamical systems. However, more quantitative results are needed before we can associate them to optical turbulence.

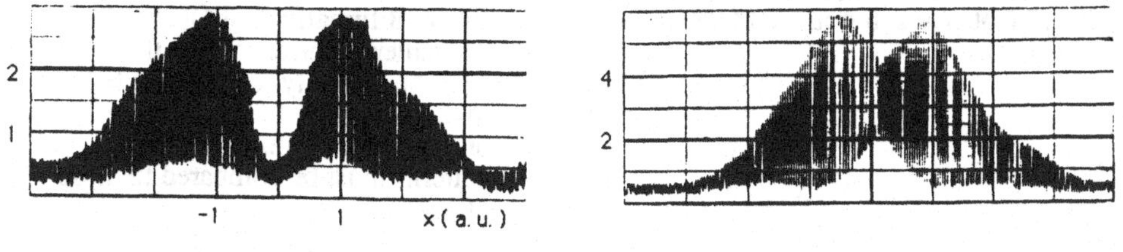

Fig.6 a:Cross section of pattern of Fig.2 d over a line trought the center in between the two maxima; b: Cross section of same pattern over a line parallel to a) trought one maximun.

IV. Qualitative theory of spontaneous symmetry breaking

An important step forward was carried out recently on the interpretation of the solutions described in section III [13]. Most of the important features of the solutions and their transitions can be described by a finite set of states and transitions described by bifurcation theory.

Our laser possess an obvious O(2) symmetry. The observed patterns after the symmetry is broken show a D_n symmetry and each bifurcation is characterized by the appearance of oscillations in time with a different frequency. This means we can classify the patterns according to the spatio-temporal symmetries of waves of the form

$$E = 1 + (z_1 e^{il\theta} + z_2 e^{-il\theta}) e^{i\Omega t} + (z_3 e^{im\theta} + z_4 e^{-im\theta}) e^{i\Omega' t}$$

where E is the electric field, z_n for $n = 1$ to 4 are functions of the radial coordinate and time, θ is the angular coordinate and Ω and Ω' are arbitrary frequencies.

Following Ref.14, we can classify the degree of symmetry of a spatial pattern in terms of a number of states labelled from 0 (fig 7) (the highest symmetry state) to 14 (the lowest symmetry.state). The arrows indicate the bifurcation sequence observed in our laser. The intensity distribution can easily be constructed from table 1. Different solutions are distinguished by their geometrical shape, time behavior of each point on the pattern, and the relative phase with respect to another point in space. With the help of measurements of these observables we have verified experimentally the results presented in Ref.14.

Isotropy subgroup	Basis	Comments
O(2) x T2	$z_1 = z_2 = z_3 = z_4 = 0$	trivial-fully symmetric
S(0,0,1) x S(1,-1,0)	$z_2 = z_3 = z_4 = 0$	l-travelling wave (l-TW)
S(0,1,0) x S(1,0,-m)	$z_1 = z_2 = z_4 = 0$	m-TW
S(0,0,1) x K x Z(π/l,π,0)	$z_1 = z_2,\ z_3 = z_4 = 0$	l-standing wave (l-SW)
S(0,1,0) x K x Z(π/m,0,π)	$z_1 = z_2 = 0,\ z_3 = z_4$	m-SW
S(0,0,1) x Z(π/l,π,0)	$z_3 = z_4 = 0$	l-TW + l-SW
S(0,1,0) x Z(π/m,0,π)	$z_1 = z_2 = 0$	m-TW + m-SW
S(1,l,m)	$z_1 = z_3 = 0$	l-TW + m-TW (Same direction)
S(1,l,-m)	$z_1 = z_4 = 0$	l-TW + m-TW (Opp. direction)
K x Z(π,lπ,mπ)	$z_1 = z_2,\ z_3 = z_4$	l-SW + m-SW
Z_k(0,π,0) x Z(π,lπ,mπ)	$z_1 = -z_2,\ z_3 = z_4$	l-SW+m-SW (m odd)
Z_k(0,0,π) x Z(π,lπ,mπ)	$z_1 = z_2,\ z_3 = -z_4$	l-SW+m-SW (m even, l odd)
Z(π/l,π,mπ/l)	$z_3 = 0$	frequencies (l≠1)
Z(π/m,lπ/m,π)	$z_1 = 0$	frequencies (m≠1)
Z(π,lπ,mπ)	$z_n \neq 0$	frequencies

Table 1: Isotropy subgroups of O(2) x T^2.

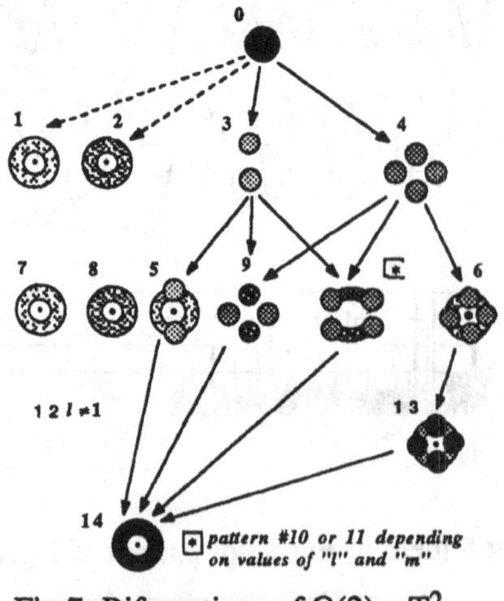

Fig.7: Bifurcations of O(2) x T^2. Thepatterns correspond to the average intensity for the case l=1, m=2. The dashed lines indicate unobservable transitions. These patterns must be compared to those of Fig. 2.

The sequence shown in Fig.2 is the experimental evidence of the spontaneous symmetry breaking of the group O(2)xT^2 as calculated from bifurcation theory. Our system has chosen only certain paths. For example patterns #1 and #2 involving only travelling waves were not observed. After the appearance of two or three frequencies the temporal behavior of the local intensity becomes usually chaotic and almost any degree of spatial symmetry tends to disappear. At this point one must go beyond the group theoretical analysis for a deeper understanding of the observations. Quantitative measurements of correlation functions and the identification of possible defects on the intensity pattern are badly needed. Numerical work on the Maxwell-Bloch equations with diffraction terms can also give insight on the physical processes that generate this complexity in space and time.

V. Numerical results

The Maxwell-Bloch equations for a two level system, including diffractive contributions in the paraaxial approximation is usually accepted as a suitable model to describe the behavior of lasers under our experimental conditions. The derivation of these equations, as given in Ref.15,yields:

$$\frac{\partial E}{\partial t} = -k\left[1 - i\,\Delta - i\,\frac{a}{2}\left(\frac{1}{4}\nabla_\perp^2 - \rho^2 + 1\right)\right]E - 2\,C\,k\,P(\rho,\phi,t)$$

$$\frac{\partial P}{\partial t} = -\gamma_\perp\left[E\,D + (1 + i\,\Delta)\,P\right] \qquad (2)$$

$$\frac{\partial D}{\partial t} = -\gamma_\parallel\left[\frac{1}{2}(E^*P + E\,P^*) + D - \chi(\rho)\right]$$

where E, P and D are the electric field, polarization and population inversion respectively; k, $\gamma_\perp$, and $\gamma_\parallel$ their relaxation rates; ρ, ϕ, and z the radial, angular and longitudinal coordinates respectively, $\chi(\rho)$ is the pump rate, C is the gain factor , Δ is the detuning between the atomic and the lowest order cavity mode (n00) resonances, and a is a constant proportional to the separation in angular frequency between transverse modes; $\nabla_\perp^2$ indicates the transverse Laplacian operator.

It is necessary to note that the derivation of Eq.2 assumes that any variation of the field with respect to z is controlled by the cavity while the medium has no influence on it. This assumption is equivalent to the well-known uniform field approximation in the plane wave models.

The numerical integration of the Maxwell-Bloch equations is performed using a modified Gallerkin method. The trial functions are the orthonormal Gauss-Laguerre polynomials. This choice is selected because they are exact solutions of Maxwell's equations with boundary conditions defined by spherical mirrors in the absence of the atomic medium. Assuming that the solution is slightly modified by the presence of the atoms, such a choice minimize the number of spatial points required to obtain a reliable precision in the numerical calculation. In particular we use 200 collocation points.

The main physical results obtained from our analysis are:

1. the appearance of structures with a broken spatial symmetry (Fig.8)

2. the presence of defects in the spatial structure (Fig.9) which, when in motion, may create travelling waves (Fig.10).

3. the creation of strongly disordered structures in space and time, which becomes radially symmetric when observed under time averaging conditions (Fig.11).

Fig 8 Results from numerical simulations a: symetric pattern; b: pattern with broken symetry. Calculated for parameter values typical of a CO2 laser. The patterns in the figure are stationary because a=0 and the detuning is also 0.

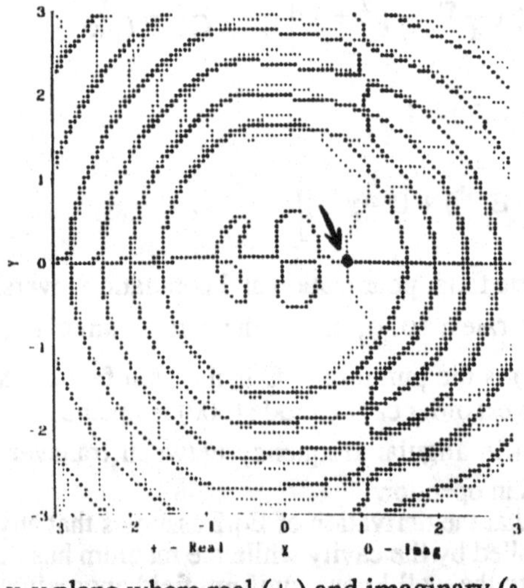

Fig.9: Locus in the x-y plane where real (+) and imaginary (•) parts of the field becomes zero corresponding to the intensity distribution of Fig.8b. The arrow shows the position of a defect .

Conclusions

One of the most important features of lasers is not only that they are able to display spatio-temporal complexity, but that it is possible to control accurately the transition from very simple and stable structures to more complex ones. Using this property we are able to use lasers to check theoretical results based on bifurcation theory, which is independent of the particular models and therefore of general interest for other branches of science. On the other hand, the existence of reliable models for the laser allows a more quantitative comparison with numerical results still using relatively simple numerical codes. The numerical results show the presence of defects wich play an essential role on the appearance of turbulence [16-17].

The temporal oscillations in CO_2 lasers have a frequency in the MHz range. Thus, frequencies are high enough in such a way that the requirements in terms of mechanical stability are easy to achieve and low enough to make measurements feasible.

In synthesis, we believe that lasers are one of the most powerful test benches in nonlinear dynamic studies and we hope the results shown above did convince you.

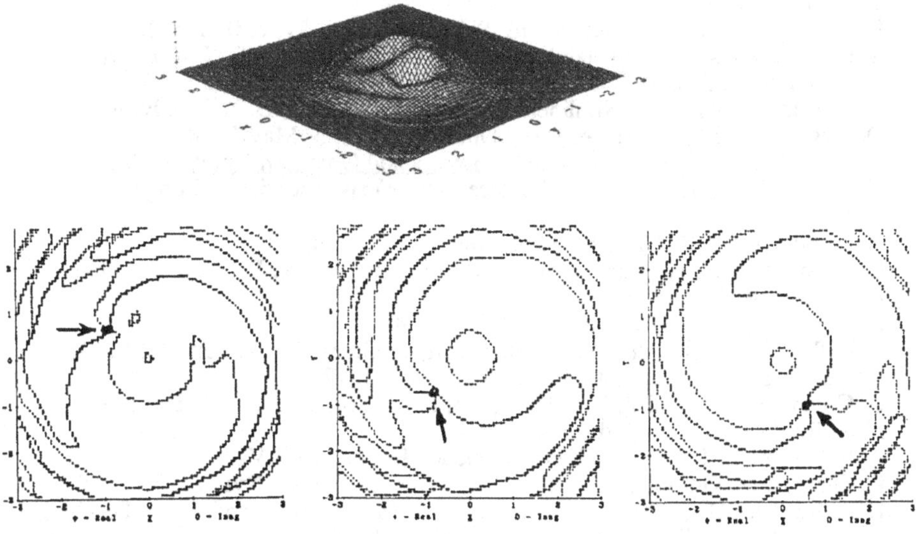

Fig.10: Idem as Fig.9 but for a=0.1. The position of the defect changes in time; it moves in a circular path around the center of the beam

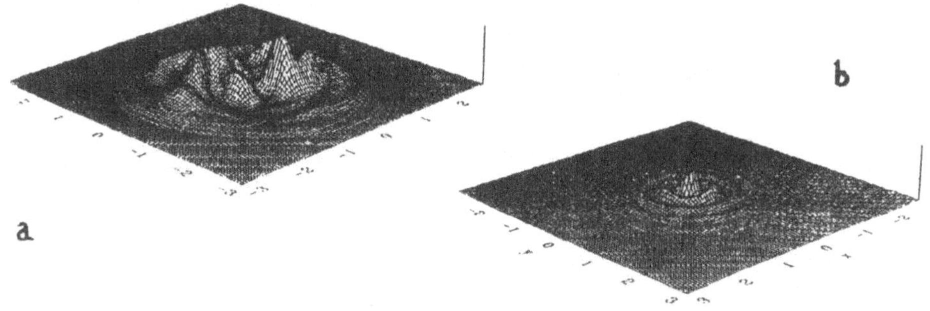

Fig.11: a: instantaneous, and b: averaged intensity distributions for a complex spatio temporal pattern (a= 0.6).

References

* Permanent address: Department of Physics, University of Strathclyde, Glasgow, U.K.

[1]. See e.g. L. A. Lugiato, L. M. Narducci, J. R. Tredicce, and D. K. Bandy in Instabilities and Chaos in Quantum Optics II , N. B. Abraham, F. T. Arecchi, and L. A. Lugiato, eds. NATO-ASI series B, Vol. 177 (Plenum Press, 1987), and references therein.

[2]. a) H. Haken, Phys. Lett.**53A**, 77 (1975); b) E. N. Lorenz, J. Atmos. Sci. **20**, 130 (1963).
[3]. F. T. Arecchi, R. Meucci, G. P. Puccioni, and J. R. Tredicce, Phys. Rev. Lett. **49**, 1217 (1982).
[4] R. S. Gioggia, and N. B. Abraham, Phys. Rev. Lett. **51**, 650 (1983).
[5] See the special issue on "Instabilities in Active Optical Media", N. B. Abraham, L.A.Lugiato, and L. M. Narducci, eds. J.Opt.Soc.Am.B, **2**, January 1985.
[6] See the special issue on "Nonlinear Dynamics of Lasers", D. K. Bandy, A. Oraevskii, and J.R. Tredicce, eds. J.Opt.Soc.Am.B, **5**, May 1988.
[7] J. R. Tredicce, and N. B. Abraham, in <u>Lasers and Quantum Optics</u>, L. M. Narducci, E. J. Quel, and J. R. Tredicce, eds. Worls Scientific (Siongapore, 1990).
[8] H. G. Solari, and R. Gilmore, Phys. Rev. A **37**, 3096 (1988).
[9] G. B. Mindlin, X. Hou, H. G. Solari, R. Gilmore, and N. B. Tufillaro, Phys. Rev. Lett. **64**, 2350 (1990).
[10]. P. W. Smith, Appl. Phys. Lett. **13**, 235 (1968).
[11]. J. R. Tredicce, E. J. Quel, A. M. Ghazzawi, C. Green, M. A. Pernigo, L. M. Narducci, and L. A. Lugiato, Phys. Rev. Lett. **62**, 1274 (1989).
[12]. C. Green, G. B. Mindlin, E. J. D'Angelo, H. G. Solari, and J. R. Tredicce, Phys. Rev. Lett. (submitted for publication).
[13]. A. E. Siegman, "Lasers", University Science Books, (Mill Valley, CA. 1986).
[14]. P. Chossat, M. Golubitsky, and B. L. Keyfitz, Dyn.Stab. Sys. **1**, 255 (1987).
[15]. L. A. Lugiato, G. L. Oppo, J. R. Tredicce, L. M. Narducci, and M. A. Pernigo, Journ. of the Opt. Soc. of America B, **7**, 1019 (1990).
[16]. P. Coullet, L. Gil, and J. Lega, Phys. Rev. Lett. **62**, 1619 (1989).
[17]. S. Rica, and E. Tirapegui, Phys. Rev. Lett. **64**, 878 (1990).

MEAN FIELD THEORY OF CANONICAL SPINODAL INSTABILITIES

M. CALVO
Centro de Física
Instituto Venezolano de
Investigaciones Científicas
IVIC
Apartado 21827
Caracas 1020A
Venezuela

Abstract. We show that the character of the spinodal descomposition line of a system described by the Landau-Ginzburg theory depends on the statistical ensemble. For the grand canonical it is well known that this line corresponds to a line of discontinuous instabilities (first order transition). For the canonical case it becomes a line of continuous instabilities (second order transitions). Our analysis is based on the mean field approximation and ignores the effect of nucleation.

E. Tirapegui and W. Zeller (eds.), Instabilities and Nonequilibrium Structures III, 249–253.

We consider, in the context of Landau-Ginzburg theory, the behavior of a system near the spinodal line, in the canonical ensemble. We obtain that the spinodal line becomes a line of continuous instabilities rather that the usual discontinuous instabilities line found for the grand-canonical ensemble [1].

We will analyze a system with a globally conserved order parameter in the vicinity of a critical point. Typical example would be a binary mixture close to the consolute point. We consider that the system is in contact with a heat reservoir at temperature T, and assume that the number of atoms in the mixture remain fixed. Landau theory [2] assumes that the partition function near criticality is given by the functional integral:

$$Z = \int D\phi \exp\{-H[\phi]\}, \tag{1}$$

where

$$H[\phi] = \int d^9x\{\frac{1}{2}(\nabla\phi)^2 + \frac{t}{2}\phi^2 + \frac{\lambda}{4!}\phi^4\} \tag{2}$$

with

$$t = \frac{T - T_c}{T_c}, \lambda > 0,$$

and $\phi(x)$ measures local deviations of the concentration. T_c is the critical temperature. The global conservation requires that

$$\int \phi(x) d^3x = c \tag{3}$$

The mean field condition for criticality requires

1. $\frac{\delta H}{\delta\phi}|_{\phi_{mf}} = 0$ (4)

Since we are imposing a constraint on ϕ, we introduce a Lagrange multiplier η and derive

$$\nabla\phi_{mf}(x) = t\phi_{mf} + \frac{\lambda}{6}\phi^9_{mf} - \eta. \tag{5}$$

2. Condition for minima requires.

$$H[\phi_{mf} + \mu] > H[\phi_{mf}] \tag{6}$$

for arbitrary but small variations μ. Expanding to order μ^2 we obtain

$$H[\phi_{mf} + \mu] - H[\phi_{mf}] = \int d^3 x \mu(-\nabla^2 + t + \frac{\lambda\phi^2_{mf}}{2})\mu, \tag{7}$$

so the positiviness of the spectrum of the operator $-\nabla^2 + t + \lambda\phi^2_{mf}/2$ guarantees minima. Criticality requires (in the canonical ensemble) that second lowest eigenvalue of this operator, ϵ_2, vanishes. (The first lowest laying eigenvalue $\epsilon_1 = 0$, is the condition for the grand canonical critical point. For the canonical ensembles variations which violate Eq. (2) are precluded). Let μ_2 be the eigenfunction associated to ϵ_2. Then condition for thermodynamical stability requires

$$H[\phi_{mf} + \mu_2] - H[\phi_{mf}] > 0, \tag{8}$$

which implies to cubic order in μ

$$\lambda \int \phi_{mf}(x)\mu_2^3(x)dx = 0. \tag{9}$$

Needless to say that the fourth order term in μ is always positive because $\lambda > 0$.

Let us apply these criteria to a system enclosed in a box of sides L_x, L_y and L_z and where the walls of this container are neutral. We conclude that the conditions for criticality become

$$\phi_{mf} = c \qquad \text{with} \qquad tc + \frac{\lambda c^2}{6} = \eta$$

and

$$\epsilon_1 = 0 = (\frac{\pi}{L_x})^2 + t + \lambda c^2/2 \tag{10}$$

Moreover since $\mu_2 = sin\frac{\pi x}{L_x}$, we conclude that condition 3 is automatically satisfied.

In conclusion we find that the line of spinodal instabilities in the large L_x limit is

$$t_{\text{spinodal}} + \lambda c^2/2 \simeq 0, \tag{11}$$

at which the solution becomes unstable continuously, that is, the local minima of $H[\phi]$ bifurcates continuously into coordinate dependent minimal field configurations. This result strongly contrast with the grand canonical case. In the latter case the corresponding solution also becomes unstable but with a finite discontinuity.

Of course in both cases the solution ϕ_{mf}, given by Eq. (10), represents a local minima of $H[\phi]$. For temperatures below the coexistence curve,

$$t_{\text{coex.}} + \frac{\lambda c^2}{6} = 0, \tag{12}$$

The homogeneous solution is metastable. That is there exists an inhomogeneous configuration which is the absolute minima of $H[\phi]$. In the metastable region the observability of the spinodal line depends in the rate of nucleation. This problem has been amply discussed in the litterature [3].

Acknowledgments

The author acknowledges illuminating discussions with Dr. R. Medina. The preparation of this work was made possible by a financial support provided by the Latin American Academy of Sciences.

References

See, J., Gunton et. al. (1983) 'Phase transitions and critical phenomena', in C. Domb and J. Lebowitz (eds.), Vol. 8, Academic Press, London.

Stanley, H. (1971) 'Introduction to phase transitions and critical phenomena', Oxford Univ. Press., Oxford.

Binder, H. (1980) 'Systems far from equilibrium', in L. Garrido (ed.), Springer-Verlag, Berlin.

PERIODIC(LAYERED) STRUCTURES UNDER CONSTRAINT*

Yves Pomeau
Laboratoire de physique statistique; URA 731 ENS 24 rue Lhomond 75231 Paris Cedex O5
Jacques Prost
Laboratoire de Physico-Chimie Théorique; ESPCI 10 rue Vauquelin 75231 Paris Cedex 05
Etienne Guyon
Physique de la Matière Hétérogène; URA 857 ESPCI 10 rue Vauquelin 75231 Paris Cedex 05
France

ABSTRACT :

Periodic, or layered, structures are met in many physical instances : they form in dissipative systems, such as Rayleigh Benard rolls, or at equilibrium in liquid crystal phases (smectics and cholesterics). The common framework for describing the long range excitations of these structures is the *phase dynamics* theory. We present here a general description of the stability of these structures when they are submitted to an external force field (flow, electric field) perpendicular to the roll axis for various boundary conditions.
The one dimensional equilibrium solution with fixed boundary conditions leads to the coexistence of compressed and dilated rolls in the same structure, an effect discovered by Pocheau and Croquette on Rayleigh Benard rolls in the presence of a transverse flow which has a known counterpart in cholesterics. Layered structures may present an undulational instability when submitted to a dilative stress : this increases the local wavenumber by adding an undulation along the roll axis. We study this instability from the point of view of linear stability and in the limit of a very large constraint. Finally we discuss the possibility of observing time dependent phenomena when the boundary conditions are such that the number of rolls may change. We discuss this time dependent behavior in the limit when the phase dynamics is much slower than the process of creation or annihilation of new rolls.
It is to be noticed that all this discussion starts from a general formulation of the problem of phase dynamics that takes into account symmetries and other simple constraints only, independent on any detailed derivation of this dynamics from the basic equations.

---

* This was presented at the Third conference on "Instabilities and non equilibrium structures" held at Valparaiso (Chile) on December 1989. More details on the questions examined here may be found in the preprint in reference 23.

E. Tirapegui and W. Zeller (eds.), Instabilities and Nonequilibrium Structures III, 255–267.

1.INTRODUCTION

When a layered system (smectic or cholesteric liquid crystals; Rayleigh Benard rolls...) is submitted to a non equilibrium vector force in a direction orthogonal to the layers, a remarkable phenomenon, called *permeation*, takes place. This phenomenon expresses the fact that an external force drives a difference between layer and barycentric velocities. In the case of cholesterics, it is associated to the rotation of the local axis as observed long ago by Lehmann in a temperature gradient at small amplitude [1,2]. A natural framework is provided by the Leslie hydrodynamic theory [3]. The connection made by Helfrich [4] between the local rotation in a cholesteric and permeation makes it possible to incorporate the Lehmann effect in the framework of the hydrodynamics of layered systems[5,6]: a permanent bulk rotation of the order parameter (i.e. local optical axis in cholesterics or the phase of the layers in the smectic case) takes place with an angular velocity proportional to the external field at small amplitude in the case of free boundary conditions. Direct observations of permeation in either case have been found extremely difficult [7] for reasons which will be discussed in the conclusion of this paper.

These experiments can naturally interpreted in the framework of a *phase dynamics equation*[11] as explained in this section 2. This equation is formally the same as the one used by Pocheau and Croquette (P.C.)[9] to analyse their experiment of "compression" of Rayleigh-Bénard (R.B.) rolls under the effect of a transverse flow: generally, the wavenumber in one dimensional layered structures is constant in space, under uniform conditions, because of the general existence of a "constant of the motion" in the direction perpendicular to the roll axis, as was discovered by Zaleski [21]. Hovever, under some specific constraint, this wavenumber may vary throughout the system even under constant external conditions. This stationary distortion is one of the possible responses of a layered system. It is only a change of the boundary conditions that change this response into a time dependent rotation of the phase, equivalent to the Lehmann rotation, and P.C. observed too this time dependent behavior.

In section 2, we derive the phase dynamics equation in its most standard form with one dimensional modulations only, perpendicular to the layers/rolls with a term representing the effect of external constraints. The elementary solutions of the resulting equations strongly depend on the boundary conditions

In section 3, we extend first the phase equation to take into account possible deformations parallel to the roll axis. The derivation of the corresponding equation requires only rather general arguments of symmetry. With the help of these equations we show that the static deformation obtained by Pocheau and Croquette may become linearly unstable if a transverse modulation is allowed. Because of the transvrse force, part of the structure is compressed and part of it is under extension. In this region the wavelength increases locally and so can optimize the energy by adding a modulation in a transverse direction. This can be analysed both from the point of view of the linear stability and in the limit of a very large constraint. A key feature of our analysis is that all interesting phenomena occur at vanishingly small velocities : This corresponds to the large box limit (i.e. a large number of periods N in the system). We study both the linear stability close to threshold ($V \sim 1 / N^2$) and the strongly non-linear regime (keeping $V \ll 1/N$). The instability takes place when the phase is fixed at both ends of the structure (It should also be possible to observe it with mixed boundary conditions).

In section 4, we show that a sustained time dependent behavior is possible with different boundary conditions and in the absence of transverse modulation of the rolls, another effect observed by Pocheau and Croquette and corresponding to the creation/destruction of rolls. More complicated phenomena can also be observed and have been analysed using different boundary conditions on both sides. In the presence of a flow which forces phase winding, new rolls have to be continuously added inside the structure; this cannot be described within the framework of the phase dynamics which excludes fast events such as the nucleation of new rolls, in the Rayleigh-Bénard "terminology" that we shall use mostly throughout this paper. However this fast dynamics may be considered as instantaneous (compared to the phase dynamics). This allows to include it in a consistent fashion into the phase dynamics equations, as shown in this last section, where we also sketch an analysis of the behavior when nucleation of new wavelengths occurs.

2.PHASE DYNAMICS EQUATION AND ITS ONE DIMENSIONAL EQUILIBRIUM SOLUTIONS

The general idea of phase dynamics is to consider an almost regular structure as represented locally in space by a phase, that measures the possible departure of the pattern from perfect regularity. This is an approximation in the sense that the dynamics of the amplitude itself is forgotten. More exactly, one assumes that this amplitude has a finite response time and so follows adiabatically the fluctuations of the phase, which have no typical time scale (because they obey a diffusion equation) and can be as slow as desired. We present below the simplest realization of this phase dynamics equation (eq. 2) in one dimension of space, but by adding terms that represents the constraint imposed to this phase by the kind of external perturbation discussed in the introduction.
In the three different physical situations (RB rolls, smectics and cholesteric liquid crystals) we have introduced so far, one may describe the systems by a complex order parameter

$$\Psi = \Psi_0 e^{i\varphi}$$

Except in the vicinity of a transition or in the core of a topological defect, the real positive amplitude Ψ_0 can be taken as constant, because it follows adiabatically the variations of the phase. In the ground state:

$$\varphi = q_0 z \qquad (1)$$

where $2\pi / q_0 = a_0$ is the period of the system and z the periodicity axis. In more general situations, φ is a function of space and time variables.

In the present section, we consider one dimensional situations φ (z,t) which correspond to translation, dilation or compression of rolls. For slowly varying phase (e.g. time long compared to the internal scales and wavelengths long compared with the periodicity), the equation for the dynamics of the phase [8,11,12,14] takes the form of a convection-diffusion equation in analogy with the smectic case [5,13]:

$$\frac{\partial \varphi}{\partial t} + V \frac{\partial \varphi}{\partial z} = \Lambda E + D \frac{\partial^2 \varphi}{\partial z^2} \qquad (2)$$

in which V is the z component of the velocity field (averaged over the thickness in the case of roll instabilities), E is the external fied alluded to in the introduction. When one considers fluctuations of the phase near equilibrium (that is in layered structures formed at equilibrium by liquid crystals), Λ is a dissipative coefficient which obeys Onsager's relations for small E.

If one considers small deviations from the ground state ($\partial u / \partial z << 1$) where the displacement u is defined by

$$\varphi = q_0 (z - u) \quad (3)$$

equation (2) can be rewritten as

$$\frac{\partial u}{\partial t} = V + \Lambda E + D \frac{\partial^2 u}{\partial z^2} \quad (4)$$

where we have made the replacements Dq_0 by D, Λ / q_0 by Λ and we have omitted the part of the convective term in (2) which is of second order in the perturbation u. This is exactly the permeation equation of layered liquid crystals[5]. The equivalent role played by a flow and by an external field is obvious from this relation although their physical origin and time reversal behavior is different. In the following we will replace $[V + \Lambda E]$ by V.

Let us consider a sample of thickness L ($0 < z < L$); we can calculate simple solutions of equation (4) for various boundary conditions which are obtained quite naturally in experiments : In liquid crystals, free boundary conditions correspond usually to free liquid surfaces and rigid B.C. to the material in contact with a solid surface with appropriate surface treatment. In R. B. rolls rigid B. C. correspond to solid lateral walls whereas free B.C. can be obtained with smoothly varying properties.

i) For free - or Neuman - boundary conditions:

$$\frac{\partial u}{\partial z}(z=0) = \frac{\partial u}{\partial z}(z=L) = 0 \quad (5)$$

we get the solution of the Lehmann rotation :

$$u = Vt \quad (6)$$

ii) For rigid - or Dirichlet- boundary conditions:

$$u\,(z=0) = 0 \;, \quad u\,(z=L) = u_0 \quad (7)$$

the solution is static :

$$u = \frac{V}{2D} z\,(L - z) + \frac{z\,u_0}{L} \quad (8)$$

It describes the P.C. effect . An equivalent effect had been predicted by Leslie[3] for cholesterics. When L is equal to an integer number of $2\pi/q_0$, the equation (8) describes a compression of rolls in the downstream region ($z < L/2$) and a dilation in the upstream region as observed in the P.C. experiments. Note that in these experiments the change in wavenumber induced by the flow was not always small.

iii) For mixed conditions

$$u\,(z=0) = 0 \qquad \frac{\partial u}{\partial z}(z=L) = 0 \quad (9)$$

a steady solution is obtained:

$$u = \frac{V}{2D} z(2L-z) \quad (10)$$

In the following paragraphs, we will study the stability of some of those simple solutions by considering :

- two dimensional perturbations in case ii)

- defect nucleation in one dimension in case iii)

3.BEYOND THE DIMENSIONAL SITUATION

We consider again the phase dynamics of a pattern of parallel "rolls" under the effect of an external constraint, but we add now the possibility of fluctuations depending on coordinates x (=coordinate along the roll axis) and z. We shall establish first the relevant equation of motion under the general constraint of isotropy of the external conditions in the absence of the forcing (that would exclude for instance Taylor-Couette rolls [22]or convection in planar nematics [15]).

3.i) Phase equation in two space dimensions

Assuming a variational form for the "elasticity" term, one finds as an equation of motion for the displacement u:

$$u_t + (\mathbf{V.\ Grad})\, u - V = -\frac{\delta F}{\delta u} \qquad (11),$$

where the terms in the left hand side are defined in equations (3) and (4), and subscripts are for derivatives : $u_t = \frac{\partial u}{\partial t}$. The right hand side represents the "elasticity" or phase diffusion effects through a Fréchet derivative of F. The form of this functional is obvious if one limits oneself to the diffusion term already introduced in equation (2). We shall add more terms to it in order to represent the following physical phenomena: first the existence of a small length scale, typically the wavelength of the structure, that is introduced through space derivatives of an order higher than in the simple phase approximation; then, as announced, the possibility of a dependence of fluctuations on coordinates others than z. All this is accomplished with the following choice for the Lyapunov functional F [5,16]:

$$F = \int dr\, (E^2(u) + \lambda^2 (\Delta u)^2) D/2 \qquad (12)$$

with $E(u) = \frac{\partial u}{\partial z} + \frac{1}{2}(\nabla u)^2$.

This form of E(u) is dictated [23] by the requirement of global rotational invariance. The form of equation (12) is the standard one for the elasticity of smectics and associates to an elasticity term of first order expressing the compression of rolls a second order -or Frank-elasticity which takes into account the curvature effect. This introduction of curvature effects and of compressibilty effects in the same dynamical equation leads inevitably to the introduction of a length λ scale, that is the ratio of these two terms, this is the microscopic de Gennes screening length [5] of the order of magnitude of the wavelength of the basic structure. The mixing of quantities of a formally different order with respect to the number of derivatives in the same expression, as in (12), might seem inconsistent. However, as verified later, this leads to a consistent expression because one may use different scalings in z and x .

The explicit form of the phase dynamic equation deduced from (11) and (12) reads:

$$\partial u/\partial t + (\mathbf{V.\ Grad})\, u - V = D\,(\partial^2 u/\partial z^2 + (1/2)(\partial/\partial z)(\nabla u)^2 + \nabla(\partial u/\partial z \nabla u) + (1/2)\nabla(\nabla u(\nabla u)^2) - \lambda^2 \nabla^4 u) \qquad (13).$$

This equation may be written formally as:

$$u_t + (\mathbf{V.\ Grad})\, u = -\frac{\delta G}{\delta u} \qquad \text{where } G = F + \int dr\, V u \qquad (14)$$

However this does not necessarily imply relaxation to a steady state since the term linear in u in G may increase or decrease indefinitely with Neuman boundary conditions for instance. This would not put a priori any restriction upon the value of u since F depends on the gradients of u only. In this case the functional G is not bounded from below and cannot play the role of a Lyapunov functional. In the case of Dirichlet b.c., G is clearly uniformly bounded as well as

$$\int dr V u = \int [u(L)-u(0)] V dz$$

and one can look for stationary solutions.

3.ii) Linear stability against undulations of the constrained structure

We study the linear stability of the solution (8) of equation (13). Let w be the small perturbation, that is added to the basic stationary solution. We shall deal with situations where the phase approximation makes sense. We consider the *large box* limit which corresponds to $\lambda/L<<1$; they are many wavelengths in the actual realization of the regular structure, as required to use phase dynamics. Anticipating the typical length scale for the most unstable perturbations to be L along the z axis-as the basic solution itself- and $(\lambda L)^{1/2}$along x, we obtain the relevant equation for w where we have omittted the terms of order λ/L smaller than the retained ones. Since we can always choose systems large enough, this limit is always meaningful.

$$w_t = D\left(\frac{\partial^2 w}{\partial z^2} + u_z \frac{\partial^2 w}{\partial x^2} - \lambda^2 \frac{\partial^4 w}{\partial x^4}\right) \quad (15).$$

Although the approximation leading to this last equation is formally consistent, it neglects the highest derivative in z that is of fourth order. In such a situation, it seems likely that a boundary layer with a thickness of order λ will take care of the additional boundary condition due to this fourth derivative. We simply assume that this does not change, at the dominant order, the solutions, obtained with a second derivative in z only, far from the boundary. The solution of the linear problem discussed in detail in reference (23) can be summarised as follows . One looks for perturbations of the form $w_0(z)\exp(iqx+\sigma t)$ so that σ becomes an eigenvalue of a linear differential operator with q as a parameter.

In the absence of external force V and if a large enough imposed dilation u_0 is applied the system gets out of equilibrium ($\sigma>0$) at a marginal state such that $u_0=-2\pi\lambda$; the sign corresponds to a *dilation* with respect to equilibrium. The instability tends to restore the optimal and shorter wavelength by building an undulation as permitted in the x direction[10,17]. The critical wavelength is the harmonic mean between the box size and the local length scale λ as this is the natural scale along x resulting from a balance between the compressive elasticity across the layers and the second order - bend- elasticity transverse to them : $q_{co}{}^2=\pi/(\lambda L)$.

We expect this to remain true as long as V is negligible. In scaled quantities, by comparing terms in M_q, we can express the condition as $V << V_c = \pi D\lambda/L^2$ or as $C<<\pi$; the dimensionless number $C=VL^2/(D\lambda)$ will be used in the following to characterise the relative efficiency of the driving force (it plays the role of a Peclet number for the problem).

On the other hand, n the absence of imposed dilation ($u_0=0$), the instability sets in for $V \cong V_c$ with $q_c \cong (\pi/\lambda L)^{1/2}$.

If $V >> V_c$, the source of instability is in the constraint V and a whole band of wavenumbers will become unstable. The border which has a arabolic shape next to the threshold V_c varies asymptotically as $V^{-1/2}$ at long wavelengths and $V^{1/2}$ at short ones..

3.iii) Undulation in the strongly nonlinear limit

We consider now in more detail the domain of strong nonlinearities, reached at large C's (but still respecting the inequality $C<<L/\lambda$). It turns out that this leads to a rather interesting mathematical structure that could have something to do with the famous von Karman conjecture on the buckling of plates for very large constraints.

The study of the nonlinear domain is made formally simpler by keeping the order of magnitudes coming from the above developments :

$u \propto \lambda$, $x \propto (\lambda L)^{1/2}$, $z \propto L$.

The strength of the constraint V is measured by the number C, and the nonlinear equation for the steady solution reads,after making a change of variables $u/\lambda \rightarrow u$, $x/(\lambda L)^{1/2} \rightarrow x$, $z/L \rightarrow z$:

$- C = (u_z+u_x{}^2/2)_z + [u_x(u_z+u_x{}^2/2)]_x - u_{xxxx}$ (16).

We restrict our analysis the boundary conditions u=0 at z=0 and 1 such that no dilation is imposed by the boundary conditions. We already established that a threshold value of C exists such that the x-independent solution of (16) bifurcates to a solution depending on x in a non trivial fashion. Moreover this equation is the Euler-Lagrange condition expressing that the functional G given by formula (14) is stationary. Thus it makes sense to look at the optimal solution with the lowest "energy". As G contains a term which is linear in u in this energy and is multiplied by the large quantity C, one expects that this optimal solution has the largest possible order of magnitude in C .

We can use a dimensional argument to evaluate the order of magnitude of the different terms in (16). More precisely we look for solutions of the form :

$u \propto C^{\alpha}$, $z \propto C^{\beta}$, $x \propto C^{\gamma}$. (17)

If we look first for extended solutions along z, the scaling of the reduced z variable as 1 should correspond to $\beta = 0$. It is possible to match the first four terms in equation (16) with $u \propto C$, $z \propto 1$ and $x \propto C^{1/2}$. This choice makes relatively negligible the last term on the right hand side which is of order C^{-1}. It also yields the parameterless equation, deduced from (16) with the scalings given by (21) but with the change of notations u in Cu , z in z , x in $C^{1/2}x$:

$-1 = (u_z+u_x{}^2/2)_z + [u_x(u_z+u_x{}^2/2)]_x$ (18),

with the b.c. u=0 at z=0 and 1. The scaling (17) with the above choice of exponents indicates that the preferred wavelength of the optimal structure increases like $C^{1/2}$ at large C's, although the amplitude of the pertubation in u varies as C. This seems to be incompatible at first with the general assumptions made in the calculations of small gradients of u; in particular, the phase fluctuation has to be small in some sense in order to permit to write the equations in the coordinates of the unperturbed system. However this can be done consistently in the limit where C goes to infinity because there is a smallness parameter independent on C, which is the ratio λ/L. The tilt of the equiphase lines remains small, as required, if the dimensionless quantity u_x (resp u_z) is small. In the large C limit this gradient scales as $(C\lambda/L)^{1/2}$ [resp $(C\lambda/L)$] and thus can remain small even with a large C as long as λ/L is smaller than C^{-1}.

Indeed it remains to prove that a nontrivial and smooth solution of (18) yields the lowest energy. This is a rather subtle question. First it can be guessed that the solution u(z)=z(1-z)/2 does not represent the minimum of the energy by looking at fluctuations near this solution. Let again w be this fluctuation with the wavenumber q in the x direction. If neutral, this fluctuation has to be the solution of:

$$w_{zz}+\frac{1}{2}q^2 w\,(z-1/2)=0 \text{ with } w=0 \text{ at } z=0 \text{ and } 1.$$

This Airy equation imposes quantized values of q^2 and those values exist, as can be shown in the WKB-limit (large q). Since the trivial solution u(z) is neutrally stable against some perturbations, a weakly nonlinear analysis would show that some small amplitude pertubation with a wavenumber close to the ones given by the above eigenvalue problem yields a solution with an energy less than the trivial one.

Let us come now to the smoothness of the solution of equation (18). This equation is the Euler-Lagrange condition for the stationarity of the functional:

$\mathcal{L}(u)=\int dr\,\{-u+[u_z+\frac{1}{2}(\frac{du}{dx})^2]^2\}$, together with the b.c. u=0 at z=0 and 1. Suppose that one has found a smooth solution U(x,z) making the functional $\mathcal{L}$ stationnary and as negative as possible with our choice of sign. Consider now the second variation of $\mathcal{L}(u)$ around this solution. Let w(x,y) be the corresponding fluctuation of u near U, then this second variation has the form:

$$d\mathcal{L}\{w\}=-\int dr\,[w^2_z+2w_z w_x U_x+\frac{1}{2}w^2_x(U_z+\frac{3}{2}U^2_x)],$$

This expression is quite remarquable in the sense that it is homogeneous as far as the derivation order of w is concerned. This implies that the sign of $d\mathcal{L}(w)$ is completely determined by the one of the quadratic form in the variable w_z and w_x that appears in the integrand of $d\mathcal{L}(w)$. This quadratic form may get positive values (and the initial solution may become unstable) if the determinant $-(U_z+\frac{1}{2}U^2_x)$ is positive somewhere in the physical domain.

Before going to the physical meaning of this last condition let us show that this determinant is certainly positive somewhere for a smooth solution U. Let us consider the following change of variables, kindly suggested by V. Hakim:

$$dz=d\tau \quad ; \quad dx=\frac{1}{2}d\tau.\,U_x$$

In term of the new variable τ, we can express the variation of U as

$$dU=U_z\,dz+U_x\,dx=d\tau\,(U_z+\frac{1}{2}U^2_x)$$

The variation of U between z =0, 1 is given in the new running variable τ as

$$\int \text{grad}U.\,d\tau=0$$

since the boundary conditions U=0 holds at both ends.

Thus, either U is zero everywhere, or it has a maximum in the interval. In such a case, the quantity $-(U_z+\frac{1}{2}U^2_x)$ must be positive somewhere in the interval z = 0,1. On the other hand, U cannot be zero everywhere on the line indexed by τ. Near z=0 (or z=1, but with a slight change of sign and of variable), the Taylor expansion of a solution of equation (18)

should include a quadratic term $-z^2/2$, and, as the line under consideration merges with the z=0 line at right angle, U cannot vanish everywhere on it.
This result may be understood as follows : the determinant of the quadratic form is precisely equal to the local gradient of the wavelength expressed in intrisic coordinates. Thus the possibility of having locally an unstable fluctuation is another manifestation that this may increase the local wavenumber by a modulation in the transverse direction. Here we have the remarkable situation that this process may be continued down to arbitrary small scales. This implies that the equation (18) cannot be uniformly valid, because the original equation (16) had actually a built-in small space scale, represented by the fourth derivative, that has precisely disappeared in (18) after the rescalings motivated by the large C limit. This is the typical situation where a boundary layer has to exist in order to match the domains where the asymptotic equation [here equation (18)] joins formal singularities of its solution. Coming back to equation (16) and looking for the balance between all terms by using the scaling forms (17), we obtain the solution
$\alpha = 0; \beta = -1/2; \gamma = -1/4.$
This choice implies first that there is a boundary layer whose thickness along z goes to zero at large C as $C^{-1/2}$. More generally, the exponents β and γ define the scaling of lower cut off lengths along the z and x direction. This scaling defines the smallest scales in the cascade of structures starting from an original large scale *zig-zag* solution as well as the adjustment of the boundary conditions at the wall. Much work remains to be done on this problem which could be approached from an asymptotic matching treatment.
However the above solution does not preclude that there are other solutions with an even smaller energy on average along the x-direction that were nonperiodic in y, but quasiperiodic or even chaotic.

4. NUCLEATION AND DESTRUCTION OF ROLLS

In this section we discuss, within the phase approximation, the nucleation or destruction of new rolls as observed in the experiments by Pocheau and Croquette [9]. Indeed, as said above, every detail cannot be discussed within the phase approximation since it imposes that the phase has a small variation within a distance of one wavelength. On the contrary, the birth (/ death) of a roll implies a localised and abrupt phase jump of 2π over a distance of order of one wavelength. The corresponding dynamics is expected to be fast, with a well defined intrinsic time scale. On the contrary, the time scale for phase dynamics increases with the size of the system and so may be much longer than any intrinsic time scale related to the destruction/creation of rolls. Accordingly, when viewed on this long time scale, the birth/death of a roll may be seen as a local and instantaneous change of phase of $+/- 2\pi$. Indeed, this is not enough to specify the nucleation/destruction process, as we have also to decide when and where it occurs. In one dimension, a situation to which we shall restrict ourselves now (the 2D case is discussed in [23]), the nucleation is due to the growth of the *Eckhaus instability* [18], triggered by a too large or too small local wavenumber, depending on which side of the stability boundary is reached. As it is well known too, this phase instability manifests itself by the vanishing of the phase diffusion coefficient, a function of u_z, contrary to what was assumed before. We shall not introduce this dependence of D, see again ref. 23 for this. Note that the nucleation does not have to be compensated in the physical domain to keep constant the number of rolls: with our choice of boundary conditions, $u_z=0$ at $z= L$, there is a possible flux of phase either to or from the outside at this boundary.

We shall assume below that the only effect of the Eckhaus instability is to put an upper bound on the phase gradient u_z that we shall denote as g. If, somewhere, u_z equals this critical value, then u jumps down by 2π and gets later a dynamics that is still described by the linear equation (4) with $(\Lambda E+V)$ replaced by V. A somewhat similar approach has been used by Langer and Fisher [19] in their description of the critical current in superfluids. We shall describe what are the consequences of this assumption in a typical 1D situation. Except at the instant of nucleation or destruction of rolls, the phase u is the solution of:

$u_t=V+Du_{zz}$ (19)

with the b.c. u=0 at z=0 and $u_z=0$ at z=L and the condition of "Eckhaus stability" $u_z<g$. The relevant steady solution of this equation is given by equation (10):

$U_0=-Vz(z-2L)/2D$,

With this solution, the maximum of u_z is at z=0 so that the above solution becomes Eckhaus unstable if $V>V_c=Dg/L$. If this is so, no convenient steady state solution of (19) exists and rolls have to be destroyed in order to insure that the condition $u_z<g$ is satisfied everywhere and all the time. We shall show first that periodic solutions exist, with the destruction of new rolls at z=0 and then compute the corresponding period both for V slightly larger than V_c and then much larger than V_c. Later we shall also prove that, within some reasonnable assumptions, those periodic solutions attracts all initial conditions.

To prove that periodic solutions exist with the rolls being always destroyed at z=0, one notices that the sought solution, because of the linearity of (19), is the sum of the steady parabolic solution U_0 plus a solution of the diffusion equation :

$u'_t=Du'_{zz}$ (20)

with the b.c. u'=0 at z=0 and $u'_z=0$ at z=L. Suppose furthermore that u'_z has a single extremum at z=0 at some instant. This will remain so at later times, and the roll destruction will occur at z=0, because the z-derivative of (U_0+u') will become first larger than g at this position. In our model, this will be the time when one will have to add (-2π) to the full u. This function does not quite satisfy the b.c. u=0 at z=0; however after a very short time it will do, getting a very large negative slope at z=0. The new function so obtained still has its largest negative derivative at z=0 and so the initial assumption will remain true.

Let us estimate now the period of these oscillations as a function of V. This can be done completely in two limits, that is for V very close to the critical value V_c and V much larger than this critical value.

If V is only slightly larger than V_c, it takes a very long time, after the destruction of a roll , to build up the derivative of the solution at z=0 to the critical value. When this happens, almost all relaxation modes of the diffusion equation (20) have relaxed almost to zero and have an amplitude negligible as compared to the less damped mode. Fourier analysis shows that then the amplitude of u' (z,t) is almost equal to:

$u'(z,t)=\pm(2a_0/\pi)\exp(-\pi D^2t/4L^2)\sin(\pi z/2L)$,

The +/- correspond to the addition /subtraction of a roll , and the coefficient $2a_0/\pi$ is such that the complete series adds up to a_0 at t=0. The origin of time is at the instant of the previous roll destruction. Thus the period T(V) is such that the derivative of this contribution u' at z=0, once added to the one of U_0 is equal to V_c, which leads to:

$$T(V)=\frac{4L^2}{D\pi^2}\ln\left[\frac{\pi^2 D a_0}{2L^2(V-V_c)}\right],$$

The expression is valid when V is only slightly larger than V_c and all scaling parameters have been written explicitely.

On the contrary, when V is much larger than V_c one expects that the period is much shorter than the phase diffusion time across the length L. At a given time, u_z at z=0 is the sum of U_0 plus all the contributions of the previous kicks.The contributions of the kicks are independent on the length L as long as the corresponding perurbation initiated at z=0 has not reached z=L through the diffusive process described by equation (24). Once this pertubation has reached L, it is damped out by the b.c. Let n* be the order of magnitude of the number of kicks in the past that have not yet been damped by this boundary effect. Thus a simple estimate of n* follows from the equality $Dn^*T(V)=L^2$. Furthermore, each kick between n =0 and n =n* has a contribution $a_0\ (DnT)^{-1/2}$ to the derivative of u at z=0. Thus the contributions of all undamped kicks to this derivative is about $\Sigma a_0\ (DnT)^{-1/2}$ with n running from zero to n*. This sum is thus of order $n^*a_0\ (Dn^*T)^{-1/2}$. This must precisely compensate the large derivative of U_0 at z=0, that is VL/D. Whence,we get the simple estimate:

$T(V)=a_0/V$.

Indeed this has a very simple physical meaning because V is a velocity and a_0 a wavelength, so that T(V) is the period of waves with those velocity and wavelength. The computation of the constant prefactor in this last formula is rather intricate and we shall not consider this question here. Note that , in this regime, except in a boundary layer of thickness of order $2(\frac{Da_0}{2V})^{1/2}$, the free boundary solution is recovered. One should be careful that T(V) should remain large compared to the microscopic nucleation time for the treatment to be valid. In anyone of the regimes discussed here , this can be achieved by using a large enough system.

In the intermediate range of values of V, that is for V/Vc of the order of a few units, the period T(V) is given implicitely by the solution of an integral equation together with the condition that u_z is exactly equal to its limit value g just before the destruction time. The details of the corresponding analysis may be found in ref.23 where we also discuss the possibility to have successive nucleations taking place at different locations. It is also possible to have more complicated asymptotic time dependence with other choices of boundary conditions, but by keeping the same phase dynamics equation and the same condition for an extreme value of the phase gradient. Numerical work is required to study these situations in detail.

5. SUMMARY AND CONCLUSIONS.

As pointed out in the introduction, the present analysis allows to give an explanation for the extreme difficulty of observing permeation in conventional flow experiment, with the noticeable exception of the experiment by Martinet et Léger [20]. This can be explained by the instability studied in section 3 of this paper. This instability has a vanishingly low threshold at large L, since the critical velocity is found to vanish like L^{-2}.Our predictions can be probably best tested in cholesterics and in Rayleigh Benard cells were very slow flow can be achieved . Furthermore the Pocheau-Croquette annular geometry provides a convenient one dimensional situation in which the considerations of section 4 can be checked. Another important domain, in which transverse flows perturb roll systems is meteorology. Our analysis may provide some hints on the reason why roll systems are never well ordered nor quite periodic in actual situations. On the theoretical level, much work remains to be done. In particular, the behavior of instabilities generated by unidirectionnal flows in basically two dimensionnal geometries deserves further investigations, in view in particular of the possibility of fractal solutions to the buckling problem described by equation (22). The existence of chaotic solutions in the nucleation of defects in one dimensional problems is the object of a present numerical investigation. Finally, we wish to stress, once more, out of this comparative study, how some physical fluid instabilities and phenomena in liquid crystals fit within an unified theoretical framework and how such a description can lead to a better mutual understanding.

References

1) O. Lehmann Ann. Phys. (Leipzig) 2 ,649 (1900)
2) N.V. Madhusudana & R. Pratiba, Mol. Cryst. & Liq. Cryst. Lett. 5 ,43 (1987)
3) F.M. Leslie, Mol. Cryst. & Liq. Cryst. 7 ,407 (1969)
4) W. Helfrich, Phys. Rev. Lett. 23 ,372 (1969)
5) P.G. de Gennes, The Physics of liquid crystals , Clarendon , Oxford (1987)
6) P. Martin, O. Parodi & P. Pershan, Phys. Rev. A 6 ,2401 (1972)
7) N.A. Clark, Phys. Rev. Lett. 40 ,1663 (1978)7)
8) A.C. Newell & J. Whitehead, J. Fluid. Mech.38 203 (1969); L.A.Segel, J.Fluid Mech. 38, 235 (1969).
9) A. Pocheau, Thèse Université Paris 7 (1987)
A. Pocheau, V. Croquette & P. Le Gal , Phys. Rev. Lett. 55 ,10 (1985)
10) M. Delaye, R. Ribotta,& G. Durand, Phys. Rev. Lett. 31 ,433 (1973),
N. Clark & R.B. Meyer, App. Phys. Lett. 30 ,3 (1973),
see also (5) page 287
11) Y. Pomeau & P.Manneville, J. de Phys. Lett. 40, L609 (1979).
12) E. Siggia & A. Zippelius, Phys. Rev. A24, 1036 (1981)
13) E. Guazzelli, E.Guyon and J.E. Wesfreid, Phil. Mag. A 48, 709 (1983).
14) H. Brand, Phys. Rev. A 35 4461 (1987)
15) E. Dubois Violette, G. Durand, E. Guyon, P. Manneville and P. Pieranski Sol. St. Phys. Supp. 14 ,447 (1978)
16) G. Greenstein & R. Pelcovitz, Phys. Rev. Lett. 47 ,856 (1981),

"From amorphous to crystalline" page 387; Editor C. Godreche Les Editions de Physique

17) J.M. Delrieu J. Chem. Phys. 60 1081 (1974)
18) A good introduction on the patterns in convective instabilities can be found in the introductory chapter of the book " Cellular structures in instabilities"J.E. Wesfreid and S. Zaleski ed. Springer Verlag (Berlin)1984

19) J.S. Langer, M. Fisher Phys. Rev. Lett. 19, 560 (1967).

20) L. Leger & A. Martinet J. de Phys. C3, 89 (1976).

21) S.Zaleski, thèse de DEA, U. de Paris 6 (1980).

22) P.Hall, Phys. Rev. A29, 2921 (1984) and references therein.

23) J. Prost, Y.Pomeau and E. Guyon, "Stability of permeative flows in roll systems", preprint, submitted for publication.

TRANSITIONS BETWEEN PATTERNS OF DIFFERENT SYMMETRIES

D. WALGRAEF†
Service de Chimie-Physique,
Université Libre de Bruxelles,
Bd du Triomphe, CP 231,
B-1050, Brussels, Belgium.

ABSTRACT. Beyond various pattern forming instabilities structures with different symmetries may be simultaneously stable. Several aspects of the transitions between such structures are studied in the framework of amplitude equations of the Ginzburg-Landau type. In particular, it is shown how boundary effects, external fields and internal fluctuations may affect these transitions. The relevance of these effects to specific experimental problems is discussed.

1. Introduction

The nucleation of spatio-temporal patterns in systems driven far from thermal equilibrium by external constraints remains the subject of intensive experimental and theoretical research [1-2]. It deals with instabilities and transitions and raises new fundamental issues related, for instance, to the spontaneous emergence of order in such diverse contexts as mechanics, hydrodynamics, chemistry, biophysics, nonlinear optics, geology, materials science and solid state physics. A general framework for the study of these phenomena has now emerged, based on bifurcation theory and normal forms, leading to amplitude equations for the spatio-temporal patterns which are reminiscent of the time dependent Ginzburg-Landau description of phase transitions [3-4].

It is now well known that, beyond pattern forming instabilities in extended systems, structures with different symmetries may be simultaneously stable (e.g.

† Senior Research Associate, National Fund for Scientific Research (Belgium).

E. Tirapegui and W. Zeller (eds.), Instabilities and Nonequilibrium Structures III, 269–282.

rolls and hexagons in non Boussinesq convection; rolls, squares and hexagons in liquid crystal instabilities and ferro-fluids, cubic and planar defect microstructures in irradiated metals and alloys, etc.). The determination of the ranges of parameters for which these structures are stable and the study of the transitions and of the transition mechanisms between these structures are thus particularly interesting. Furthermore in nonequilibrium pattern formation, there is usually no free energy argument nor variational principle to tell us when and where such transitions should occur. Hence, no universal transition mechanism may be expected and the aim of this paper is to show that besides fluctuations, boundary effects, or external fields, or defects may play a determinant role in the transitions between different types of nonequilibrium patterns.

Three examples will be discussed: the effect of boundaries on the hexagon-rolls transition which is essential for the understanding of recent experimental results in convective systems; the effect of external fields on the transition to spatio-temporal structures with special emphasis on waves induced by Hopf bifurcations where strong resonances may occur between the unstable modes and fields which do not have the same symmetry; finally, it will be shown, in the incommensurate transition of quartz, how thermal fluctuations affect the phase diagram and the transitions between the different structures.

2. Boundary Effects on the Transition between Hexagons and Rolls

Hexagonal and roll patterns may be induced by hydrodynamical instabilities of the Bénard-Marangoni or and Rayleigh-Bénard type, for example in non-Boussinesq fluids or under temporal temperature modulation [5-13]. The existence of these structures has also been predicted in reaction-diffusion systems beyond Turing instabilities [14-15]. The transition between rolls and hexagons have been experimentally studied in the Rayleigh-Bénard convection of water near 4^0C where non-Boussinesq effects are important [7]. More recently other experiments of this type have been performed and detailed comparisons with existing theories have been made, showing some discrepancies between predicted and observed transition thresholds [8]. As shown below, this problem may be understood by considering finite size and boundary effects.

From the theoretical point of view these transitions may be described by the Haken model [16]:

$$\tau_0 \partial_t \sigma(\vec{x},t) = [\epsilon - \frac{\xi_0^2}{4q_c^2}(q_c^2 + \nabla^2)^2]\sigma(\vec{x},t) + v\sigma^2(\vec{x},t) - u\sigma^3(\vec{x},t) \tag{1}$$

where $\epsilon = (b - b_c)/b_c$, b being the bifurcation parameter and b_c its critical value. For $b > b_c$, the patternless solution is unstable and new stable solutions of this Landau-Ginzburg type of dynamics appear which correspond, in unbounded two-dimensional systems to :

(1) roll structures or unidirectional modulations of the order parameter. They are described by the usual amplitude equation:

$$\tau_0 \dot{A} = [\epsilon + \xi_0^2(\partial_x - \frac{i}{2q_c}\partial_y^2)^2]A - |A|^2 A \tag{2}$$

where $\sigma(\vec{x}, t) = [A(\vec{x}, t)e^{iq_c x} + c.c.]/\sqrt{3u}$. Hence, steady rolls of amplitude $|A| = A_r = \sqrt{\epsilon}$ may appear via a second orderlike transition, or supercritical bifurcation at $b = b_c$.

(2) hexagonal or triangular structures appearing via a first orderlike transition (subcritical bifurcation) and described by the following amplitude equations:

$$\tau_0 \dot{A}_i = [\epsilon + \frac{4\xi_0^2}{q_c^2}(\vec{q}_i.\vec{\nabla})^2]A_i + vA_{i-1}^* A_{i+1}^* - |A_i|^2 A_i - \gamma \sum_{j \neq i} |A_j|^2] A_i \tag{3}$$

where $\sigma(\vec{x}, t) = [\sum_{i=1}^3 A_i(\vec{x}, t)e^{i\vec{q}_i \vec{x}} + c.c.]/\sqrt{3u}$, with $|\vec{q}_i| = q_c$, and $\vec{q}_1 + \vec{q}_2 + \vec{q}_3 = 0$. Note that in the Haken model u is constant and $\gamma = 2$. When u is space dependent as in the Proctor-Sivashinsky model [3] or in non-Boussinesq convection [6], γ may take different values. In the following I will assume that $\gamma > 1$ since otherwise, rolls would always be unstable contrary to square patterns for example. This case will be discussed elsewhere.

The corresponding steady states are given by

$$|A_i| = A_\pm = \frac{1}{2(1+2\gamma)}[v \pm \sqrt{v^2 + 4(1+2\gamma)\epsilon}] \quad . \tag{4}$$

The linear stability analysis shows that the rolls are stable for

$$\epsilon > \frac{v^2}{(1-\gamma)^2} = \epsilon_r \tag{5}$$

while A_+ is the only stable hexagonal pattern in the range :

$$-\frac{v^2}{4(1+2\gamma)} < \epsilon < \frac{v^2(2+\gamma)}{(1-\gamma)^2} = \epsilon_h \tag{6}$$

Hence, as shown in figure 1, there is an hysteresis loop between hexagons and rolls in the range $\epsilon_r < \epsilon < \epsilon_h$. In the purely deterministic case, a transition from hexagons to rolls occurs, on increasing the constraint, at $\epsilon = \epsilon_h$, while, on decreasing the constraint, a transition from rolls to hexagons takes place at $\epsilon = \epsilon_r$. In the presence of fluctuations, however, since the dynamics (1) is potential, these transitions should occur when the potential is the same for rolls and hexagons, i.e. for

$$\epsilon = \epsilon_s = \frac{v^2(\sqrt{2(1+\gamma)} - 1)}{(1+2\gamma)(\sqrt{2(1+\gamma)} - 2)^2} \quad . \tag{7}$$

However, the noise intensity is usually too small in nonequilibrium systems for this transition being experimentally observable. Nevertheless the condition (7) gives us the value of the bifurcation parameter for which the two structures may

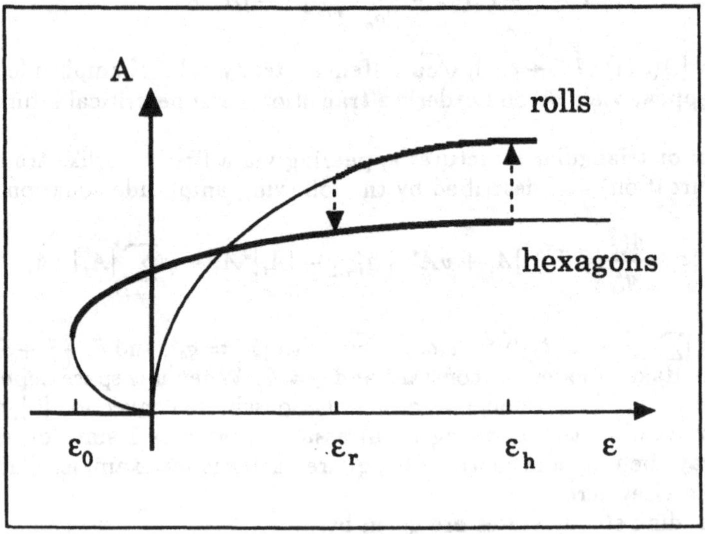

Figure 1 : Bifurcation diagram for the Haken model showing the hysteresis loop between rolls and hexagonal planforms (heavy lines represent stable states while thin lines correspond to unstable states).

coexist. Effectively, a front separating rolls and hexagons in an unbounded system will move towards the rolls for $\epsilon < \epsilon_s$ and towards the hexagons for $\epsilon > \epsilon_s$. At $\epsilon = \epsilon_s$ it will remain stationnary . This effect may be relevant for systems where there is frustration at the boundaries: for example when a stable structure is incompatible with the boundary conditions or when a structure is forced at the boundary. This is the case in convective instabilities where hexagonal planforms are usually incompatible with the geometry of the system. This is for example the case in the experiments of Ciliberto et al. [8] where hexagonal planforms cannot satisfy the boundary conditions imposed by the cylindrical geometry of the container. Hence, even in regimes where hexagons are selected, rolls are forced at the boundaries in a layer of width $l_c \propto \xi_0/\sqrt{\epsilon}$. These rolls may be considered as defects and they thus act as nucleation centers which trigger the transition from hexagonal to roll patterns [17]. The corresponding transition threshold $\bar{\epsilon}_h$ may be computed with the potential associated to the dynamics and one finds :

$$\bar{\epsilon}_h = \epsilon_s + \frac{3\xi_0^2}{l_c\Gamma} \tag{8}$$

For the experimental conditions discussed here (water at 28.3^0C, Prandtl number P=2, aspect ratio $\Gamma = 18$), the theory [6] predicts the following values for the parameters of the amplitude equation (3) : $v = 0.1031$, $\gamma = 1.44$ and $\xi_0^2 = 0.132$ (and $\epsilon_r = 5.4910^{-2}$, $\epsilon_h = 0.189$). These values lead to a threshold value $\bar{\epsilon}_h = 0.093$

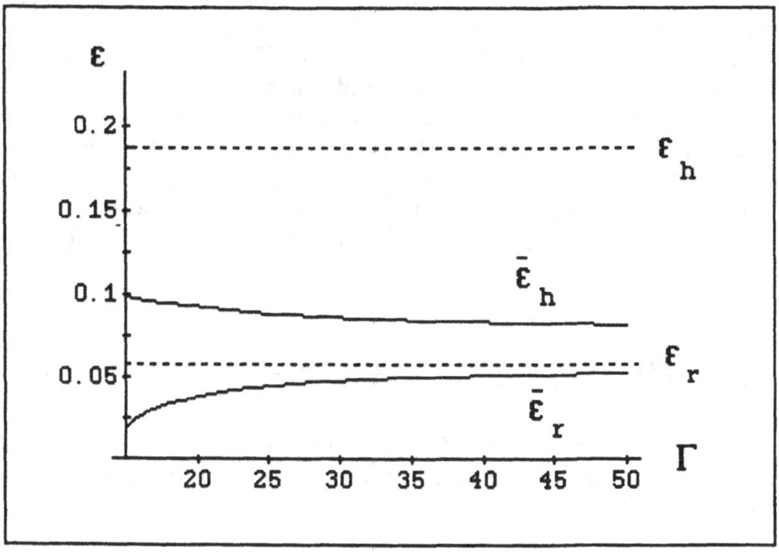

Figure 2 : Aspect ratio dependence of the transition thresholds for rolls and hexagons in the case of Bénard-Marangoni instability in finite containers (all the symbols are defined in the text).

which is close to the experimental value 0.09.
The situation is quite different on decreasing the constraint: since the noise intensity is negligeable and since they are forced at the boundaries, roll are stable until the lower hysteresis limit. However, this limit is moved as a result of the finite size of the system [4]. In the present case, one finds that the stability limit $\bar{\epsilon}_r$ for critical rolls is given by

$$(\gamma - 1)\epsilon_r - v\sqrt{\epsilon_r} + \xi_0^2(\frac{\pi}{L})^2 = 0 \qquad (9)$$

or $\bar{\epsilon}_r = 0.0326$ (experimentally $\bar{\epsilon}_r = 0.03$). The aspect ratio dependence of the thresholds $\bar{\epsilon}_r$ and $\bar{\epsilon}_r$ is shown in fig.2. From this discussion we see that finite size effects may be essential in the localization of the transitions between different structures. In particular, forced nucleation may occur as a consequence of geometrical frustration at the boundaries.

3. External Forcing of 2D Wave Patterns

As shown recently, pure spatial or temporal modulations imposed on Hopf bifurcations leading to wave patterns modify the selection and stability properties of the resulting spatio-temporal structures [**18-21**]. For example, it was shown that, in one-dimensional systems, beyond spatio-temporal Hopf bifurcations, left and right traveling waves are linearly coupled by uniform oscillations or steady spatial modulations provided the frequency of the oscillations or the wavenumber of the modulations are close to two times the critical ones. Hence, a spatially uniform forcing may restore the left-right symmetry and transform traveling waves into standing waves in regimes where the latter are otherwise unstable.

In two-dimensional systems, new possibilities occur and I will discuss two examples here. Consider an isotropic system invariant under space and time translations and parity transformations which undergoes a Hopf bifurcation with finite wavenumber k_c and frequency ω_c. The growth rate of the unstable modes is written as :

$$\omega(\vec{k}) = \epsilon - \xi_0^2(k^2 - k_c^2)^2 + i(\omega_c + \omega_1(k^2 - k_c^2) + \omega_2(k^2 - k_c^2)^2) + O((k^2 - k_c^2)^3) \quad (10)$$

and the slowly varying envelopes of left and right travelling waves propagating in, say, the x-direction, satisfy, after appropriate scaling, the following amplitude equations [22-24] :

$$\begin{aligned} \dot{A} + c\nabla_x A &= \epsilon A + (1 + i\alpha)[\nabla_x - \frac{i\nabla_y^2}{2k_0}]^2 A + i\eta[\nabla_x^2 + \frac{\nabla_y^2}{\chi}]A - (1 + i\beta)A|A|^2 \\ &\quad - (\gamma + i\delta)A|B|^2 \\ \dot{B} - c\nabla_x B &= \epsilon B + (1 - i\alpha)[\nabla_x + \frac{i\nabla_y^2}{2k_0}]^2 B - i\eta[\nabla_x^2 + \frac{\nabla_y^2}{\chi}]B - (1 - i\beta)B|B|^2 \\ &\quad - (\gamma - i\delta)B|A|^2 \end{aligned} \quad (11)$$

When $\gamma > 1$, only travelling are stable, while for $\gamma < 1$ standing waves are the only stable structure, and I will consider the first case in the following. Of course, patterns built on travelling waves which propagate in different directions may also be expected. The stability of these structures also depends on the coefficient of the nonlinear couplings between the different waves. On assuming that this coefficient is scalar and written as $\kappa + i\lambda$ with $\kappa > 1$, we are in a situation where the only stable structure of the system corresponds to waves travelling in one direction.

The effect of a purely spatial forcing of hexagonal symmetry has been discussed in [21]. It was shown that, when the wavenumber of the forcing is close to $3k_c$, travelling waves with wavevectors parallel to the basic vectors of the imposed hexagonal modulation are nonlinearly coupled. This leads to a stabilization mechanism for two-dimensional wave patterns. Similar effects may also be obtained with purely temporal forcings. For example, temporal modulations of frequency ω_0 such that $n\omega_0 \neq 3\omega_c$ couple triplets of waves travelling in directions making $2\pi/3$ angles (e.g. $A expi(\vec{q}_1\vec{r} - \omega_c t)$, $B expi(\vec{q}_2\vec{r} - \omega_c t)$, $C expi(\vec{q}_3\vec{r} - \omega_c t)$, with $\vec{q}_1 + \vec{q}_2 + \vec{q}_3 = 0$ and $|\vec{q}_1| = |\vec{q}_1| = |\vec{q}_1| = q_c$) . The corresponding uniform amplitude equations are:

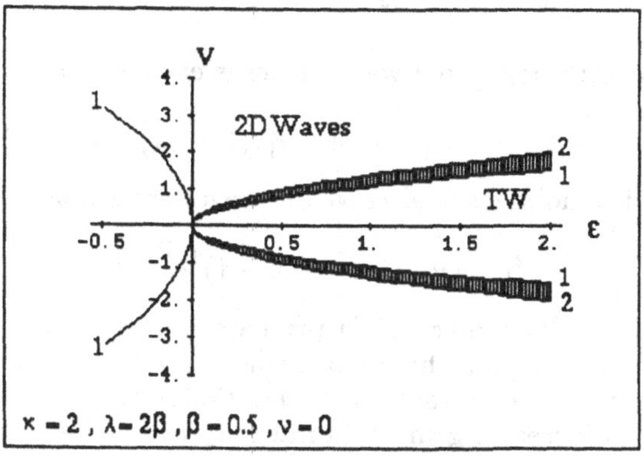

curve 1 : stability limit of 2D wave pattern
curve 2 : stability limit of 1D traveling wave

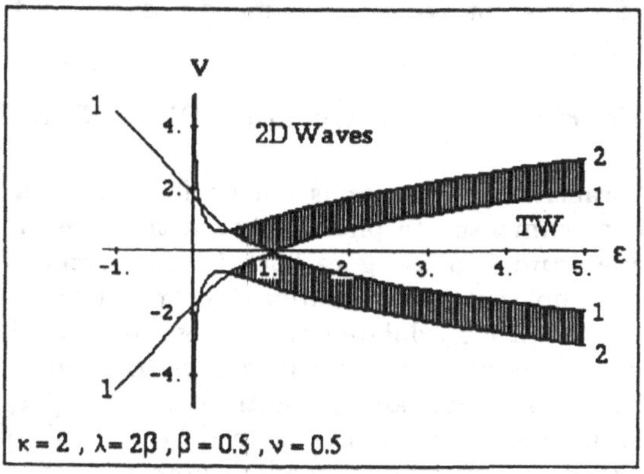

Figure 3 : Phase diagram for a Hopf bifurcation forced by temporal modulations of frequency close to three times the critical frequency, at zero and non zero detuning.

$$\dot{A} = (\epsilon + i\nu)A + v\bar{B}\bar{C} - (1 + i\beta)A|A|^2 - (\kappa + i\lambda)A(|B|^2 + |C|^2)$$
$$\dot{B} = (\epsilon + i\nu)B + v\bar{A}\bar{C} - (1 + i\beta)B|B|^2 - (\kappa + i\lambda)B(|A|^2 + |C|^2)$$
$$\dot{C} = (\epsilon + i\nu)C + v\bar{B}\bar{A} - (1 + i\beta)C|C|^2 - (\kappa + i\lambda)C(|A|^2 + |B|^2) \qquad (12)$$

where $v \propto h^n$, h being the amplitude of the external modulation, and $\nu = \frac{n\omega_0}{3} - \omega_c$, the frequency detuning between the external field and the waves .

Hence, from the fixed point condition ($A = Rexpi\Phi_A, B = Rexpi\Phi_B, C = Rexpi\Phi_C$) :

$$[(1+2\kappa)^2+(\beta+2\lambda)^2]R^4-[2\epsilon(1+2\kappa)+2\nu(\beta+2\lambda)+v^2]R^2+(\epsilon^2+\nu^2)=0 \quad (13)$$

it may be deduced that hexagonal wave patterns exist when

$$v^4+4v^2[\epsilon(1+2\kappa)+\nu(\beta+2\lambda)]-4[\nu(1+2\kappa)-\epsilon(\beta+2\lambda)]^2>0 \quad (14)$$

On the other hand, unidimensional traveling waves are stable when

$$v^2\epsilon-(\nu-\lambda\epsilon)^2-(\kappa-1)^2\epsilon^2>0 \quad (15)$$

The corresponding phase diagram is presented in fig.3 for zero and non-zero frequency detuning ν, showing the regions where 2D and 1D wave structures are individually or simultaneously stable. One sees that, either on increasing ϵ at fixed field intensity, or on increaising the field intensity at fixed ϵ, one crosses a region where the only stable structure corresponds to hexagonal wave patterns.

Hence we have here another example where the stabilization of a pattern by an external field of a different symmetry is made possible by the non-variational character of the dynamics.

4. Fluctuation Effects on Incommensurate Phase Transitions in Quartz

While the $\alpha-\beta$ transition in quartz is known for almost a century [**25**], the observation of an incommensurate phase between the low-temperature α phase and the high temperature β phase goes back to the seventies [**26-29**]. This is one of the many examples where transitions occur between normal crystalline phases and incommensurate modulated structures [**30**]. Its interest lies in the fact that modulated structures of different symmetries may be simultaneously stable. Effectively, experimental and theoretical analysis [**31**] concluded to the existence of striped modulated structures defined by one pair of wavevectors (1-q phase) or of modulated structures of triangular symmetry defined by three pairs of wavevectors separated by $2\pi/3$ angles (3-q phase). It turns out that the stablest incommensurate phase corresponds to 3-q structures of about 10 nm wavelength which persist in a small temperature range (1.4 K) around 846 K. According to the mean field Landau description, this phase transition should be second order [**31**]. Furthermore, in the presence of uniaxial stresses, a striped 1-q modulated phase appears between the β and the 3-q phases [**32**].

On the other hand, it is well known that critical fluctuations can strongly affect such a transition behavior and, according to renormalization group calculations, the transition point may even be unstable [**33**]. Hence, fluctuations should be incorporated in the analysis of the system's behavior near the critical temperature where the β, 3-q and 1-q phases meet, and this is the purpose of the following discussion.

According to the phenomenological description of the $\alpha-\beta$ transition of quartz [**34**] , the free energy can be expanded in a power series of the Fourier components $\sigma_{\vec{q}}$ of the order parameter (which corresponds to the displacement of the Si atoms

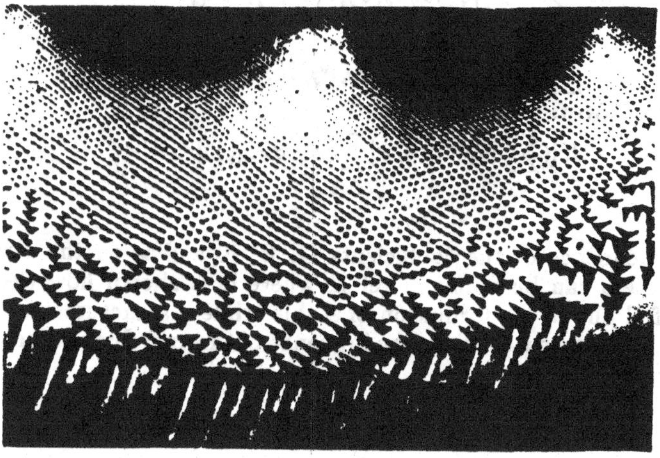

Figure 4: Incommensurate structure of triangular symmetry in quartz (courtesy of C.Leroux).

from their equilibrium position in the β phase) and of the strain field. After elimination of the elastic degrees of freedom, the quadratic part of the free energy may be written, in scaled variables, as :

$$\mathcal{F}_2 = \sum_{\vec{q}} (-\epsilon + d(q^2 - q_0^2)^2 + \Delta q^2 cos^2(3\phi))|\sigma_{\vec{q}}|^2 \tag{16}$$

where $\epsilon = (T_i - T)/T_i$, T_i being the β-INC transition temperature; q_0 is the length of the critical modulation wavevectors and ϕ the angle between $\vec{q}$ and the two-fold x axis of the α phase. The corresponding marginal stability surface which limits the stability domain of the β phase in the $(\epsilon, \vec{q})$ space is shown in fig.5. One sees that 6 unstable modes are expected at $T = T_i$, which correspond to modulation wavevectors of length q_0 and of orientation $\phi = (2n+1)\pi/6$. Hence single- and triple-q structures may in principle be nucleated for $T < T_i$ but their stability is determined by the higher order terms of the Landau free energy which may be written as [**31**] :

$$\mathcal{F}_{>2} = \sum_{\vec{q},\vec{q}_1,\vec{q}_2} G(\{\vec{q}\})\sigma_{\vec{q}}\sigma_{\vec{q}_1}\sigma_{\vec{q}_2}\delta(\vec{q}+\vec{q}_1+\vec{q}_2)$$
$$\sum_{\vec{q},\vec{q}_1,\vec{q}_2,\vec{q}_3} H(\{\vec{q}\})\sigma_{\vec{q}}\sigma_{\vec{q}_1}\sigma_{\vec{q}_2}\sigma_{\vec{q}_3}\delta(\vec{q}+\vec{q}_1+\vec{q}_2+\vec{q}_3)$$
$$- K[(\sum_{\vec{q}} |\sigma_{\vec{q}}|^2)^2 + \nu|\sum_{\vec{q},\vec{q}_1} \sigma_{\vec{q}_1}\sigma_{\vec{q}-\vec{q}_1}|^2] \tag{17}$$

The detailed dependence of the coefficients G, H, K and ν on the elastic constants of the system can be found in ref.31 where it is shown how the angular q-dependence of G allows the formation of stable triple-q structures via a second order transition. Let us now consider the influence of thermal fluctuations on these various structures.

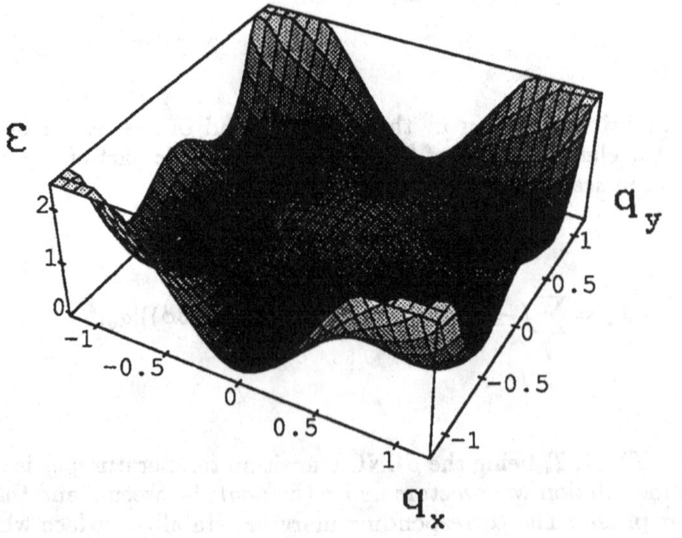

Figure 5: Representation, in the (ϵ, $\vec{q}$) space, of the surface which limits the stability domain of the β phase of quartz, according to the free energy (16) (d=1, q_0 =1 , Δ = 0.5).

4.1. Single-q or stripe phase

The existence and stability of a single-q structure of wavevector $\vec{q}$ of length q_0 and oriented along one of the directions defined by $\phi = (2n+1)\pi/6$ may be deduced from the Landau free energy (16-17) [31]. If one takes into account the contribution of thermal fluctuations, the equation of state may be written, at the Hartree approximation [15][35] :

$$0 = \epsilon < \sigma_{q_0} > -u < \sigma_{q_0} >^3 -u < \sigma_{q_0} > \int d\vec{k} < \sigma_{\vec{k}}\sigma_{-\vec{k}} > \tag{18}$$

The correlation function $< \sigma_{\vec{k}}\sigma_{-\vec{k}} >$ may be evaluated at the same approximation through the self-consistent scheme :

$$< \sigma_{\vec{k}}\sigma_{-\vec{k}} > = \frac{1}{r + d(k^2 - q_0^2)^2 + \Delta k^2 cos^2(3\phi)}$$

$$r = -\epsilon + 2u < \sigma_{q_0} >^2 + u \int d\vec{k} < \sigma_{\vec{k}}\sigma_{-\vec{k}} > \tag{19}$$

As a result, the equation of state may be rewritten as :

$$\epsilon = r_1 + \frac{2u}{\sqrt{d}} \frac{1}{\sqrt{r_1 + \Delta q_0^2}} K(\frac{\Delta q_0^2}{r_1 + \Delta q_0^2}) \tag{20}$$

where $K(m)$ is the complete elliptic integral of the first kind and with

$$r_1 < \sigma_{q_0} > -u < \sigma_{q_0} >^3 = 0 \tag{21}$$

Therefore the corresponding modulated structure arises with a finite amplitude (i.e. via a *first-order* transition) :

$$< \sigma_{q_0} > = \sqrt{\frac{r_{1c}}{u}}$$

at

$$\epsilon_1 = r_{1c} + \frac{2u}{\sqrt{d}} \frac{1}{\sqrt{r_{1c} + \Delta q_0^2}} K(\frac{\Delta q_0^2}{r_{1c} + \Delta q_0^2}) \tag{22}$$

where $\epsilon_1 = \epsilon(r_{1c})$ corresponds to the minimum of the curve $\epsilon(r_1)$ defined by equation (20). In the strong anisotropy limit ($\Delta q_0^2 >> r_1$), one has :

$$\epsilon_1 = \frac{3u}{\sqrt{d\Delta q_0^2}} ln(\Delta q_0^2) \tag{23a}$$

while in the weak anisotropy limit ($\Delta q_0^2 << r_1$), one has :

$$\epsilon_1 = 3(\frac{2u}{\sqrt{d\Delta q_0^2}})^{2/3} \tag{23b}$$

and we have here another example of the so-called Brazovskii effect [36] where fluctuations modify qualitatively the character of the transition.

4.2. Triple-q or triangular phase

A similar discussion may be performed for the 3-q structures. In this case, the cubic term of the free energy couples the amplitudes of the three modulations and behaves as $\sigma_1\sigma_2\sigma_3 cos(3\phi)$. The minimization process then shows that $cos(3\phi)$ is proportional to σ (leading to the observed rotation of the triangular structure), so that the cubic term effectively behaves as σ^4 [31]. The transition is thus second order as for 1-q structures, and the equation of state may be written as [35]:

$$0 = \epsilon < \sigma_{q_i} > -u_d < \sigma_{q_i} >^3 -u_{nd} < \sigma_{q_i} > \sum_{j\neq i} < \sigma_{q_j} >^2$$

$$-u < \sigma_{q_i} > \int d\vec{k} < \sigma_{\vec{k}}\sigma_{-\vec{k}} > \qquad (24)$$

On evaluating the correlation function at the same approximation level as in the 1-q case, one finds [35]:

$$\epsilon = (1+2\gamma)r_3 + \frac{2u}{\sqrt{d}}\frac{1}{\sqrt{r_3+\Delta q_0^2}}K(\frac{\Delta q_0^2}{r_3+\Delta q_0^2}) \qquad (25)$$

where $\gamma = \frac{u_{nd}}{u_d}$ and

$$r_3 < \sigma_{q_0} > -u < \sigma_{q_0} >^3 = 0 \qquad (26)$$

Therefore the 3-q modulated structure arises with a finite amplitude

$$< \sigma_{q_0} > = \sqrt{\frac{r_{3c}}{u}} \qquad (27)$$

and a finite rotation angle ϕ_c, since

$$\cos(3\phi_c) \propto \sqrt{\frac{r_{3c}}{u}} \qquad (28)$$

at

$$\epsilon_3 = (1+2\gamma)r_{3c} + \frac{2u}{\sqrt{d}}\frac{1}{\sqrt{r_{3c}+\Delta q_0^2}}K(\frac{\Delta q_0^2}{r_{3c}+\Delta q_0^2}) \qquad (29)$$

where $\epsilon_3 = \epsilon(r_{3c})$ corresponds to the minimum of the curve $\epsilon(r_3)$ defined by equation (25). In the strong anisotropy limit, one has :

$$\epsilon_3 = \frac{3u}{\sqrt{d\Delta q_0^2}} ln(\Delta q_0^2(1+2\gamma)^{2/3}) \qquad (30a)$$

while in the weak anisotropy limit, one has :

$$\epsilon_3 = 3(1+2\gamma)^{1/3}\left(\frac{2u}{\sqrt{d}}\right)^{2/3} \tag{30b}$$

Hence, the thresholds for 1-q and 3-q structures are different, and, since

$$\epsilon_3 > \epsilon_1, \tag{30}$$

($\epsilon_3 - \epsilon_1 = \frac{4u}{\sqrt{d\Delta q_0^2}} ln(1+2\gamma)$ in the strong anisotropy limit), a 1-q modulated structure should occur between the β and 3-q phases, even in the absence of uniaxial stresses in agreement with recent experimental observations [37]. To conclude, we see that critical fluctuations modify qualitatively the incommensurate phase transitions of quartz, which should be effectively first order. They also provide an intrinsic symmetry breaking effect which lifts the degeneracy of the mean field transition point between the β, 1-q and 3-q phases . The quantitative aspects of the problem and the effect of fluctuations on the transition lines in the presence of uniaxial stresses will be discussed in forthcoming publications.

Acknowledgments. Fruitful discussions with P.Borckmans, P.Coullet, G.Dewel, C.Leroux, C.Perez-Garcia and J.E.Wesfreid are gratefully acknowledged.

References

1. P.Coullet and P.Huerre, "New Trends in Nonlinear Dynamics and Pattern Forming Phenomena : The Geometry of Nonequilibrium.," Plenum, New York, 1990.
2. F.Busse and L.Kramer, "Nonlinear Evolution of Spatio-temporal Structures in Dissipative Continuous Systems," Plenum, New York, to appear 1990.
3. D.Walgraef, in "Nonlinear Phenomena in Materials Science," G.Martin and L.P.Kubin, eds., Transtech, Aedermannsdorf (Switzerland), 1988, p. 77.
4. P.C.Hohenberg and M.C.Cross, in "Fluctuations and Stochastic Phenomena in Condensed Matter," Lecture Notes in Physics 268, L.Garrido ed., Springer, New York, 1987.
5. E.Palm, J.Fluid Mech. **8** (1960), p. 183.
6. F.Busse, J.Fluid Mech. **30** (1967), p. 625.
7. M.Dubois, P.Bergé and J.E.Wesfreid, J.Physique **39** (1978), p. 1253.
8. S.Ciliberto, E.Pampaloni and C.Perez-Garcia, Phys.Rev.Lett. **61** (1988), p. 1198.
9. C.Normand, Y.Pomeau and M.Velarde, Rev.Mod.Phys. **49** (1977), p. 581.
10. M.Besterhorn and C.Perez-Garcia, Europhys.Lett. **4** (1987), p. 1365.
11. M.N.Roppo, S.H.Davis and S.Rosenblatt, Phys.Fluids **27** (1984), p. 796.
12. P.C.Hohenberg and J.Swift, Phys.Rev. **A35** (1987), p. 3855.
13. C.W.Meyer, G.Ahlers and D.Cannell, Phys.Rev.Lett. **59** (1987), p. 1577.

14. L.M.Pismen, J.Chem.Phys. **72** (1980), p. 1900.
15. D.Walgraef, G.Dewel and P.Borckmans, Adv.Chem.Phys. **49** (1982), p. 311.
16. H.Haken, "Advanced Synergetics," Springer, Berlin, 1983.
17. P.Coullet, L.Gil and D.Repaux, in "Instabilities and Nonequilibrium Structures II," E.Tirapegui and D.Villaroel, eds., Kluwer Acad. Publ., Dordrecht, 1989, p. 189.
18. H.Riecke, J.D.Crawford and E.Knobloch, Phys.Rev.Lett. **61** (1988), p. 1942.
19. D.Walgraef, Europhys.Lett. **7** (1988), p. 485.
20. I.Rehberg, S.Rasenat, J.Fineberg, M.De La Torre Juarez and V.Steinberg, Phys.Rev.Lett. **61** (1988), p. 2449.
21. P.Coullet and D.Walgraef, Europhys.Lett. **10** (1989), p. 525.
22. M.C.Cross, Phys.Rev.Lett. **57** (1986), p. 2935.
23. P.Coullet, S.Fauve et E.Tirapegui, J.Physique (Paris) **46** (1985), p. 787.
24. H.R.Brand, P.S.Lomdhal et A.C.Newell, Phys.Lett. **118A** (1986), p. 67.
25. H.Le Chatelier, C.R.Acad.Sci.(Paris) **108** (1889), p. 1046.
26. G.Van Tendeloo, J.Van Landuyt and S.Amelinckx, Phys.Status Solidi **A33** (1976), p. 723.
27. J.P.Bachheimer, J.Phys.Lett. **41** (1980), p. L559.
28. G.Dolino, J.P.Bachheimer, B.Bergé, C.Zeyen, G.Van Tendeloo. J.Van Landuyt and S.Amelinckx, J.Phys.(Paris) **45** (1984), p. 901.
29. E.Snoeck, C.Roucau and P.Saint Gregoire, J.Phys.(Paris) **47** (1986), p. 2041.
30. R.Blinc and A.P.Levanyuk, "Incommensurate Phases in Dielectrics," North Holland, Amsterdam, 1986.
31. T.A.Aslanyan, A.P.Levanyuk, M.Vallade and J.Lajzerowicz, J.Phys.C : Solid State Phys. **17** (1983), p. 6505.
32. G.Dolino, P.Bastie, B.Bergé, M.Vallade, J.Bethke, L.P.Regnault and C.Zeyen, Europhys.Lett. **3** (1987), p. 601.
33. O.Biham, D.Mukamel, J.Joner and X.Zhu, Phys.Rev.Lett. **59** (1987), p. 2439.
34. T.A.Aslanyan and A.P.Levanyuk, Solid State Commun. **31** (1979), p. 547.
35. P.Borckmans, G.Dewel and D.Walgraef, preprint (1990).
36. S.A.Brazovskii, Sov.Phys. (J.Exp.Theor.Phys.) **41** (1975), p. 85.
37. P.Bastie, F.Mogeon and C.Zeyen, Phys.Rev. **B38** (1988), p. 786.

DIRECT SIMULATION OF THREE-DIMENSIONAL TURBULENCE

M. E. Brachet
CNRS, Laboratoire de Physique Statistique
Ecole Normale Supérieure
24 rue Lhomond 75231 Paris Cedex 05
FRANCE

ABSTRACT. The three- dimensional Navier-Stokes equations are numerically integrated by a spectral method. Using the symmetries of the initial data a resolution of 864^3 and high Reynolds numbers ($R_\lambda = 140$) are obtained. Visualisations of the resulting turbulent flow show that the vorticity is spatially more concentrated than the energy dissipation. Consequently the turbulent activity is strongly correlated with low pressure zones. Measures of the exponent of fractal codimension confirm these results by showing more intermittency for vorticity: $\mu = 0.86$, than for energy dissipation: $\mu = 0.38$.

Turbulence is called developed when the scales transporting energy and those in which dissipation occurs are widely separated. This necessitates Reynolds that are at least on the order of several thousand. Much higher Reynolds numbers (on the order of several million) are obtained experimentally. Nevertheless, the current experimental methods measure only the velocity, and thus knowledge of the small-scale structures characterized by large velocity gradients remains fragmentary [1]. To study developed turbulence numerically, a large range of spatial scales, and hence high resolution, is essential. One is quickly limited by the size of the computer. An idea that comes to mind is to simplify the geometry of the flow, for example by using periodic boundary conditions [2].

The Taylor-Green vortex [3] is the three-dimensional flow that develops from the initial data:

$$V_x = sin(x)cos(y)cos(z)$$
$$V_y = -cos(x)sin(y)cos(z)$$
$$V_z = 0,$$

following the Navier-Stokes equations:

$$\partial_t \mathbf{v} + (\mathbf{v} \cdot \nabla)\mathbf{v} = -\nabla p + \nu \nabla^2 \mathbf{v}$$
$$\nabla \cdot \mathbf{v} = 0$$

periodic boundary conditions.

This is perhaps the simplest system in which to study the generation of excitation at small scales and the resulting turbulence. Since the initial conditions are products

E. Tirapegui and W. Zeller (eds.), Instabilities and Nonequilibrium Structures III, 283–289.

of trigonometric functions, we can use spectral methods, which are both simple to implement and accurate [4].

Compared to flows which are simply periodic, the Taylor-Green vortex displays additional symmetries. By taking advantage of these additional symmetries in the spectral integration algorithm for the Navier-Stokes equations, it is possible to gain factors of 64 in memory, of 32 in number of operations. and thus a factor of 4 in the separation of scales for a given computational power. In practice it is possible to run on a Cray-2 with a resolution of 864^3 (at the cost of tens of CPU hours per turnover time).

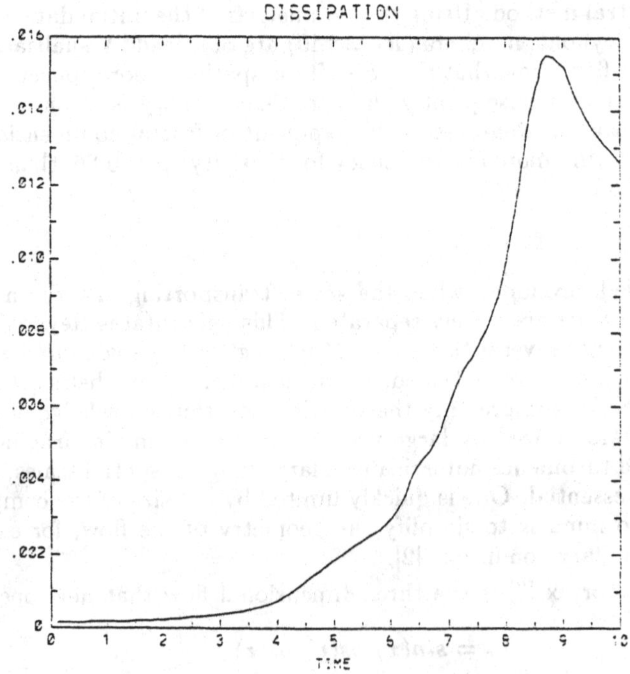

Figure 1. Energy dissipation $\epsilon(t) = \nu/2 \int dr \Sigma_{ij} (\partial_i v_j + \partial_j v_i)^2$ versus time. The maximum is reached, after a few eddy turnover times, around $t = 9$.

The evolution at large Reynolds numbers of the Taylor-Green vortex follows essentially two phases [5]. During the first phase, the viscous effects can be neglected, and small-scale structures are generated which are well-organized and laminar. During the second phase, viscous diffusion plays an important role in the dynamics and distordered dissipative structures are created. The energy dissipation attains its maximum at a late stage of the viscous phase. In the simulation which we present here, with a resolution of 864^3 and with a Reynolds number (defined as the inverse of the kinematic viscosity) of 5000, this maximum takes place at $t = 9$ (see Figure1). Figure 2 shows the energy spectrum, defined here as the kinetic energy per unit volume and per wavenumber (averaged over angle) at the moment of maximum energy dissipation. Note that, over more than a decade, an inertial range is present over which the spectrum follows a power law with exponent close to the value of $-5/3$ predicted by Kolmogorov [6], followed by a dissipative zone. At this instant in time, the Reynolds number R_λ defined using the Taylor scale is about 140.

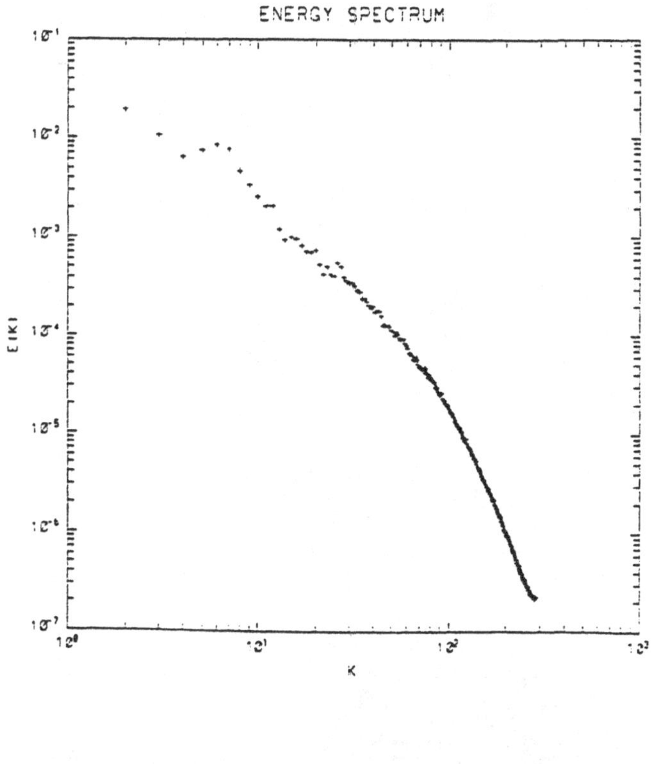

Figure 2. Energy spectrum $E(k)$ at $t = 9$. Note that the slope is close to Kolmogorov's value $-5/3$ beteween $k = 6$ and $k = 60$.

The chaotic and highly intermittent appearance of the small-scale excitations is shown in Figure 3 which represents, in a planar section at $y = \pi/4$, the velocity field **v**. Figure 4 represents in the same fashion the local pressure p. Note that the energy dissipation $\sigma^2 = \nu/2\Sigma_{ij}(\partial_i v_j + \partial_j v_i)^2$ when visualized (data not shown), appears somewhat less spatially concentrated than the vorticity $\omega^2 = (\nabla \times v)^2$. This is confirmed by examination of figure 4, which shows that the zones of turbulent

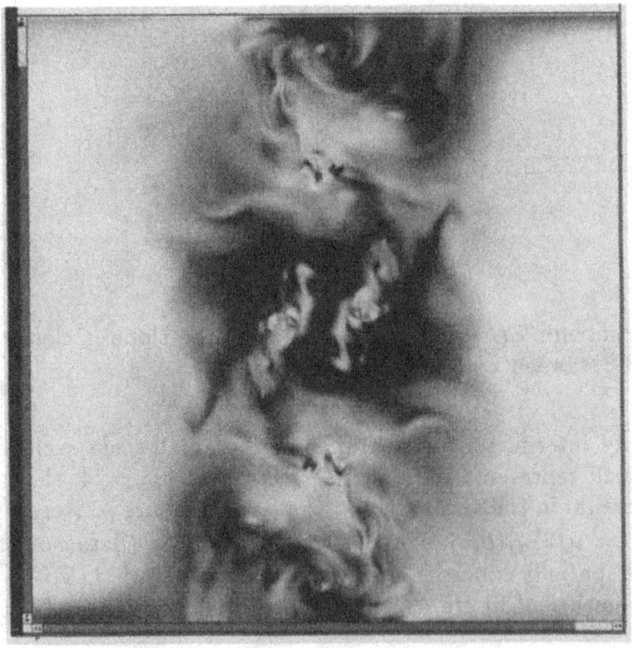

Figure 3. Raster visualisation of the magnitude of velocity in the plane $y = \pi/4$ at $t = 9$. Because of the symmetries, only the aera $0 < x < \pi, 0 < z < \pi$ is shown.

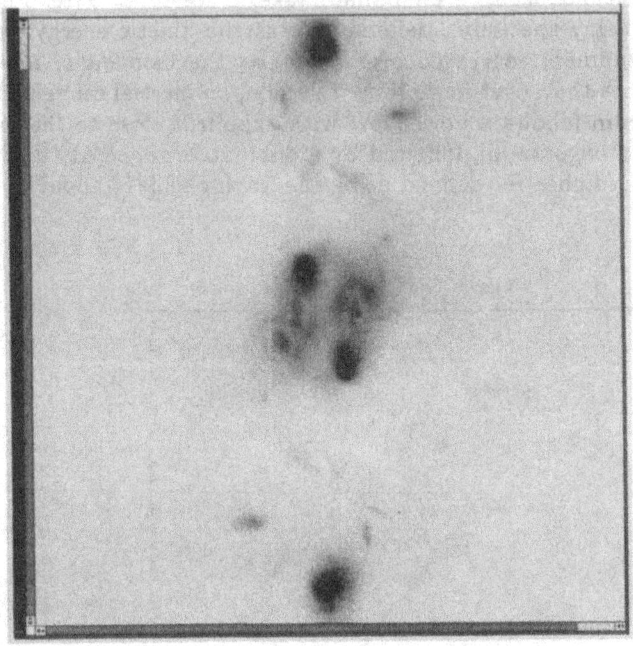

Figure 4. Raster visualisation of the pressure field in the plane $y = \pi/4$ at $t = 9$. Because of the symmetries, only the aera $0 < x < \pi, 0 < z < \pi$ is shown.

activity are in low-pressure regions. Indeed, by taking the divergence of the Navier-Stokes equations, we find that the pressure obeys the equation:

$$\nabla^2 p + 1/4\Sigma_{ij}(\partial_i v_j + \partial_j v_i)^2 - 2\omega^2 = 0.$$

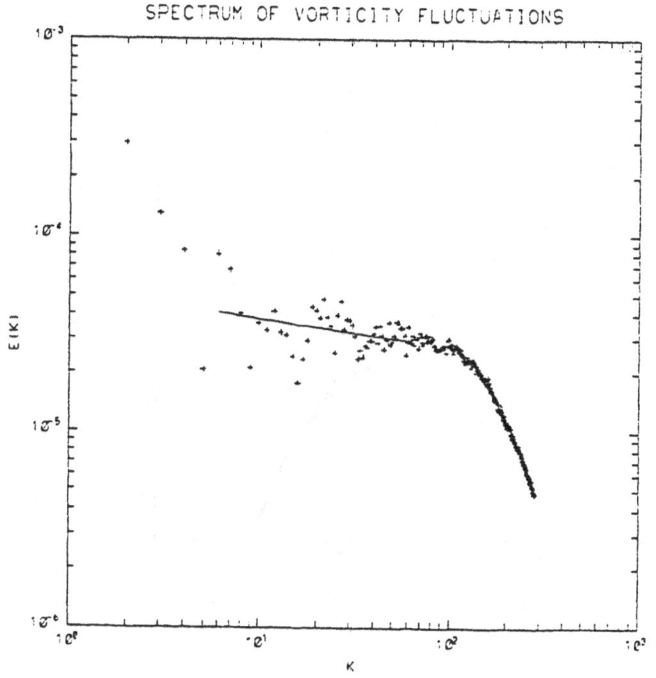

Figure 5. Spectrum of vorticity fluctuations. The continuous line is a least square fit between $k = 6$ and $k = 60$ of the form $k^{-1+\mu}$ yielding $\mu = 0,86$.

It is therefore natural to establish an analogy to electrostatics, with the pressure corresponding to the potential resulting from negative and positive charges distributed according to the vorticity and the energy dissipation, respectively. The vorticity concentrations thus act like sources of low pressure and their greatest relative concentration is the cause of the spatial correlation observed between turbulent activity and low-pressure regions. We can quantify this relative concentration difference by measuring the intermittency of the vorticity and of the energy dissipation via the spectra of their local fluctuations. Figures 5 and 6 show that these spectra follow power laws in $k^{1-\mu}$ with $\mu = 0.38$ for the energy dissipation and $\mu = 0.86$ for the vorticity. The exponent μ which was first introduced by Kolmogorov in 1962 [7] can also be interpreted as the Fourier fractal dimension of the dissipation in the limit of infinite Reynolds number [8].

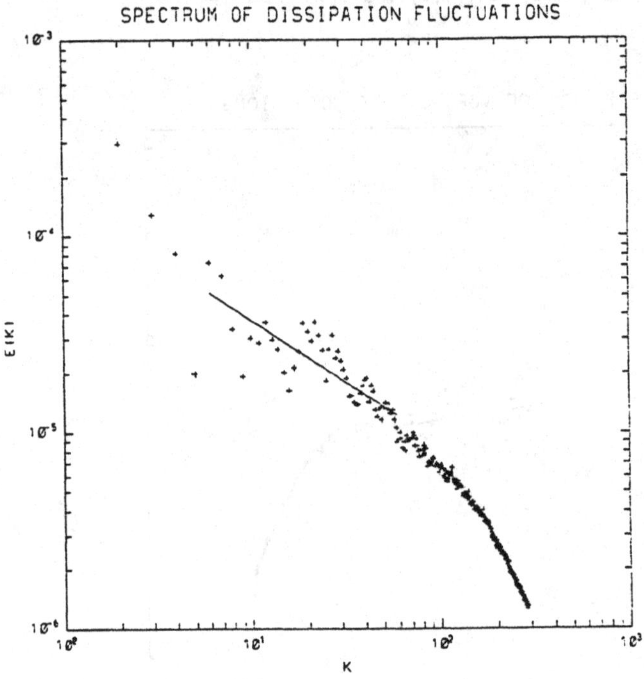

Figure 6. Spectrum of energy dissipation fluctuations. The same fit as in figure 7 yields $\mu = 0,38$.

Note that the non-uniqueness of the exponent μ that we observe here is taken into account in the multifractal theories of intermittency [9]. It may be possible to detect experimentally the high-vorticity low-pressure zones of fully developed turbulence by running a Wilson-type bubble chamber containing a suspension of active particles [10].

Acknowledgments

The computations were done on the CCVR Cray-2 at Palaiseau, using the Fast Fourier Transforms of C. Temperton. The raster visualisations were performed with the help of DRET contract 87/1483.

References

-1. A.S. Monin et A.M. Yaglom, Statistical Fluid Mechanics, 2, M.I.T. Press, 1975.

-2. S.A. Orszag, Les Houches, 1973, in R. Ballian et J.L. Peube ed., Gordon and Breach, 1977, p. 235.

-3. G.I. Taylor et A.E. Green, Proc. Roy. Soc. Londres, **A 158**, 1937, p. 499.

-4. D. Gottlieb et S.A. Orszag, Numerical Analysis of Spectral Methods: Theory and Applications (Soc. Indurstr. Appl. Math., Philadelphie, 1977).

-5. M. Brachet, D. Meiron, S.A. Orszag, B. Nickel, R. Morf et U. Frisch, J. Fluid Mech. , **130**, 1983, p. 411.

-6. A.N. Kolmogorov, C.R. Acad. Sci.URSS, **30**, 1941, p. 301.

-7. A.N. Kolmogorov, J. Fluid Mech., **13**, 1962, p. 82.

-8. B. Mandelbrot, Turbulence and Navier Stokes Equation, R. Teman, ed. (Lecture Notes in Math., Springer, **565**, 1976, p. 121).

-9. G. Parisi et U. Frisch, Turbulence and Predictability in Geophysical Fluid Dynamics and Climate Dynamics, M. Ghil et G; Parisi eds, (North-Holland, Amsterdam, 1985, p.71).

-10. Y. Couder and Y. Pomeau, private communication.

TRANSITION BETWEEN DIFFERENT SYMMETRIES IN CONVECTION

C. Pérez-García (*), S. Ciliberto (‡) and E. Pampaloni (‡)

(*) Departamento de Física, Universidad de Navarra,
31080 Pamplona, Navarra, Spain

(‡) Istituto Nazionale di Ottica, Largo E. Fermi 6,
50125 Arcetri-Firenze, Italia

ABSTRACT. The transition between hexagons a rolls in convective patterns have been studied. The transition thresholds and changes in the Nusselt number are discussed theoretically in terms of calculations made with weakly nonlinear analyses. The influence of defects on these transition is also discussed.

1. INTRODUCTION

Convection provides a good example of dynamical pattern- forming systems. In this system the rest (conducting) state is replaced by motions which organize themselves to form patterns. Usually these patterns are formed by rolls. However, by adding some complexity (poor conducting boundaries, a binary mixture, non-Boussinesq conditions, etc.) other symmetries are also possible. Squares are stable ([1]) in convective cells with insulating walls. The same planform arises in binary mixtures in the Soret driven regime ([2]). On the other hand, hexagons can appear: i) with an upper surface free and with a temperature-dependent surface tension (Bénard- Marangoni instability) ([3-4]); ii) when the transport coefficients are temperature- dependent (non-Boussinesq conditions) ([5-7]), iii) when the mean temperature varies (increases or decreases) linearly in time ([8]) and iv) by means of a temperature modulation in the cell ([9-10]).

At secondary thresholds a transition between different symmetries is also possible. In these nonequilibrium transitions the state with higher symmetry (hexagons, squares) is replaced by one with lower symmetry (rolls). A transition between squares and rolls can be observed in binary mixtures in Soret driven regime ([2]). Some weakly nonlinear theoretical analyses predict that hexagons can become unstable against rolls after some secondary threshold in simple liquids under non-Boussinesq conditions ([5]). Experiments give support to this result ([6-7]). More recently a similar transition have been observed in temperature-modulated convection ([10]). In Bénard-Marangoni convection there is also some indications that such a transition can occur ([4b,11]). The transition between hexagons and squares have been observed

E. Tirapegui and W. Zeller (eds.), Instabilities and Nonequilibrium Structures III, 291–300.

in convection under strong non-Boussinesq conditions, probably also due to the complex characteristics of the polymer used ([12]).

Heat flow measurement can characterize globally those transitions, because this quantity is sensitive to the symmetry variations ([13]). In particular, the slope of the Nusselt number (the ratio between the total heat flow and the conductive flow) as a function of the supercritical heating changes for different symmetries ([13-14]). Standard calorimetric techniques allows for the determination of the Nusselt number and, therefore, to determine global changes in the pattern. Other kind of techniques used to characterize these transitions are laser-Doppler anemometry ([6b]) and an optical technique based on the deflections of a laser beam that crosses the convecting fluid. With this last technique one can determine the temperature field integrated in the direction of the laser beam. This allows to reproduce the pattern of motions and also global parameters like the Nusselt number. An extensive description of this technique can be found in refs. ([15]).

Usually an important difference between theory and experiments in pattern forming systems arise as a consequence of the finite extent of real systems. Theoretical studies deal with an infinite aspect ratio Γ (the ratio between a characteristic horizontal length and the layer thickness), while in experiments lateral boundaries introduce important effects. The first effect is the modification of the temperature threshold for which convection starts.

The second effect is the induction of defects that, although local, lead to global changes in the pattern ([16]). The unavoidable presence of these defects in spontaneous patterns introduces disorder even near threshold.

2. Convection under non-Boussinesq conditions

Usually the convective motions are studied under an approximation due to Oberbeck and Boussinesq (OB). In this approximation temperature dependence of the fluid parameters is neglected, except for the thermal expansion effects responsible for buoyancy forces. Moreover, the viscous dissipation is neglected in comparison with the conductive term in the energy balance equation. There are some theoretical investigations on the effects of departures from OB approximation ([5]). The work of Busse (19[illegible]) is the most complete because he considers the effect of the temperature variations with all the fluid parameters and the effect of the Prandtl number in a coherent manner, although this study is limited to small departures from the OB approximation and near the convective threshold. Experiments partially confirm some of the predictions of this theoretical work ([6]). The works of Ahlers ([6c]) and Walden & Ahlers ([14]) are the most complete, because these have been made in liquid He under very precise conditions and a complete comparison between experimental and theoretical predictions is given.

We recall briefly here the main results of the Busse's analysis ([5b]). The departures from the OB approximation is quantitatively given by means of a parameter $\mathcal{P}$, defined in formulas (8.11) in ref (5b). The departures from the OB approximation give rise to the following consequences: i) near threshold the pattern is hexagonal; ii) subcritical motions are possible and iii) an hysteretic transition between hexagons and rolls is predicted, when the heating increases. Therefore the bifurcation is **transcritical**, and several threshold values can be distinguished

$$\varepsilon_a \le 0 \le \varepsilon_r \le \varepsilon_h \qquad (2.1)$$

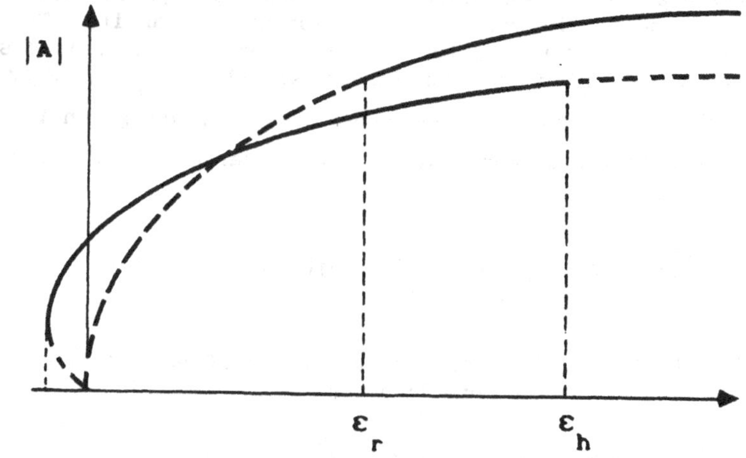

Figure 1. Bifurcation diagram and stability threshold in convection under non-Boussinesq convection

where ε represents the supercritical heating ($\varepsilon = \Delta T - \Delta T_c / \Delta T_c$ where ΔT_c indicates the critical temperature difference across the convective layer). For $\varepsilon_a \le \varepsilon \le 0$ hexagons and no-convective state are possible solutions. In the interval $0 \le \varepsilon \le \varepsilon_r$ only hexagons are stable. Between $\varepsilon_r \le \varepsilon \le \varepsilon_h$, hexagons and rolls can coexist. For $\varepsilon \ge \varepsilon_h$ only rolls are stable. These thresholds depend on $\mathcal{P}$ as given in ref (5b). The sign of the parameter $\mathcal{P}$ determine the direction of circulation in hexagonal cells: for negative $\mathcal{P}$ convective motions rise in the center of the cell and descend by the sides.

3. Amplitude equations and het flow for hexagonal patterns

Now we present a different approach that allows to obtain the transition threshold between hexagons and rolls, and the corresponding expressions for the heat flow, based on the so-called **amplitude equations**. The solutions of the hydrodynamics equations can be written in the form ([17])

$$\mathbf{V}(\mathbf{r},z,t) = \begin{pmatrix} u(\mathbf{r},z,t) \\ w(\mathbf{r},z,t) \\ \theta(\mathbf{r},z,t) \end{pmatrix} \approx \left[\frac{R_c}{\langle w_c \theta_c \rangle_z}\right]^{1/2} \begin{pmatrix} \nabla\Psi(\mathbf{r},t)u_c(z) \\ \Psi(\mathbf{r},t)w_c(z) \\ \Psi(\mathbf{r},t)\theta_c(z) \end{pmatrix} \tag{2.2}$$

where $\mathbf{r}$ is the vector in the horizontal plane and $u(\mathbf{r},z,t)$, $w(\mathbf{r},z,t)$ and $\theta(\mathbf{r},z,t)$ denote the horizontal and the vertical components of the velocity field and the temperature field, respectively. The subscript c indicate the linear solution of the corresponding quantity. The order parameter $\Psi(\mathbf{r},t)$ is the projection of the hydrodynamical variables onto the slowest unstable mode. (The linear solutions $u_c(z)$, $w_c(z)$ and $\theta_c(z)$, as well as the critical wavenumber $k_c = 3.1$, are given in ([20])). The development in normal modes of this order parameter can be written as

$$\Psi(\mathbf{r},t) = \sqrt{2}\ \mathrm{Re}\left[\sum_{\mathbf{k}} \psi_{\mathbf{k}} \exp\,(i\mathbf{k}\cdot\mathbf{r})\right] \tag{2.3}$$

The multiplicative factors in the r.h.s. of eq. (2.2) have been chosen in order to have the normalization

$$\mathcal{N} \equiv \frac{(N-1)R}{R_c} = \frac{1}{S}\int dxdy\ \Psi^2 = \sum_{\mathbf{k}} |\psi_{\mathbf{k}}|^2 \tag{2.4}$$

where $N = q_{tot}/q_{cond}$ is the Nusselt number, the ratio between total heat flow $q_{tot} = q_{cond} + q_{conv}$ and the conductive heat flow q_{cond}, and S the horizontal area of the layer. Therefore, the normalized Nusselt number $\mathcal{N}$ is the sum of the intensities of all the modes present in the system.

Experimentally one can see that the convective patterns are very ordered, with a predominant geometry (rolls, hexagons, etc.). This means that, the modes in the development (2.3) are concentrated near some sharp peaks (two for rolls and six for hexagons). Therefore it is useful to consider the following development for the order parameter

$$\Psi(\mathbf{r},t) = \sqrt{2}\ \mathrm{Re}\left[\sum_{i=1}^{3} A_i(\mathbf{r},t) \exp\,(i\mathbf{k}_i\cdot\mathbf{r})\right] \tag{2.5}$$

where we consider as reference the hexagonal form, which is formed by the superposition of three sets of straight rolls characterized by wavenumbers that obey $\Sigma\mathbf{k}_i = 0$, $|\mathbf{k}_i\cdot\mathbf{k}_j| = k^2/2$ $(i\neq j)$ and $i = 1,2,3$, and by amplitudes A_i defined as

$$A_i(\mathbf{r},t) = \sqrt{2}\ \mathrm{Re}\left[\sum_{\mathbf{k}\simeq\mathbf{k}_i} \psi_{\mathbf{k}} \exp\left(i(\mathbf{k}-\mathbf{k}_i)\cdot\mathbf{r}\right)\right] \qquad (2.6)$$

The sum is over the modes inside a peak in Fourier space with wavenumbers around $\mathbf{k}_i$.

Convective patterns can be described by means of evolution equations for these amplitudes, known as **amplitude equations**, that are obtained either by introducing (2.2) and (2.5) into the hydrodynamic equations ([19]) or simply from symmetry arguments ([20]). The advantage of this procedure is that the final nonlinear equations are simpler than the initial hydrodynamic ones. In the case of a hexagonal pattern they take the form

$$\tau_0 \frac{\partial A_i}{\partial t} = \left\{\varepsilon + \xi_0^2\left[\frac{\partial}{\partial x_i} - \frac{i}{2k_c}\frac{\partial^2}{\partial y_i^2}\right]^2\right\} A_i + a\, A^*_{i+1} A^*_{i+2} -$$

$$- b\left[\sum_{i\neq j} |A_j|^2\right] A_i - c\,|A_i|^2 A_i \quad (i = i \bmod 3) \qquad (2.7)$$

The relaxation time τ_0, the correlation length ξ_0 and the critical Rayleigh number R_c and wavenumber k_c are obtained from a linear stability analysis of the full hydrodynamic equations of the system. The superscript * denotes the complex conjugate and $\partial/\partial x_i$ and $\partial/\partial y_i$ are the spatial derivatives parallel and perpendicular to the vector $\mathbf{k}_i$, respectively. The physical meaning of the nonlinear coefficients a, b and c, will be specified later on.

These equations can be written in a variational form

$$\tau_0 \frac{\partial A_i}{\partial t} = - \frac{\delta \mathcal{F}}{\delta A_i^*} \qquad (2.8)$$

where $\mathcal{F}$ is a functional, analogous to a potential, defined as

$$\mathcal{F} = \int dxdy \left\{ \sum_{i=1}^{3} \left[-\varepsilon |A_i|^2 + \left| \xi_0 \left(\frac{\partial}{\partial x_i} - \frac{i}{2k_c} \frac{\partial^2}{\partial y_i^2} \right) A_i \right|^2 \right] + \right. \qquad (2.9)$$

$$\left. - a \prod_{i=1}^{3} |A_i| + \frac{1}{2} \sum_{i=1}^{3} \left[b \sum_{j\neq i} |A_j|^2 |A_i|^2 + \frac{c}{2} |A_i|^4 \right] \right\} + f(\varepsilon, a)$$

provided that on the boundaries $A_i = \partial A_i/\partial y_i = 0$.

The Lyapunov functional $\mathcal{F}$ provides an interesting tool to determine the dynamics of the systems, but it is not directly accessible from experiments. Some elaborations, based on local measurements must be made in order to exploit such an analysis ([16b]). Moreover in some cases (convection in liquid ^{4}He) optical inspection of the cell is very difficult. Stationary solutions are obtained by the relation $\delta\mathcal{F}/\delta A_i^* = 0$. In the homogeneous case these are i) a pattern of rolls $|A_1| \neq 0$, $|A_2| = |A_3| = 0$ and ii) a hexagonal pattern $|A_1| = |A_2| = |A_3|$. The linear stability of these solutions can be determined by the matrix $\delta^2\mathcal{F}/\delta A_i \delta A_j^*$ linearized around the stationary solutions. This allows to determine that rolls are the unstable solutions for $0 \leq \varepsilon \leq \varepsilon_r$, where ε_r is given by

$$\varepsilon_r = \frac{a^2 c}{(b-c)^2} \tag{2.10}$$

Hexagons are stable in the interval $\varepsilon_a \leq \varepsilon \leq \varepsilon_h$ with

$$\varepsilon_a = -\frac{a^2 c}{4(2b+c)} \qquad \varepsilon_h = \frac{a^2(b+2c)}{(b-c)^2} \tag{2.11}$$

It is interesting to relate the local description of amplitude equations with the global heat flow measurements which are experimentally accessible. The comparison can be made taking into account that, by substitution of eq. (2.5) in eq. (2.4), we obtain the relation

$$\mathcal{N} \equiv \frac{(N-1)R}{R_c} = \frac{1}{S}\int dxdy \sum_{i=1}^{3} |A_i|^2 = \sum_{i=1}^{3} \langle |A_i|^2 \rangle_{x,y} \tag{2.12}$$

where the bracket $\langle\,\rangle_{x,y}$ indicates the average on the horizontal plane. For hexagons the normalized Nusselt number $\mathcal{N}$ gives

$$\mathcal{N}_h = 3|A_h|^2 = \frac{3}{C}\left\{\frac{a^2}{4C} + \varepsilon + a\sqrt{a^2 + 4C\varepsilon}\right\} \tag{2.13}$$

where $C = c+2b$. For rolls the expression is simply

$$\mathcal{N}_r = |A_r|^2 = \varepsilon/c \tag{2.14}$$

As a consequence the slope of $\mathcal{N}_r(\varepsilon)$ gives a direct information about the coefficient c, while the determination of b and a from N_h requires a more delicate fitting of the dependence on ε and $\varepsilon^{1/2}$.

Of course, for a comparison of (2.13)-(2.14) with heat flow measurements one must take into account lateral effects, as well as the contributions of defects to the heat transport. This have been in some recent works ([21,22]). (The role of the lateral walls on the transition thresholds is treated also in the contribution of Dr. Walgraef in these proceedings).

4. Stationary defects in hexagonal patterns

The influence of the defects on the transition threshold between different symmetries is clearly shown in experiments. However, their influence on the heat flow is very small, because this parameter is

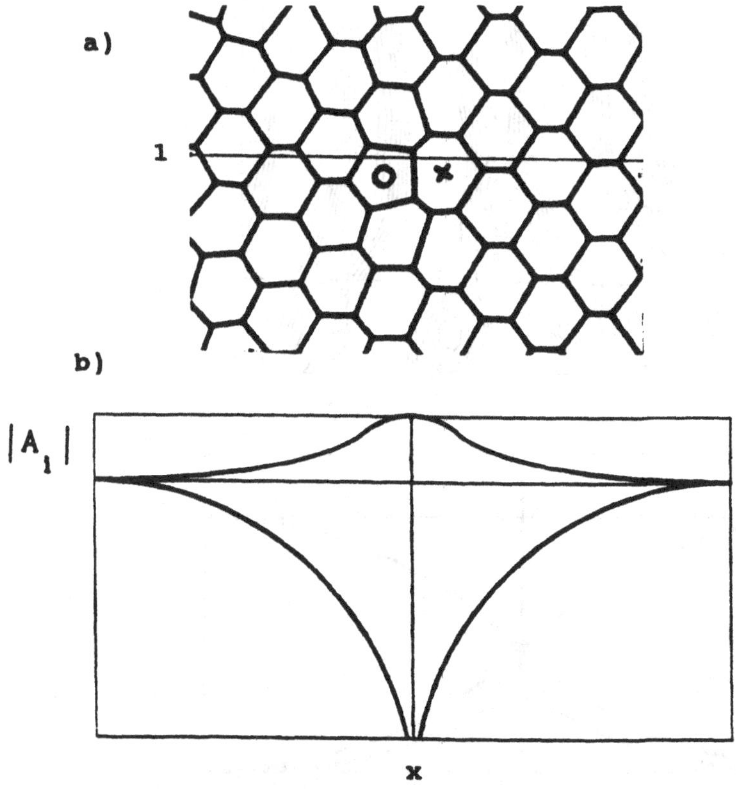

Fig. 2. a) Reconstruction of a hexagonal pattern in Bénard- Marangoni convection, showing a penta-hepta pair ([24]). b) Variation of the amplitudes of the three modes following the line 1 is Fig 2a. (schematic)

global and defects are localized in space. In this paragraph we only intend to give a qualitative view of the two types of stationary defects that appear in when hexagons and rolls can compete.

In hexagonal patterns the stable defects is the pentagon-heptagon pair ([23]). We show in Fig. 2 a reconstruction of a hexagonal pattern in Bénard-Marangoni convection ([24]). Following the center of the hexagons, one can realise that this defect corresponds to dislocations in two of the three directions that form the hexagonal lattice. As suggested recently ([22]) this defect corresponds to a local transition between an hexagonal pattern and a pattern of rolls, because at the core of these pairs only one of the three systems of rolls that form the hexagonal symmetry survives. Therefore the amplitude of the three modes that form the hexagonal pattern may have a spatial variation as sketched in Fig. 2 b, following the line labeled 1 in Fig. 2a.

When the rolls are dominant the typical defect is not a pure dislocation, but a grain boundary. In the core of such a grain boundary

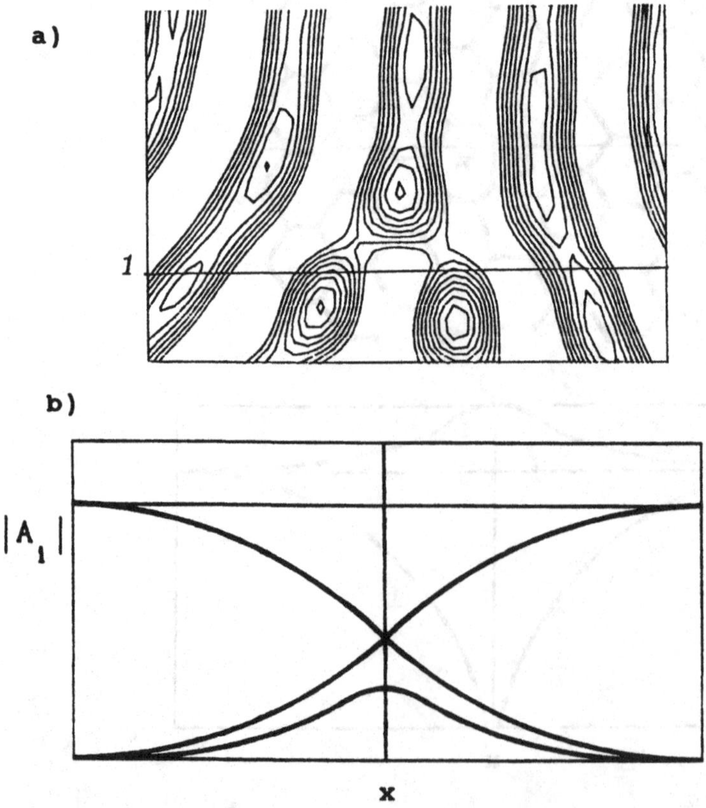

Fig. 3. a) Temperature field in an experiment in water under non-Boussinesq conditions (ε = 0.14) ([7]). b) Scheme of the amplitude variation of the different modes following the line 1 in Fig.3a.

hexagonal cells are still present. This confirms also the conjecture that the unstable "phase" reappear in the defects of the stable one. In Fig. 3 we show the reconstruction of the temperature field in a experiment in water under non-Boussinesq conditions ([7]). A schematic view of the behaviour of the amplitude of the modes following the line marked in Fig. 3a is shown in Fig. 3b. As one can deduce from (2.7), when two oblique modes are present the third one must also be present.

Although these results are preliminary, the main conclusion is that defects can play an important role similar to that of condensation nuclei in first order phase transitions in equilibrium a point that must be analysed in more detail in future works.

REFERENCES

(1) Busse F.H. & Riahi, N., 1980 J. Fluid Mech. **96**, 243
Jenkins, D.R. & Proctor, M.R.E. 1984 J. Fluid Mech. **139**, 461

(2) Moses, E., & Steinberg, V. 1986 Phys. Rev. Lett. **57**, 2016
Le Gal, P., Pocheau, A. & Croquette 1985 Phys. Rev. Lett **55**, 10
Bigazzi P., Ciliberto S. & Croquette V, J. Physique, submitted

(3) Bénard, H., 1900 Rev. Gen. Sci. Pur & Apli. **11**, 1261
Pearson, J.R.A. 1958 J. Fluid Mech. **4**, 489

(4) Pantaloni, J., Bailleux, R., Salan, J. & Velarde, M.G. 1979 J. Nonequilib. Thermodyn **4**, 201
Cerisier P., Jamond C., Pérez-García C. & Pantaloni J. 1987 Phys. Fluids **30**, 954

(5) Palm E., 1960 J. Fluid Mech. **8**, 183
Busse, F.H., 1967 J. Fluid Mech. **30**, 625

(6) Hoard C.Q., Robertson C.R. & Acrivos A. 1970 Int. J. Heat Mass Transfer **13**, 849
Dubois M., Bergé P. & Wesfreid J.E. 1978 J. Physique **39**, 1253
Ahlers, G. 1980 J. Fluid Mech. **98**, 137

(7) Ciliberto S., Pampaloni E. & Pérez-García C. 1988 Phys. Rev. Lett. **61**, 1198

(8) Krishnamurti R. 1968 J. Fluid Mech. **33**, **445**; **457**

(9) Roppo M.N., Davis S.H. & Rosenblatt S. 1984 Phys. Fluids **27**, 796
Hohenberg P.C. & Swift J. 1987 Phys. Rev. **A35**, 3855

(10) Meyer, C.W., Ahlers, G. & Cannell, D. 1987 Phys. Rev. Lett. **59**, 1577

(11) Bestehorn M. & Pérez-García C., 1987 Europhys. Lett. **4** ,1365

(12) Oliver, D.S. & Booker, J.R. 1983 Geophys. Astrophys. Fluid Dynamics **27**, 73
White D.B. 1988 J. Fluid Mech. **191**, 247

(13) Schlüter, A., Lortz, D. & Busse, F.H. 1965 J. Fluid Mech. **23**, 129

(14) Walden R.W. and Ahlers G. 1981 J. Fluid Mech. **109**, 89

(15) Ciliberto S., Francini F. & Simonelli, F. 1985 Opt. Commun. **54**, 381
Rubio, M.A., Ciliberto S & Albavetti, L. J. Fluid Mech (to appear)

(16) Steinberg, V., Ahlers, G & Cannell, D.S. 1985 Phys. Script. **T9**. 97
Heutmaker M.S. & Gollub J.P. 1987 Phys Rev **A35**, 242

(17) Ahlers G., Cross M., Hohenberg P.C. & Safran, S., 1981 J. Fluid Mech., **110**, 297

(18) Cross, M 1980 Phys Fluids **23**, 1727
(19) Newell, A.C. & Whitehead, J.A. 1969 J. Fluid Mech. **38**, 279
Segel, L.A. 1969 J. Fluid. Mech. **38**, 203
(20) Caroli, B., Caroli, C. & Roulet, B. 1984 J. Crystal Growth **68**, 677
(21) Pérez-García, C., Pampaloni, E. & Ciliberto, S. 1990 Europhys. Lett. **12**, 51
(22) Ciliberto, S., Coullet, P., Lega, J., Pampaloni, E. & Pérez-García, C., preprint
(23) Pantaloni, J. & Cerisier, P. 1983 **Cellular Structures in Instabilities** (E. Wesfreid & S. Zaleski, eds.), p. 197, Springer, Berlin.
(24) Cerisier P., Jamond C., Pérez-García C. & Pantaloni J. 1987 Phys. Rev. **A 35**, 1949

SHALLOW SURFACE WAVES IN A CONVECTING FLUID

C. M. Alfaro, H. Aspe & M. C. Depassier
Facultad de Física
Universidad Católica de Chile
Casilla 6177, Santiago 22, Chile

ABSTRACT. We consider the interaction between surface waves in shallow fluids and convection. We show that the presence of a thermal gradient compensates viscous losses which would otherwise damp the waves. Propagating solitary waves are found with their amplitude determined by the excess of the Rayleigh number above its critical value.

Introduction

In the absence of dissipation surface waves whose nonlinear evolution is given by Korteweg-de Vries equation may propagate on a shallow layer of fluid. This equation is integrable and solitons are among its solutions. When viscous dissipation is taken into account the waves are damped. They can be maintained if an additional source of energy is available to compensate viscous losses. One such mechanism is to let the fluid flow along an inclined plane, depending on the inclination of the plane the surface waves are damped or amplified. Here we have considered a different mechanism, extraction of energy from the gravitational field through buoyancy.

We consider then a fluid which is bounded below by a plane stress-free surface held at constant temperature and above by a free surface at constant pressure on which the heat flux is fixed. Neglecting surface tension, the linear stability theory for this problem shows that oscillatory as well as convective instabilities are present.[1] Our interest is on the surface waves so we shall concentrate on the oscillatory solution. The long wavelength oscillatory solution is found at a critical Rayleigh number $R_c = 30$ independently of the value of the Prandtl number or of the Galileo number and the frequency at the onset coincides with that of ordinary shallow gravity waves. Here we describe the nonlinear theory for the evolution of the waves. The main effect of the coupling between convection and the surface wave is the appearance of instability and dissipation, the combined effect gives rise to solitary waves with the amplitude determined by the value of the Rayleigh number.[2,3]

E. Tirapegui and W. Zeller (eds.), Instabilities and Nonequilibrium Structures III, 301–305.

Mathematical Formulation

The equations that govern the evolution of the fluid, in the Boussinesq approximation are

$$\nabla \cdot \mathbf{v} = 0$$

$$\frac{d\mathbf{v}}{dt} = -\nabla p + \sigma \nabla^2 \mathbf{v} - \sigma^2 G \rho \hat{\mathbf{z}}$$

$$\frac{dT}{dt} = \nabla^2 T$$

$$\rho = 1 - \frac{R}{\sigma G}(T - T_o)$$

where $d/dt = \partial/\partial t + \mathbf{v} \cdot \nabla$ is the convective derivative; $\mathbf{v} = (u, 0, w)$ is the fluid velocity, p is the pressure, and T is the temperature. T_o is a reference temperature. On the upper free surface $z = 1+\eta(x,t)$ the boundary conditions are the continuity of the normal and tangential stresses and a kinematical equation for the surface displacement η:

$$\eta_t + u\eta_x = w$$

$$p - p_a - \frac{2\sigma}{N^2}[w_z + u_x\eta_x^2 - \eta_x(u_z + w_x)] = 0$$

$$(1 - \eta_x^2)(u_z + w_x) + 2\eta_x(w_z - u_x) = 0$$

and

$$T_z - \eta_x T_x = -N$$

Subscripts x and z denote derivatives with respect to the horizontal and vertical coordinates respectively. Here $N = (1+\eta_x^2)^{1/2}$, and p_a is a constant pressure exerted on the upper free surface. On the lower surface $z = 0$ the boundary conditions are

$$w = u_z = 0, \qquad \text{and} \qquad T = T_b$$

We have chosen the depth of the static fluid as the unit of length, the static temperature difference between the upper and lower boundaries as unit of temperature, the value of the static density on the free surface as unit of density and the heat diffusion time as time unit. The dimensionless parameters involved in the problem are the Prandtl number σ, the Rayleigh number R and the Galileo number G.

Asymptotic Expansion

The nonlinear evolution of the instability is obtained by means of an asymptotic expansion. After introducing new variables

$$\xi = \epsilon(x - ct), \qquad \tau = \epsilon^3 t$$

we look for a solution of the form

$$\eta(\xi,\tau) = \epsilon^2(\eta_0 + \epsilon\eta_1 + \epsilon^2\eta_2 + \ldots)$$

$$u(\xi,z,\tau) = \epsilon^2(u_0 + \epsilon u_1 + \epsilon^2 u_2 + \ldots)$$

$$w(\xi,z,\tau) = \epsilon^3(w_0 + \epsilon w_1 + \epsilon^2 w_2 + \ldots)$$

$$p(\xi,z,\tau) = p_s(z) + \epsilon^2(p_0 + \epsilon p_1 + \epsilon^2 p_2 + \ldots)$$

$$T(\xi,z,\tau) = T_s(z) + \epsilon^3(\theta_0 + \epsilon\theta_1 + \epsilon^2\theta_2 + \ldots).$$

where the quantities with the subscript s represent the static solutions. In addition, the critical Rayleigh number is slightly above its critical value, so we let $R = R_c + \epsilon^2 R_2$ and proceed to solve as usual to each order in ϵ. We find that the solution is given by

$$\eta = \epsilon^2\left(\frac{f(\xi,\tau)}{c} + \epsilon\frac{g(\xi,\tau)}{c} + O(\epsilon^2)\right)$$

$$u = \epsilon^2\left(f + \epsilon g + O(\epsilon^2)\right)$$

$$w = \epsilon^3\left(-f_\xi z - \epsilon g_\xi z + O(\epsilon^2)\right)$$

$$p = p_s + \epsilon^2\left(\frac{\sigma^2 G f}{c} + \epsilon(\frac{\sigma^2 G g}{c} - 2\sigma f_\xi + \sigma R_c f_\xi \frac{z^4 - 6z^2 + 5}{24}) + O(\epsilon^2)\right)$$

and

$$T = T_s + \epsilon^3\left((f_\xi + \epsilon g_\xi)\frac{z^3 - 3z}{6} - \epsilon c f_{\xi\xi}\frac{z^5 - 10z^3 + 25z}{120} + O(\epsilon^2)\right).$$

The functions $f(\xi,\tau)$ and $g(\xi,\tau)$ are arbitrary functions, their evolution equations are determined as solubility conditions at higher order. One finds that f satisfies

$$f_\tau + \delta_1 f f_\xi + \lambda_3 f_{\xi\xi\xi} = 0$$

and g satisfies

$$g_\tau + \delta_1 (fg)_\xi + \lambda_3 g_{\xi\xi\xi} + \lambda_2 f_{\xi\xi} + \lambda_4 f_{\xi\xi\xi\xi} + \delta_2 (f f_\xi)_\xi = 0.$$

where the parameters δ_i, λ_3 and λ_4 are functions of the Prandtl and Galileo numbers. The coefficient λ_2 is proportional to the Rayleigh number, $\lambda_2 = \sigma(R - R_c)/(15\epsilon^2)$.

An improved equation correct to order ϵ for $u = f + \epsilon g$ can be obtained by adding the evolution equations for f and g above. The final evolution equation that we obtain for the horizontal velocity u or equivalently for the surface deformation $\eta = u/c$ is finally

$$u_\tau + \delta_1 u u_\xi + \lambda_3 u_{\xi\xi\xi} + \epsilon(\lambda_2 u_{\xi\xi} + \lambda_4 u_{\xi\xi\xi\xi} + \delta_2 (u u_\xi)_\xi) = 0.$$

This equation with $\delta_2 = 0$ has been studied numerically.[4] For any initial condition, for periodic boundary conditions, regular arrays of soliton like pulses emerge. The additional nonlinearity when δ_2 is different from zero has an additional destabilizing effect. The amplitude of the pulses may be estimated by means of an asymptotic expansion around the solution of the leading order KdV equation. We let then

$$\frac{\partial}{\partial \tau} = \frac{\partial}{\partial t} + \epsilon \frac{\partial}{\partial s}$$

and

$$u = u_0 + \epsilon u_1 + \cdots.$$

In leading order u_0 satisfies the KdV equation the solution of which we take to be

$$u_0 = N(s)\text{sech}^2\left[\left(\frac{\delta_1 N(s)}{12\lambda_3}\right)^{1/2} (\xi - \frac{\delta_1 N(s)}{3} t)\right]$$

The slow time dependence of the amplitude is determined from the solubility condition[5] of the solution of order ϵ:

$$\frac{\partial u_1}{\partial t} + \delta_1 \left(u_0 \frac{\partial u_1}{\partial \xi} + u_1 \frac{\partial u_0}{\partial \xi}\right) + \lambda_3 \frac{\partial^3 u_1}{\partial \xi^3} = -\frac{\partial u_0}{\partial s} - \lambda_2 \frac{\partial^2 u_0}{\partial \xi^2}$$
$$-\lambda_4 \frac{\partial^4 u_0}{\partial \xi^4} - \delta_2 \frac{\partial}{\partial \xi}\left(u_0 \frac{\partial u_0}{\partial \xi}\right).$$

The solubility condition yields

$$\frac{dN}{ds} = \frac{\delta_1 N^2}{315\lambda_3^2}(21\lambda_2\lambda_3 - \delta_1(5\lambda_4 - 12\delta_2)N).$$

In this problem the coefficient $(5\lambda_4 - 12\delta_2)$ is positive, therefore we expect to see pulses with an amplitude

$$N = \frac{21\lambda_2\lambda_3}{\delta_1(5\lambda_4 - 12\delta_2)}.$$

To sum up, in a shallow viscous layer of fluid subject to an external temperature gradient, undamped solitary waves may propagate, their amplitude being determined by the excess of the Rayleigh number above its critical value.

Acknowledgments

This work was partially supported by FONDECYT.

References

1. R. D. Benguria & M. C. Depassier, Phys. Fluids 30, 1678 (1987).
2. C. M. Alfaro & M. C. Depassier, Phys. Rev. Lett. 62, 2597 (1989).
3. H. Aspe & M. C. Depassier, Phys. Rev. A 41, 3125 (1990).
4. T. Kawahara, Phys. Rev. Lett. 51, 381 (1983).
5. E. Ott & N. Sudan, Phys. Fluids 12, 2388 (1969).

AN EXACT MODEL OF RESISTIVE CONVECTION IN A CYLINDRICAL PLASMA

L. GOMBEROFF
Department of Physics
Faculty of Sciences
University of Chile
Casilla 653, Santiago
Chile

ABSTRACT. By using non-ideal magnetohydrodynamic fluid equations, it is shown that resistivity and thermal conductivity can lead to stationary convection in a current-carrying cylindrical plasma in an external magnetic field. Convection takes place when $(\eta/\kappa) = 8\pi/3$, where η is the resistivity and κ the thermal conductivity. When (B_θ/B_z) is much larger and much smaller that one, convection occurs for all values of the azimuthal wavenumber m. However, when (B_θ/B_z) is of order one convection occurs only for $m \gg 1$. The convective cells are helically twisted tubes and the number of convection cells is equal to 2m. The condition for convection, i.e., $R > R_{crit}$, where R is the Rayleigh number, is the same for all cases.

1. INTRODUCTION

Large scale stationary convection has been the subject of several studies. The first by Simon (1968) showed that in partially ionized plasmas, convection can be driven by the toroidal curvature with the neutrals providing the required dissipation. Later, Kadomtsev and Pogutse (1970) showed that the resistive rippling mode leads to convection cells. Okuda and Dawson (1973) showed the existence of thermally excited convection caused by driftlike modes. The formation of a quasimode of large spatial extent has been proposed by Roberts and Taylor (1965). Subsequently, Wobig (1972) and Maschke and Paris (1975) showed that when the plasma is unstable to a magnetohydrodynamics interchange, viscosity and resistivity lead to convection in a shearless magnetic field. Dagazian and Paris (1979) showed the existence of stationary convection-like modes in a plasma slab with magnetic shear.

More recently, Gomberoff and Hernández (1983, 1984) showed the existence of stationary convection in a current carrying cylindrical plasma as a consequence of viscosity and thermal conductivity. The convective modes occur for large wavenumbers and shearless magnetic fields satisfying $(B_\theta/B_z) \ll 1$.

Later, Gomberoff (1984) showed that resistivity and thermal

E. Tirapegui and W. Zeller (eds.), Instabilities and Nonequilibrium Structures III, 307–316.

conductivity also lead to convection for low wavenumbers and $(B_\theta/B_z) \ll 1$. In this case, convection was shown to take place also in the opposite limit, i.e., for large wavenumbers and $(B_\theta/B_z) \gg 1$ (Gomberoff, 1985).

Finally, Gomberoff (1983) and Gomberoff and Palma (1984) showed that the combined effect of viscosity, resistivity, and thermal conductivity can also lead to large scale steady convection.

As mentioned above, in the case of resistive convection, convective states can occur either for low wavenumbers and $(B_\theta/B_z) \ll 1$ or for large wavenumbers and $(B_\theta/B_z) \gg 1$. It turns out that in both cases the condition for convection is the same. This fact suggests that convection can take place for any (B_θ/B_z).

By solving the model in a closed form, it will be shown that convection can occur for any value of B_θ/B_z provided that the azimuthal wavenumber m is much larger than 1. If m is small convection takes place only in the aforementioned limits.

In Section 2 the model is described and solved. In Section 3 the dispersion relation is derived. In Section 4 the nature of the marginal modes is analysed. In Section 5 a description of the flow pattern is given and in Section 6 the main results are discussed.

2. THE MODEL

The model consists of a current-carrying cylindrical plasma of length L and radius a. The plasma is assumed to be surrounded by perfectly conducting walls.

The system is described by the following equations:

$$\beta\left(\frac{\partial\vec{v}}{\partial t} + \vec{v}\cdot\nabla\vec{v}\right) = \vec{J}\times\vec{B} - \vec{\nabla}p, \tag{1}$$

$$\frac{\partial\rho}{\partial t} + \vec{\nabla}\cdot(\rho\vec{v}) = 0, \tag{2}$$

$$\frac{\partial p}{\partial t} + \vec{v}\cdot\vec{\nabla}p - \frac{2}{3}\kappa\nabla^2 p - \frac{2}{3}\eta|\vec{J}|^2 - S_o = -\gamma p\vec{\nabla}\cdot\vec{v}, \tag{3}$$

$$\vec{E} + \vec{\nabla}\times\vec{B} = \eta\vec{J}, \tag{4}$$

$$\vec{\nabla}\cdot\vec{B} = 0, \tag{5}$$

$$\vec{\nabla}\times\vec{B} = 4\pi\vec{J}, \tag{6}$$

$$\frac{\partial\vec{B}}{\partial t} = -\vec{\nabla}\times\vec{E}, \tag{7}$$

where κ is the thermal conductivity, η the resistivity, γ the adiabacity coefficient, and S_o is a constant heat source required to mantain the equilibrium pressure profile.

The equilibrium is characterized by a magnetic field given by

$$B_z^{(0)} = B_0,\ B_\theta(0) = B_I(r/a), \tag{8}$$

where B_o and B_I are constants.

The equilibrium velocity is zero and from Eq. (1) it then follows that

$$p^{(0)} = P_o - (B_I/4\pi)(r/a)^2 \ , \tag{9}$$

where P_o is a constant.

The rotational transform q is constant and therefore the magnetic field is shearless:

$$q = \frac{2\pi r B_z{}^{(0)}}{1\ B_0{}^{(0)}} = \frac{2\pi a}{L}\frac{B_o}{B_I} \ , \tag{10}$$

Assuming the density to be nearly constant $\rho \simeq \rho_0$, the motion incompressible, and that all perturbed quantities behave like

$$f^{(1)}(r,\theta z) = f^{(1)}(r)\exp(im\theta + ikz + \Omega t) \ , \tag{11}$$

Eq. (1) reduces to

$$[\Omega^2 + \hat{\Omega}\hat{\eta}\beta^2 = (m - nq)^2]\vec{\xi} = \vec{\nabla}\hat{p}^{(1)} - -2i|m - nq|\ (\xi_\theta \hat{e}_r - \xi_r\ \hat{e}_\theta), \tag{12}$$

where

$$\vec{\xi} = (\frac{1}{a + \eta\beta^2/4\pi})\ \vec{v}^{(1)}, \quad \hat{\eta} = (\frac{\rho_o}{4\pi a^2 B_I^2})^{1/2}\eta \ ,$$

$$\hat{\Omega} = (4\pi\rho_o^2/B_I^2)\Omega \ , \tag{13}$$

and

$$\hat{p}^{(1)} = \frac{4\pi}{B_I^2}\{p^{(1)} + [(\vec{B}\cdot\vec{B})^{(1)}/8\pi]\} \ , \tag{14}$$

is the total perturbed pressure.

The integer n in Eq. (12) comes from taking periodic boundary conditions at the end of the cylinder:

$$k = -\ 2\pi n/L, \quad n = 0,1,2,... \tag{15}$$

In terms of the components of $\vec{\xi}$, Eq. (12) reduces to

$$\xi_r = \frac{a^2}{\Lambda}\left(\frac{\partial p^{(1)}}{\partial r} + \frac{m}{r}\sigma\hat{p}^{(1)}\right),$$

$$\xi_\theta = \frac{ia}{\Lambda}\left(\sigma\frac{\partial \hat{p}^{(1)}}{\partial r} + \frac{m}{r}\hat{p}^{(1)}\right), \tag{16}$$

$$\xi_z = (ika^2/\Lambda)(1 - \sigma^2)\hat{p}^{(1)},$$

where

$$\sigma = 2(m - nq)/[\hat{\Omega}^2 + \hat{\Omega}\hat{\eta}\beta^2 = (m - nq)^2], \tag{17}$$

and

$$\Lambda = [\hat{\Omega}^2 + \hat{\Omega}\hat{\eta}\beta^2 + (m - nq)^2](\sigma^2 - 1) . \tag{18}$$

The perturbed magnetic field has been assumed to be a force-free field, i.e., $\nabla x B^{(1)} = \beta B^{(1)}$. The proporcionality constant β can be determined by taking the curl of the following relation:

$$\Omega\,\vec{B}^{(1)} = (i\,B_I/a)(m - nq)\,v^{(1)} - (\eta/4\pi)\nabla x(\nabla x\vec{B}^{(1)}). \tag{19}$$

The last Equation follows from the linearized Eqs. (4) - (7) and the incompressibility assumption, $\nabla\cdot v = 0$.

Thus, taking the curl of Eq. (19) and using Eqs. (16) yields

$$\beta = k\sigma . \tag{20}$$

3. THE DISPERSION RELATION

The dispersion relation can be obtained by taking the divergence of Eqs. (16) and setting it equal to zero. The result is

$$\nabla^2\hat{p}^{(1)} + k^2\sigma^2\hat{p}^{(1)} = 0 . \tag{21}$$

This is Bessel's equation whose regular solution at the axis of the cylinder, $r = 0$, is

$$\hat{p}^{(1)} = \alpha\, J_m(k(\sigma^2 - 1)^{1/2}\, r) = 0, \tag{22}$$

where α is a constant.

The assumption of perfectly conducting walls implies $\xi_r\ (r = 0) = 0$. Thus, from the first Eq. (15) it follows that

$$m\,J_m(k(\sigma^2 - 1)^{1/2}a) + $$
$$+ ka(\sigma^2 - 1)^{1/2}\,J'_m(k(\sigma^2 - 1)^{1/2}a) = 0 \qquad (23)$$

Using the recurrence relation

$$x\,J'_m(x) = x\,J_{m-1}(x) - m\,J_m(x), \qquad (24)$$

leads to

$$J_{m-1}(k(\sigma^2 - 1)^{1/2}a) = -\left[m(\sigma - 1)^{1/2} / k\,a(\sigma + 1)^{1/2}\right]$$
$$\times J_m(k(\sigma^2 - 1)^{1/2}a) \ . \qquad (25)$$

This equation can be written in a more convenient form by using

$$(x/2)J_{m-1} - (x/2)\,J_{m+1} = m\,J_m \ . \qquad (26)$$

Thus Eq. (25) reduces to

$$J_{m-1}(k(\sigma^2 - 1)^{1/2}a) = -\,|(\sigma - 1)/(\sigma + 1)|J_{m+1}(k(\sigma^2 - 1)^{1/2}. \qquad (27)$$

This equation has a first zero for some value of the argument, $x = x_o$, and some value of σ, $\sigma = \sigma_o$. For these values

$$\sigma_o{}^2 = 1 + (x_o^2/k^2a^2) \ . \qquad (28)$$

This relation fixes the wavenumber k. Notice that $1 < \sigma_o < \infty$, where the extremes, i.e., the values 1 and infinity are reached for $ka \to \infty$ and $ka \to 0$, respectively.

The growth rate can now be obtained from Eq. (17) by setting $\sigma = \sigma_o$. The results is

$$\hat{\Omega} = -\frac{1}{2}\hat{\eta}\beta^2 + \left[\frac{1}{4}\hat{\eta}^2\beta^2 + (2/\sigma_o)|m - nq| - (m - nq)^2\right]^{1/2}. \qquad (29)$$

The marginal modes are given by

$$mq_{1,2} = m \mp (2/\sigma_o). \qquad (30)$$

In figure 1 the growth rate $\hat{\Omega}$ as a function of $nq = 2|k|\,a\,(B_O/B_I)$ is shown. It can be seen that there is another mode characterized by $m = nq$ for which $\hat{\Omega} = 0$. However, in contrast to the modes given by Eq. (3), this modes is compressible (see Gomberoff and Hernández, 1984).

Eqs. (1) - (7) have a linear incompressible solution provided that γ in Eq. (3) is infinite (Gomberoff and Maschke, 1981). Under some conditions, however, the modes given by Eq. (23) satisfy Eq. (3) for arbitrary finite γ.

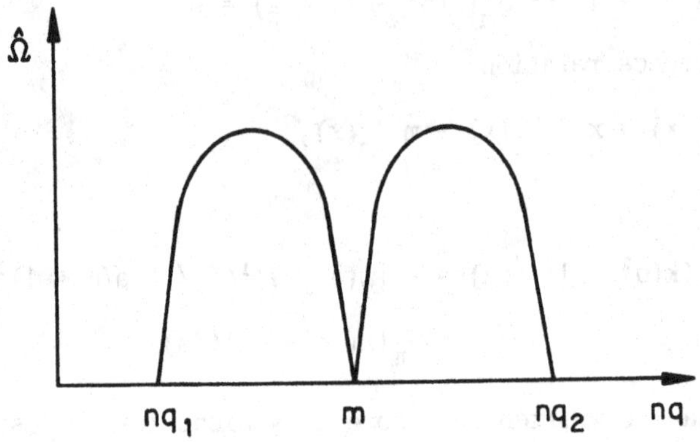

Figure 1. Growth rate-$\hat{\Omega}$ vs nq for $\eta \neq 0$ and $\kappa \neq 0$.

4. PHYSICAL MODES WITH FINITE ADIABATICITY COEFFICIENT

In order to determine the condition under which the modes given by Eq. (23) are physical, let us write the linearized Eq. (3) for $\hat{\Omega} = 0$:

$$v_r^{(1)} \frac{dp^{(0)}}{dr} = \frac{2}{3} \kappa \nabla^2 p^{(1)} + \frac{4}{3} \eta J^{(0)} \cdot J^{(1)}, \tag{31}$$

From Eq. (14) and the fact that the states given by Eq. (23) are incompressible, it follows that (see Gomberoff and Palma, 1984)

$$p^{(1)} = \frac{B_I \sigma_o}{4\pi(m - nq)(\sigma^2 - 1)} (m \sigma_o \hat{p}^{(1)} + r \frac{\partial \hat{p}^{(1)}}{\partial r}). \tag{32}$$

Taking the Laplacian of the last equation and using Eq. (21) (see Appendix B of Gomberoff and Palma, 1984)

$$p^{(1)} = \frac{B_I^2 \sigma^2 k^2}{4\pi(m - nq)(\alpha^2_{(0)} - 1)} ([m \sigma_o{}^3 + 2(\sigma_o{}^2 - 1)]\hat{p}^{(1)} + $$

$$+ \sigma_o{}^2 r \frac{\partial \hat{p}^{(1)}}{\partial r}) . \tag{33}$$

The last term in Eq. (31) can be calculated from Eqs. (6), (19), and the forcefree field condition. The result is

$$\vec{J}^{(0)} . \vec{J}^{(1)} = [2 B_I{}^2 (m - nq) k^2 \sigma_o / (4\pi)^2 \Lambda] (\sigma_o{}^2 - 1) \hat{p}^{(1)} . \tag{34}$$

Inserting Eqs. (33) and (34) into Eq. (3) yields

$$(\frac{\eta}{\kappa} - \frac{8\pi}{3}\frac{\sigma_o{}^3 m + 2(\sigma_o{}^2 - 1)}{\sigma_o{}^3 m + \frac{4}{3}(\sigma_2{}^2 - 1)}\hat{p}^{(1)} = \frac{1}{m\sigma_o}(\frac{\eta}{\kappa} - \frac{8\pi}{3})r\frac{\partial\hat{p}^{(1)}}{\partial r} . \quad (35)$$

This equation is satisfy throughout the plasma column provided that $(\eta/\kappa) = 8\pi/3$. From Eq. (28) it follows that such condition holds for $\sigma_o \to 1$ and $\sigma_o \to \infty$, i.e., for very large and very small wavenumbers. From Eq. (35) it follows that for intermediate wavenumbers the condition $(\eta/\kappa) = 8\pi/3$ can also be achieved provided that m is sufficiently large.

Thus, the states given by Eq. (34) are imcompressible and satisfy the complete set of linearized Eqs. (1) - (7) for arbitrary finite γ.

5. FLOW PATTERN

Since $\vec{\nabla}\cdot\vec{v} = 0$, $\vec{v}$ can be written in terms of the flux function $\vec{G}$ in the following form

$$\vec{v} = \vec{\nabla}x\vec{G}. \quad (36)$$

Without lost of generality one can take $G_r = 0$ and by making use of the helical symmetry of the perturbation, the flux surfaces are given by

$$\phi(r,\theta,z) = m\ G_z - k\ r\ G_\theta = \text{const.} \quad (37)$$

It can be seen easily that ϕ can be written in terms of v_r as follows:

$$\phi = \mathrm{Re}(-\ irv_r) = A\{J_m(k(\sigma_o{}^2 - 1)^{1/2}r) +$$

$$+ (k(\sigma_o{}^2 - 1)^{1/2}/m)J'(k(\sigma_o{}^2 - 1)^{1/2}r)\ \sin(m\theta + kz)\}. (38)$$

The flux surfaces ϕ = const. define cells which have the form of helically twisted tubes as shown in Figure 2 for the case m = 1.

The center of each tube is determined by the extremes of ϕ, i.e.,

$$\frac{\partial\phi}{\partial} = \frac{\partial\phi}{\partial\theta} = 0. \quad (39)$$

Hence,

$$m\theta + kz = (2n + 1)(\pi/2), \quad n = 0,1,2,... \quad (40)$$

and the components of $\vec{v}$ on the center of each tube are given by

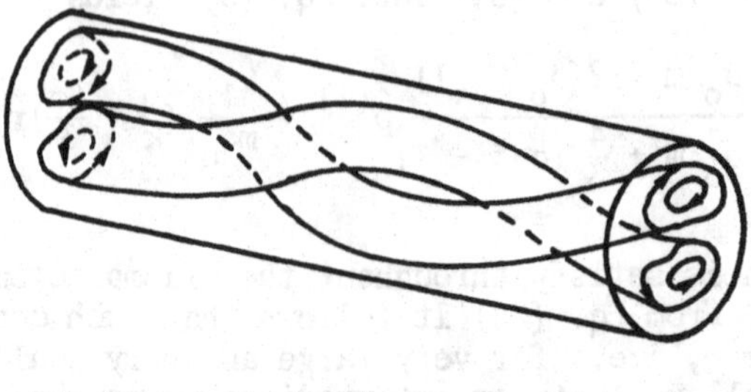

Figure 2. Helical convection cells for $m = 1$

$$v_r = 0,\ v_\theta = (B_\theta/B_I)v_z,$$

$$v_z = (-1)^{n+1}\ AkaJ_m(k(\sigma_o^{\,2} - 1)^{1/2}r). \tag{41}$$

6. DISCUSSION

It has been shown that the marginal states given by $nq_{1,2} = m \mp 2/\sigma$ are incompressible and satisfy the linearized Eqs. (1) - (7) provided that $\eta/\kappa = 8\pi/3$ and $(B_\theta/B_z) \gg 1$ or $B_\theta/B_z \ll 1$. In the case when $(B_\theta/B_z) \simeq 1$, the marginal states $nq_{1,2}$ become physical only if the azimuthal wavenumber m is sufficiently large.

The states characterized by $nq_{1,2} = m \mp 2/\sigma_o$ are not only marginal but also stationary, i.e., they satisfy $R_e\Omega = I_m\Omega = 0$. These states are the analog in a plasma of the well known stationary convection in Hydrodynamics (Maschke and Paris, 1975; Gomberoff and Maschke, 1981).

The critical Rayleigh number is given by (Gomberoff, 1984)

$$R_{crit} = (16\pi/3)(B_I^{\,2}/B_o)\ . \tag{42}$$

Since $nq_{1,2} = m \mp 2/\sigma_o$ because of Eqs. (10) and (15) it follows that

$$(B_I/B_o) \simeq |k|a/m\ . \tag{43}$$

This equation shows the relation between the external magnetic field and the wavenumber of the perturbation.

The treatment for $R = R_{crit}$ is a condition for marginal stability. The linear theory does not say anything about the behaviour for $R > R_{crit}$. However, the states under consideration correspond to what is called "exchange of instabilities" by Chandrasekhar (1961), where it is shown that for $R > R_{crit}$, the complete set of nonlinear equations possesses stationary convection solutions that bifurcate from the equilibrium solution.

Thus the states under consideration lead to large scale steady convection in a shearless current-carrying cylindrical plasma. The number of convective cells is equal to 2m (see Gomberoff and Palma, 1984). Therefore, since $B_I/B_0 \simeq 1$ only large m values can lead to convection, the plasma column will, in such cases, break up into several convection cells.

ACKNOWLEDGMENTS

This work has been supported by FONDECYT, Grant N° 90-1008, and by Departamento Técnico de Investigación de la Universidad de Chile, Grant N° E 2812/8812.

REFERENCES

Chandrasekhar, S. (1961) 'Hydrodynamics and Hydromagnetic Stability', Clarendon, Oxford.

Dagazian, R.Y. and Paris, R.B. (1977) 'Stationary convection-like modes in a plasma slab with magnetic shear', Phys. Fluids 20, 917-927.

Gomberoff, L. and Maschke, F.F. (1981) 'Non-ideal effects on the stability of a cylindrical current-carrying plasma", in E. Tirapequi (ed.), Field Theory, Quantization, and Statistical Physics, D. Reidel Publishing Company, Dordrecht, pp. 123-145.

Gomberoff, L. and Hernández, M. (1983) 'Stationary convection in a cylindrical plasma', Phys. Rev. A 27, 1244-1246.

Gomberoff, L. (1983) 'Resistive and viscous convection in a cylindrical plasma', Phys. Rev. A 28, 3125-3127.

Gomberoff, L. (1984) 'Resistive convection in a cylindrical plasma', J. Plasma Phys. 31, 29-37.

Gomberoff, L. and Hernández, M. (1984) 'Large-scale stationary convection in a cylindrical current-carrying plasma', Phys. Fluids 27, 392-398.

Gomberoff, L. and Palma, G. (1984) 'Stationary convection due to resistivity, viscosity, and thermal conductivity in a cylindrical plasma', Phys. Fluids 27, 2022-2027.

Gomberoff, L. (1985) 'Resistive convection in a cylindrical plasma. Part 2', J. Plasma Phys. 34, 299-303.

Kadomtsev, B. and Pogutse, O. (1970) 'Turbulence in toroidal plasmas' in M. Leontovich (ed.), Reviews of Plasma Physics, Plenum, New York, Vol. 5, pp. 249-498.

Maschke, E.K. and Paris, R.B., (1975) in Proceedings of the 5th. International Conference on Plasma Physics and Controlled Fusion, Tokyo, (IAEA, Vienna), Vol. 1, p. 205.

Okuda, H. and Dawson, J.M. (1973) 'Theory and numerical simulations in plasma diffusion across a magnetic field, Phys. Fluids, 16, 408-426.
Simon, A. (1968) 'Convection in a weakly ionized plasma in a non-uniform magnetic field', Phys. Fluids 11, 1186-1191.
Roberts, H. and Taylor, J.B. (1965) 'Gravitational resistive instability of an incompressible plasma in a sheared magnetic field', Phys. Fluids 8, 315-322.
Wovig, H. (1972) 'Convection of a plasma in a gravitational field', Plasma Phys. 14, 403-416.

ENERGY METHOD IN THE STABILITY OF VISCOELASTIC FLUID

W.Zeller and R.Tiemann
Instituto de Física
Universidad Católica de Valparaíso
Casilla 4059, Valparaíso, Chile.

ABSTRACT. The energy method is applied to investigate the stability theory of a viscoelastic fluid. Assuming separation of time and spatial dependence on the velocity perturbation's field, the formal presentation may be placed as in the Newtonian case.

INTRODUCCION

Viscoelastic fluids made up by macromolecules show unusual behaviours in comparison with the ones of the Newtonian fluids, because they present elastic properties that are typical of solids. These properties introduce the presence of a memory function, such as a relaxation module F, which plays an important role in the dynamics of the fluid. In fact, the stress tensor $\overleftrightarrow{T}$ depends on the history of the strain of the viscoelastic medium, although there is a retardation effect that flags rapidly, if one turns back in time.

The basic flow is ruled by the equations of momentum and continuity, besides the frontier conditions. The stress tensor in Maxwell's viscoelastic model is obtained from the equation:

$$\overleftrightarrow{T} + \lambda_1 \partial\overleftrightarrow{T}/\partial t = \eta\dot{\overleftrightarrow{\gamma}}(\vec{U}) = \eta[\nabla\vec{U} + (\nabla\vec{U})^+], \qquad (1)$$

being U(x,t) the velocity of the fluid, λ_1 the relaxation time and the dynamic viscosity. In general terms, the stress tensor can be noted down as:

E. Tirapegui and W. Zeller (eds.), Instabilities and Nonequilibrium Structures III, 317–320.

$$\overleftrightarrow{T} = \int_{-\infty}^{t} dt' \; F(t-t') \dot{\overleftrightarrow{\gamma}}(\vec{U},t') \tag{2}$$

and, in particular, for Maxwell's model one has the relaxation module $F = (\eta/\lambda_1) \exp[-(t-t')/\lambda_1]$.

ENERGY EVOLUTION EQUATION

When introducing disturbances $\vec{u}=\vec{U}-\vec{U}'$ and p=P-P' within the fields of velocity and pressure of the basic state respectively, one gets the following guidance equations:

$$\partial\vec{u}/\partial t + (\vec{U}\cdot\nabla)\vec{u} + (\vec{u}\cdot\nabla)\vec{U} + (\vec{u}\cdot\nabla)\vec{u} +-\nabla p + \nabla\cdot\overleftrightarrow{T} \tag{3}$$

$$\nabla\cdot\vec{u} = 0 \tag{4}$$

$$\overleftrightarrow{T}(\vec{u}) = \int_{-\infty}^{t} dt' \; F(t-t') \dot{\vec{\gamma}}(\vec{u},t'). \tag{5}$$

The density supposed to be included in $\overleftrightarrow{T}$ and p has not been explicited. Starting from the former equations, it is possible to build up the following evolution relation of the disturbance of the mean kinetic energy:

$$d\varepsilon(t)/dt = d\langle \tfrac{1}{2}u^2\rangle/dt = -\langle\vec{u}\cdot\overleftrightarrow{D}\cdot\vec{u}\rangle - \langle\vec{u}\cdot[\nabla\cdot\overleftrightarrow{T}(\vec{u})]\rangle, \tag{6}$$

in which the $D_{ij} = \tfrac{1}{2}[\partial U_i/\partial x_j + \partial U_j/\partial x_i]$, with (i,j=1,2,3), are the components of $\overleftrightarrow{D}$. It is easy to prove that:

$$\langle\vec{u}\cdot[\nabla\cdot\overleftrightarrow{T}(\vec{u},t)]\rangle = \int_{-\infty}^{t} dt' \; F(t-t') \langle\nabla\vec{u}:\nabla\vec{u}\rangle, \tag{7}$$

being $\nabla\vec{u}:\nabla\vec{u} = (\partial u_i/\partial x_j)(\partial u_j/\partial x_i) = |\nabla\vec{u}|^2$. The evolution equation of energy permits to find a record regarding the stability of the system. The dissipation speed of energy $\langle u\cdot\nabla\cdot T\rangle$ and the generation of energy coupled to the basic flow $\langle u\cdot\overleftrightarrow{D}\cdot u\rangle$ play opposite roles. The stability of the fluid will be assured if $d\varepsilon/dt \leq 0$; on the contrary, it is unstable. At the limit, one has marginal or neutral stability.

If one assumes that it is possible to make a division between the space and temporary variables of the disturbance field of velocities $\vec{u}(\vec{x},t) = \vec{v}(\vec{x})f(t)$, the energy evolution equation gets the shape of:

$$d\varepsilon(t)/dt = <|\nabla\vec{w}|^2> \{-(<\vec{v}\cdot\overleftrightarrow{D}\cdot\vec{v}>/<|\nabla\vec{w}|^2>)\ f^2(t) - \nu(t)\}, \tag{8}$$

where $\nu(t) = \int_{-\infty}^{t} dt'\ F(t-t')\ f^2(t')$ and $\overleftrightarrow{D} = \nabla\vec{U}(\vec{x},t,\nu)$. Using Poincaré's one finds:

$$(\ell^2/\alpha)<|\nabla v|^2> \geq <|v|^2>, \tag{9}$$

in which ℓ is a characteristic length and α has to be determined by a variational problem or by one of kinematically admissible maxima. One can define a ν_ε, so that:

$$\nu_\varepsilon(\nu,t) = \text{máx}(-<\vec{v}\cdot\overleftrightarrow{D}\cdot\vec{v}>/<|\nabla\vec{w}|^2>)\ f^2(t) > (-<\vec{v}\cdot\overleftrightarrow{D}\cdot\vec{v}>/<|\nabla\vec{w}|^2>) f^2(t) \tag{10}$$

In consequence, replacing in (8) and integrating, starting from the instant t_0 in which the disturbançe begins, there is a generalization for a viscoelastic fluid of Serrín's theorem of stability:

$$\varepsilon(t) \leq \varepsilon(t_0)\ \exp[-(2\alpha/\ell)\int_{t_0}^{t} dt'\ (1/f^2(t')\{\nu - \nu_\varepsilon\}]. \tag{11}$$

The solution of (9) for a constant decay of gives an estimated limit of the stability. With $|D_m| = \text{mín}\ [D_{11}, D_{22}, D_{33}]$, one proves that: $-<\vec{v}\cdot\overleftrightarrow{D}\cdot\vec{v}> < |D_m|<|\vec{v}|^2>$,

$$\nu\varepsilon(\nu,t) \leq |D_m|(\ell^2/\alpha)\ f^2(t) \tag{12}$$

has to be fulfilled.

Finally, it is important to point up that the inequality (9), time independent, has the same shape of the one gotten for Newtonian fluids. Nevertheless, in relation (11), which corresponds to the disturbance of energy, there appears an exponential function in which one can fin the viscoelastic properties of the fluid. In spite of this fact, there is kept a certain degree of similarity with the ones corresponding to a simply viscous fluid. The relation between ν and ν_ε is, indeed, determinant for the stability of the fluid.

REFERENCES

1. R.B.Bird et al., Dynamics of Polimetric Fluids, Vol. 1, Fluid Mechanics. Ed. J. Wiley & Sons (1977).

2. S.Chandrasekhar, Hydrodynamics and Magnetohydromagnetic Stability. Oxford UP (1961).

3. D.D.Joseph, Stability of Fluid Motions. Springer (1976).

4. B.J.A.Zielinska and Y.Demay, Phys. Rev. A38, 897 (1988).

5. M.Sokolov and R.T. Tanner, phys. Fluids 15, 534 (1972).

6. S.Carmi and M.Sokolov, Phys. Fluids 17, 544 (1974).

7. J.Martínez and C.Pérez-García, Liquids 2, 1281 (1990).

8. J.Serrín, Arch. Rat. Mech. Anal. 3, 1 (1959).

SOLITARY WAVES, TOPOLOGICAL DEFECTS, AND THEIR INTERACTIONS IN SYSTEMS WITH TRANSLATIONAL AND GALILEAN INVARIANCE

Christian Elphick
Departamento de Física
Universidad Técnica Federico Santa María
Casilla 110-V. Valparaíso-Chile

ABSTRACT. We present a method, based on dynamical symmetry groups, to study the interactions of solitary waves and topological defects (vortices) in systems with translational and Galilean invariance.

§I. INTRODUCTION

Let $\mathfrak{S}$ be a system of partial differential equations involving n independent variables $x = (x_1 = t, x_2, \ldots, x_n)$, and m dependent variables $q = (q_1, \ldots, q_m)$. Let X (resp. Q) be the space representing the independent (resp. dependent) variables. A symmetry group of the system $\mathfrak{S}$ is a local group of transformations G acting on an open subset $M \subset X \times U$ such that if $q = f(x)$ is a solution of $\mathfrak{S}$, then for any $g \in G$, $q = gf(x)$ is also a solution of $\mathfrak{S}$ [1].

Specially interesting are those solutions of $\mathfrak{S}$ with particle-like behaviour. These special solutions are called localized structures (LS). Well known examples of LS are the solitary waves, and the topological defects. It follows that if $\mathfrak{S}$ admits a symmetry group, then a given localized structure configuration is never unique, but is labelled by the symmetry group parameters. More generally, we regard the parameters as collective coordinates or collective fields (for parameters associated with internal symmetries). Once one knows a particular LS solution of $\mathfrak{S}$ it is natural to ask whether one can construct multi-LS solutions (in analogy with multi-soliton solutions of integrable systems), and how single-LS solutions interact among themselves? We address in this work these questions for a system $\mathfrak{S}$ whose symmetry group is that of space-time translations and Galilean transformations. Rather than being general we shall consider two particular systems $\mathfrak{S}$. In Section 2 we study the Kuramoto-Korteweg de Vries equation whose single-LS solution is a solitary wave, and in Section 3 we study the Gross-Pitaevskii equation whose single-LS solution is a vortex line (topological defect).

E. Tirapegui and W. Zeller (eds.), Instabilities and Nonequilibrium Structures III, 321–329.

Before moving to Section 2 let us give a rough idea of how we shall proceed. Let $q = q_{LS}$ be a single-LS solution of $\mathfrak{S}$ parametrized by its corresponding collective coordinates and fields. By varying them we obtain an orbit $\mathfrak{O}$ of G. To construct a multi-LS we make a suitable ansatz which, roughly speaking, will be a certain functional of $\mathfrak{O}$. Next, one has to specify how to the symmetry group parameters vary. To do this one projects the infinite dimensional configuration space Γ of the system onto ker $\mathcal{L}^{\dagger}$, $\mathcal{L}$ being the Fréchet derivative of $\mathfrak{S}$ at q_{LS} (the "mass matrix" in field theory language). When $\mathcal{L}$ is self-adjoint Γ gets projected onto a subspace whose tangent vectors are the Goldstone bosons[1] (zero modes of the "mass matrix") of the theory. The number of these zero modes is the number of transformations $g \in G$ that do not leave q_{LS} invariant (Goldstone theorem).

§II. Interacting Solitary Waves in the Kuramoto-Korteweg de Vries (KKdV) Equation

The KKdV equation [2-3] written in conservation law form is

(1a) $$\partial_t q + \partial_x j = 0$$

with current

(1b) $$j(q) = \nu \partial_x q + \partial_{xx} q + \partial_{xxx} q + \frac{1}{2} q^2$$

The symmetry group of eqs. (1a, 1b) corresponds to the following three one-parameter groups $G_1 : (t, x, q) \rightarrow (t + t_0, x, q)$ (time translations), $G_2 : (t, x, q) \rightarrow (t, x + x_0, q)$; (space translations), and $G_3 : (t, x, q) \rightarrow (t, x - vt, q + v)$ (Galilean boosts). Global invariants of G_1 and G_2 (with parameter $x_0 = ct_0$) are $z = x - ct$, and $p = q$. So a group invariant solution of (1a, 1b) is of the form $q = p(x - ct)$. Introducing this form in (1a, 1b) and requiring $p \rightarrow 0$ as $|z| \rightarrow \infty$ we obtain the o.d.e.

(2) $$\frac{d^3p}{dz^3} + \frac{d^2p}{dz^2} + \nu \frac{dp}{dz} - cp + \frac{1}{2} p^2 = 0$$

There exists a curve $\nu = \nu(c)$ such that (2) admits an homoclinic orbit connecting biasymptotically the fixed point $q = 0$ with itself. Accordingly the solution of (1a, 1b) corresponds to a solitary wave (sw) moving with velocity c. Its profile looks like

[1]Strictly speaking a Goldstone boson is described by a field whose value at each x in X may be identified with a local coordinate along the flow generated by the corresponding tangent vector. If some of the $q's$ are odd Grassmann variables one should speak about Goldstone bosons and Goldstone fermions (described by odd Grassmann valued fields).

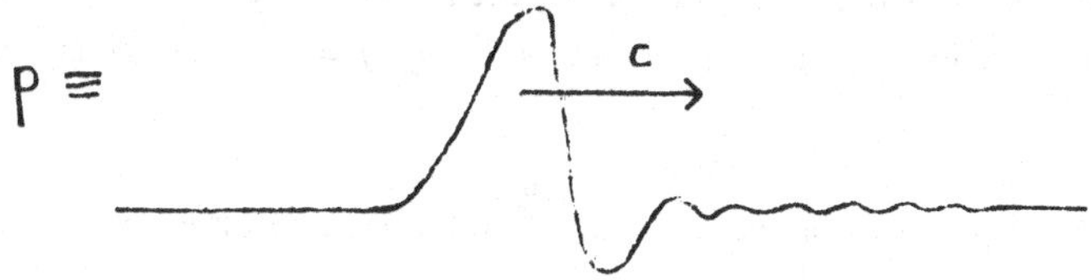

To construct a multi-*sw* solution and study the interaction of *sw's* we make use of the symmetry group of the KKdV eq., and assume that q can be written as a linear superposition of *sw* solutions of (1a, 1b). To wit we make the following ansatz for q near the ith *sw* ($|z - x_i| < \sigma$, $\sigma = sw$ width, $p_j(z) = p(z - x_j)$)

$$q = \{[p_i(z) + v_i] + p_{i-1}(z) + p_{i+1}(z)\} + R(z - x_i, t) \tag{3}$$

In writing this expression we have assumed that the ith *sw* interacts only with its nearest neighbours. We also let the group parameters x_j, v_j become collective coordinates of the jth *sw*, $\dot{x}_j = v_j$, $v_j = v_j(t)$. Finally, R stands for a small correction measuring the deviation of the terms in curly trackets from an exact solution. We suppone $|\partial_t R| \ll |R|$.

Let us write (1a) in terms of z, t. We obtain

$$\partial_t q + \partial_z J = 0 \tag{4}$$

where $J(q) = -cq + j(q)$ is the current in a frame moving with velocity c. In what follows our main assumption will be that the net "momentum flux" due to R acrross the boundary of $[x_i - \sigma, x_i + \sigma]$ vanishes, i.e.

$$\hat{J}(R)\Big|_{x_i - \sigma} - \hat{J}(R)\Big|_{x_i + \sigma} = 0 \tag{5}$$

where $\hat{J}(R) = J(p_i + R) \approx \frac{\delta J}{\delta p_i} R \equiv LR$.

Introducing (3) in (4) we obtain (neglecting $\mathfrak{O}(R^2)$ terms)

$$\mathcal{L}R = -\dot{v}_i - \partial_z\left((v_i - v_{i-1})p_{i-1} + (v_i - v_{i+1})p_{i+1} + p_i(p_{i-1} + p_{i+1})\right) \tag{6}$$

where $\mathcal{L} = \partial_z \frac{\delta J}{\delta p_i} \equiv \partial_z L$. Due to translational invariance $\mathcal{L}$ has a zero mode $e_1 = \frac{dp_i}{dz}$, $\mathcal{L}e_1 = 0$, and due to Galilean invariance $\mathcal{L}^2$ has a zero mode $e_2 = 1$, $\mathcal{L}^2 e_2 = 0$. The matrix representation of $\mathcal{L}$ in the basis $\{e_1, e_2\}$ is the Jordan matrix of a codimension two singularity $\mathsf{J} = \begin{pmatrix} 0 & 1 \\ 0 & 0 \end{pmatrix}$.

Integrating now (6) on $[x_i - \sigma, x_i + \sigma]$ and using the no flux condition (5) we arrive at the following equations for the group parameters x_j, v_j, $j = 1, \cdots N$, N being the number of $sw's$.

$$(7) \quad \begin{pmatrix} \dot{x}_i \\ \dot{v}_i \end{pmatrix} = \mathsf{J} \begin{pmatrix} x_i \\ v_i \end{pmatrix} + \begin{pmatrix} 0 \\ F(x_i - x_{i-1}, x_i - x_{i+1}) + (v_i - v_{i+1})G(x_i - x_{i+1}) + (v_i - v_{i-1})G(x_i - x_{i-1}) \end{pmatrix}$$

where

$$F(x_i - x_{i-1}, x_i - x_{i+1}) = \frac{1}{2\sigma}[p(-\sigma)(p(x_i - x_{i-1} - \sigma) + p(x_i - x_{i+1} - \sigma)) \\ -p(\sigma)(p(x_i - x_{i-1} + \sigma) + p(x_i - x_{i+1} + \sigma))]$$

and

$$G(x) = \frac{1}{2\sigma}[p(x - \sigma) - p(x + \sigma)].$$

Eqs. (7) are not the end of the story. It follows from (4) that $\int_{-\infty}^{\infty} q(z,t)dz$ is a conserved quantity. From our ansatz (3) this is equivalent to require $\sum_{j=1}^{N} \dot{v}_j = 0$, condition that is violated by (7). To get out from this impasse we introduce a new global collective coordinate y associated with G_2, and change $x_j \to x_j + y$ in (3). Hence eqs. (7) get modified according to $F \to F - \dot{y}G(0)$, $v_i - v_{i\pm1} \to v_i - v_{i\pm1} - \dot{y}$. The equation for $\dot{y}$ follows from $\sum_{j=1}^{N} \dot{v}_j = 0$.

Let us remark that if we add to the r.h.s. of (1a) a G_2, G_3 symmetry breaking perturbation, then its leading order effect will be to unfold the codimension two singularity, that is we obtain eqs. (7) with $\mathsf{J} \to \begin{pmatrix} 0 & 1 \\ \mu & \nu \end{pmatrix}$, where μ, ν are parameters characterizing the perturbation.

We close this section by giving the equations for x_j, v_j, $j = 1, \cdots, N$ when $|x_{j+1} - x_j| \gg \sigma$. Using that $p \sim e^{-\alpha x}\cos(\beta x + \phi)\; x \rightsquigarrow +\infty$, and $p \sim e^{\gamma x}\; x \rightsquigarrow -\infty$, we obtain

$$(8a) \quad \dot{x}_i = v_i$$

$$(8b) \quad \dot{v}_i = f_R e^{-\gamma(x_{i+1} - x_i)} + f_L e^{-\alpha(x_i - x_{i-1})}\cos(\beta(x_i - x_{i-1}) + \phi_L) + \\ (v_i - v_{i+1}) g_R e^{-\gamma(x_{i+1} - x_i)} + (v_i - v_{i-1}) g_L e^{-\alpha(x_i - x_{i-1})}\cos(\beta(x_i - x_{i-1}) + \psi_L)$$

where f_R, f_L, g_R, g_L, ϕ_L and ψ_L are known constants.

§III. Vortex Interactions in Galilean Invariant Systems: A Simple Example

A simple model for superfluid $HeII$ is that of a weakly interacting Bose gas where the interaction potential is approximated by a repulsive delta function [4]. The second quantized Hamiltonian is

$$(9) \quad H = \int d^3x \left(\frac{\hbar^2}{2m_4} |\nabla\psi|^2 + \frac{g}{2}|\psi|^4 \right)$$

where ψ is a bosonic field, m_4 the mass of the ^4He atom, and g a coupling constant. The evolution equation for ψ follows from Heisenberg's equation $i\hbar\partial_t\psi = [\psi, H]$. Splitting $\psi = Z + \Phi$, Z being the ground state expectation value of ψ, and Φ a "small operator correction" one obtains that Z satisfies the $1+3-d$ NLS equation

$$i\hbar\partial_t Z = -\frac{\hbar^2}{2m_4}\nabla^2 Z + g|Z|^2 Z \tag{10}$$

The symmetry group of the $1+3-d$ NLS equation corresponds to the following transformations: space-time translations $\underline{x} \to \underline{x}+\underline{x}_0$, $t \to t+t_0$, scale transformation $(\underline{x}, t, Z) \to (\lambda\underline{x}, \lambda^2 t, \lambda Z)$, rotations $\underline{x} \to R\underline{x}$, $R \in SO(3)$, $U(1)$ transformations $Z \to e^{i\phi}Z$, and Galilean boosts $(\underline{x}, t, Z) \to (\underline{x} - \underline{v}t, t, Ze^{i/\hbar(m_4\underline{v}\cdot\underline{x} - \frac{m_4}{2}v^2 t)})$.

Standard topological arguments ($\Pi_1(U(1)) \cong \mathbf{Z}$) lead to the existence of nontrivial topological solutions of (10) characterized by an integer n (the charge of the solution). In fact, writing $Z = Z_n(\underline{x})e^{-\frac{i}{\hbar}E_0 t}$ we obtain the Gross-Pitaevskii equation [[4]-[5]]

$$\frac{\hbar^2}{2m_4}\nabla^2 Z_n + E_0 Z_n - g|Z_n|^2 Z_n = 0 \tag{11}$$

with topological solutions representing vortex lines where Z_n vanishes and its phase turns by $2\pi n$. A solution representing a straight vortex line has the form ($E_0 = gn_0$) $Z_n = \sqrt{n_0}e^{in\theta}f_n(r)$ where r, θ are polar coordinates, $f_n \sim r^{|n|}$ for r small, and $f_n \sim \sqrt{n_0}\left(1 - \frac{n^2}{2}\left(\frac{r_0}{r}\right)^2\right)$ for r large, $r_0 = \hbar(2m_4 g n_0)^{-1/2}$ being the core radius. The energy of a vortex state with charge n behaves as n^2. Hence, vortices with $|n| > 1$ have a finite lifetime, and eventually decay into vortices with $|n| = 1$.

To study vortex interactions we construct a multi-vortex state according to the following ansatz

$$Z = \frac{e^{-\frac{i}{\hbar}E_0 t}}{(\sqrt{n_0})^{N-1}} \prod_{j=1}^{N} Z_{n_j}(\underline{r} - \underline{r}_j) \tag{12}$$

with the property that it vanishes at $\underline{r} = \underline{r}_j$, $j = 1, \cdots N$, $|Z|^2 \sim n_0$ for r large, and its charge is $\sum_{j=1}^{N} n_j$.

Due to the form of (12) one is forced to freeze the Galilean and $U(1)$ groups. As the amplitude of Z is fixed at infinity the scaling group is automatically broken. Time translations and $SO(2)$ rotations can be included in a $U(1)$ transformation. We are left with the space translational group that appear in (12) through the parameters $\underline{r}_j$ that are promoted to collective coordinates $\underline{r}_j = \underline{r}_j(t)$. It remains to specify how the $\underline{r}_j$ vary. This follows by demanding that (12) be a solution of (10). A short cut to arrive at the equation for $\underline{r}_j$, $j = 1, \cdots, N$, is to work near, say, $\underline{r} = \underline{r}_i$, use the $U(1)$ group, and write the ansatz

$$Z = e^{-\frac{i}{\hbar}E_0 t} e^{i\psi}(Z_{n_i}(\underline{r} - \underline{r}_i) + R) \tag{13}$$

where ψ is a singular collective field ($\oint_c \nabla\psi \cdot d\underline{\ell} = 2\pi \sum_{j\neq i} n_j$, c being an enough large circle), and R is a small correction. Introducing (13) in (10) we obtain

$$\mathcal{L}R = i\hbar\dot{\underline{r}}_i \cdot \nabla Z_{n_i} - \frac{i\hbar^2}{m_4}\nabla\psi \cdot \nabla Z_{n_i} - \frac{i\hbar^2}{2m_4} Z_{n_i} \cdot \nabla^2\psi + \frac{\hbar^2}{2m_4} Z_{n_i}(\nabla\psi)^2 \tag{14}$$

where the linear operator $\mathcal{L}$ is

$$\mathcal{L} = \frac{\hbar^2}{2m_4}\nabla^2 + E_0 - 2g|Z_{n_i}|^2 - gZ_{n_i}^2 C \tag{15}$$

C being the charge conjugation operator, $CR = \overline{R}$. The existence of a continous symmetry group implies that $\mathcal{L}$ is a singular operator. Hence, R will exist provided the r.h.s. of (14) belongs to Ran$\mathcal{L}$. A short calculation shows that $\mathcal{L}$ is selfadjoint, $\mathcal{L}^+ = \mathcal{L}$, with respect to the scalar product $<\varphi_1, \varphi_2> = \mathrm{Re}\int \varphi_1\overline{\varphi}_2$. Therefore, the r.h.s. of (14) must be orthogonal to ker $\mathcal{L}$. Using the symmetry group of the NLS equation [6], and defining $\tilde{Z}_n = \lambda Z_n(\lambda\underline{r})e^{i\frac{m_4}{\hbar}\underline{v}\cdot\underline{r}}$ we obtain

(16.a) $\quad \mathcal{L}\nabla\tilde{Z}_n|_{\lambda=1,\underline{v}=0} = 0 \quad$ (translational invariance)

(16.b) $\quad \mathcal{L}i\tilde{Z}_n|_{\lambda=1,\underline{v}=0} = 0 \quad$ ($U(1)$ invariance)

(16c) $\quad \mathcal{L}\frac{\partial\tilde{Z}_n}{\partial\underline{v}}|_{\lambda=1,\underline{v}=0} = i\hbar\nabla\tilde{Z}_n|_{\lambda=1,\underline{v}=0} \quad$ (Galilean invariance)

(16.d) $\quad \mathcal{L}\frac{\partial\tilde{Z}_n}{\partial\lambda}|_{\lambda=1,\underline{v}=0} = -2E_0\tilde{Z}_n|_{\lambda=1,\underline{v}=0} \quad$ (Scale invariance)

Defining $e_1 = -\hbar\partial_x\tilde{Z}_n$, $e_2 = \frac{\partial\tilde{Z}_n}{\partial v_x}$, $e_3 = -\hbar\partial_y\tilde{Z}_n$, $e_4 = \frac{\partial\tilde{Z}_n}{\partial v_y}$, $e_5 = -2E_0 i\tilde{Z}_n$, $e_6 = \frac{\partial\tilde{Z}_n}{\partial\lambda}$, ($e_1, \cdots, e_6$ are evaluated at $\lambda = 1$, $\underline{v} = 0$), we obtain that the matrix representation of $i\mathcal{L}$ in the basis $\{e_1, \cdots, e_6\}$ is the Jordan matrix associated with a triple codimension two singularity, i.e., $i\mathcal{L} = \mathrm{diag(J, J, J)}$, $\mathrm{J} = \begin{pmatrix} 0 & 1 \\ 0 & 0 \end{pmatrix}$.

The equations for $\dot{\underline{r}}_i$ and ψ follow by requiring that the r.h.s. of (15) be orthogonal to e_1, e_3 and e_5. We obtain

$$-\rho_{\underline{K}_i} \wedge \dot{\underline{r}}_i = -\frac{\hbar}{m_4}\rho_{\underline{K}_i} \wedge \nabla\psi|_{\underline{r}_i} \tag{17.a}$$

$$\nabla^2\psi = 0 \tag{17.b}$$

where $\rho = m_4 n_0$,and $\underline{\kappa}_i$ is the circulation vector of the $i^{\underline{\text{th}}}$ vortex, it is parallel to the vortex line, and its magnitud $\kappa_i = |n_i|\frac{h}{m_4}$ (h/m_4 is called the quantum of circulation). The solution of (17.b) subjet to the condition just below (13) is

$$\psi = \sum_{j \neq i} n_j tan^{-1}\left(\frac{y - y_j}{x - x_j}\right) \tag{18}$$

Using (18) in (17a) we arrive at the classical result [7]

$$-\rho\underline{\kappa}_i \wedge \dot{\underline{r}}_i = -\sum_{j \neq i} \frac{\rho}{2\pi}\kappa_i\kappa_j \left(\frac{\underline{r}_i - \underline{r}_j}{|\underline{r}_i - \underline{r}_j|^2}\right) = -\frac{\partial}{\partial \underline{r}_i}\mathfrak{P} \tag{19}$$

where the potential $\mathfrak{P} = \sum_{j\neq i} \frac{\rho}{2\pi}\kappa_i\kappa_j \ln|\underline{r}_i - \underline{r}_j|$. Equation (19) gives the balance between the Magnus force acting on the $i^{\underline{\text{th}}}$ vortex (l.h.s. of(19)) and the external force, due to the other vortices, acting on the $i^{\underline{\text{th}}}$ vortex. As (19) is a "force versus velocity relation" it follows that a particle moving in a plane has twice the degrees of freedom of a straight vortex line with two dimensional motion [7]. This behaviour of a vortex line was expected as we froze the Galilean group from the beginning.

We give now a brief discussion of curved vortex lines. First, consider an isolated vortex with circulation vector $\underline{\kappa}_i$, and let the group parameter $\underline{r}_i$ become a collective field, $\underline{r}_i = \underline{r}_i(z,t)$ (we use cylindrical coordinates r, θ, z). Proceeding as before we obtain

$$-\rho\underline{\kappa}_i \wedge \dot{\underline{r}}_i = \rho\frac{\kappa_i^2}{4\pi}\left(\ln\frac{\Lambda}{r_0}\right)\frac{\partial^2 \underline{r}_i}{\partial z^2} \tag{20}$$

with Λ an upper cutoff, and r_0 the vortex core radius. Equation (20) gives the balance between the Magnus force, and the restoring force tending to straighten the vortex. This equation admits solutions of the form $\underline{r}_i \sim e^{-i(\omega t - kz)}$, representing circularly polarized plane waves, the Kelvin waves, propagating along the vortex line. The dispersion relation is $\omega = \frac{\kappa_i}{4\pi}\left(\ln\frac{\Lambda}{r_0}\right)k^2$. What is the value of Λ? If the vortex line is slightly bent, we may estimate $\Lambda \sim k^{-1}$, hence

$$\omega \cong \frac{\kappa_i}{4\pi}\left(\ln\frac{1}{kr_0}\right)k^2 \tag{21}$$

As a consequence of the Galilean invariance of the NLS equation, the above estimate for ω turns out to be exact. To get this result one must go to higher orders in perturbation theory, and make a judicious use of (16c). An elegant derivation of (21) was given long time ago in [4]. It is worth remarking that (26) in [4] follows from (16.c).

The equations describing a system of interacting curved vortex lines follow from (17.a, 17.b), (20), and (21). We get

$$-\rho \underset{\sim}{\kappa}_i \wedge \dot{\underset{\sim}{r}}_i = -\frac{\hbar}{m_4} \rho \underset{\sim}{\kappa}_i \wedge \nabla\psi|_{\underset{\sim}{r}_i} + \frac{\rho\kappa_i^2}{4\pi} \mathfrak{D} \underset{\sim}{r}_i \tag{22.a}$$

$$\Delta\psi = 0, \qquad \oint_C \nabla\psi \cdot d\underset{\sim}{\ell} = 2\pi \sum_{j \neq i} n_j \tag{22.b}$$

where $\mathfrak{D}$ is a differential operator whose representation in Fourier space is $k^2 \ln kr_0$, and Δ is the $3-d$ Laplacian operator. The solution of (22.b) follows from the Biot-Savart formula

$$\nabla\psi(\underset{\sim}{x}) = \sum_j \frac{n_j}{2} \int d\underset{\sim}{\rho}_j \wedge \frac{(\underset{\sim}{x} - \underset{\sim}{\rho}_j)}{|\underset{\sim}{x} - \underset{\sim}{\rho}_j|^3} \tag{23}$$

where $\underset{\sim}{\rho}_j = z_j \hat{z} + \underset{\sim}{r}_j(z_j, t)$.

§IV. Concluding Remarks

We have presented a group-theoretic approach to interacting localized structures in systems with translational and Galilean invariance. In Section II we found that $sw's$ interact via a short range force (see eqs. (8a, 8b)). Roughly speaking, this result is a consequence of the fact that there is no spontaneous symmetry breaking in one dimensional systems. An equivalent way to understand this result, is to notice that the fixed point $q = 0$ (the asymptotic value of p) is translationally invariant. Excitations just above $q = 0$ are "massive" with "masses" $m_L = \gamma$, and $m_R = \alpha \pm i\beta$. This causes that the force between two $sw's$ is short-ranged and violates Newton's third law. Violation of the action-reaction law forced us to introduce the global collective variable y.

In Section III we dealt with interacting vortices. Of course eq. (19) is a well known result [[7],[8]], rederived here in a way that seemed nice to us. As the vortex solution of (11) is not $U(1)$- invariant, the interaction between far apart vortices is long-ranged. This interaction is mediated by the "zero mass" phase field ψ (see (17a)). On the other hand, the interaction between two nearby vortices is mediated by the "massive" amplitude field. For vortices with charge $\pm n$, $n > 0$ this interaction is of order ρ^{2n-1}, ρ being their relative distance. Using the words of Konopleva and Popov [9], vortices behave very much like protons: at large distances they interact via photons, and at small distances the interaction is mediated by pions.

References

1. P.J. Olver, "Applications of Lie Groups to Differential Equations," Springer-Verlag, New York, 1986.

2a. J. Topper and T. Kawahara, J. Phys. Soc. Japan **44** (1978;), p. 663.

2b. T. Kawahara, Phys. Rev. Lett. **51** (1984), p. 381.

2c. S. Toh and T. Kawahara, J. Phys. Soc. Japan **54,** (1985), p. 1257.

3. C. Elphick, G. Ierley, O. Regev and E. Spiegel, *Columbia Univ. Preprint.*

4. L.P. Pitaevskii, Sov. Phys. JEPT **13** (1961), p. 451.

5. E.P. Gross, Nuovo Cimento **20** (1961), p. 454.

6. C. Elphick and E. Meron, Phys. Rev. A. **40** (1989), p. 3226.

7. E.B. Sonin, Rev. Mod. Phys. **59 # 1** (1987), p. 87.

8. A.L. Fetter, Phys Rev. **151** (1966), p. 100.

9. N.P. Konopleva and V.N. Popov, "Gauge Fields," Harwood, New York, 1981.

DYNAMICS OF NONLINEAR SCHRÖDINGER DEFECTS IN THREE DIMENSIONS

FERNANDO LUND
Departamento de Física
Facultad de Ciencias Físicas y Matemáticas
Universidad de Chile
Casilla 487-3, Santiago, Chile

ABSTRACT. The equation of motion of defect solutions to the nonlinear Schrödinger equation in three dimensions is derived from a variational principle.

1. Introduction

The complex Ginzburg-Landau equation describes the behaviour of many systems far from equilibrium, and it has recently been suggested[1] that defect solutions to that equation may drive a transition from ordered to disordered behaviour of systems so described. In this context, the dynamics of two dimensional defects has attracted considerable attention[2]. A special case of the complex Ginzburg-Landau equation is the nonlinear Schrödinger equation

$$i\partial_t\psi = -\nabla^2\psi + |\psi|^2\psi. \tag{1}$$

where ψ is a complex scalar field. We shall here study the dynamics of defect solutions to this equation in three dimensions.

When written in terms of amplitude and phase

$$\psi = \sqrt{\rho}e^{i\phi}, \tag{2}$$

the nonlinear Schrödinger equation becomes

$$\partial_t\rho = -2\nabla(\rho\nabla\phi) \tag{3}$$

$$\partial_t\phi = -\frac{1}{4\rho^2}(\nabla\rho)^2 + \frac{1}{2\rho}\nabla^2\rho - (\nabla\phi)^2 - \rho. \tag{4}$$

These equations can be obtained as extrema of the action functional

$$S = \int dtdx \left(\rho\partial_t\phi + \rho(\nabla\phi)^2 + \frac{1}{4\rho}(\nabla\rho)^2 + \frac{1}{2}\rho^2\right) \tag{5}$$

E. Tirapegui and W. Zeller (eds.), Instabilities and Nonequilibrium Structures III, 331–336.

with respect to variations of ρ and ϕ.

In three dimensions, defect (vortex) solutions to (1) have a multivalued phase ϕ and are defined over all space except at closed (or infinite) lines. The amplitude ρ vanishes at the location of those singularities and grows to a constant value away from them. The gradients of the phase blow up at the location of the singularities, and the region nearby, where the amplitude cannot be taken as constant, is termed the vortex core. It is easy then to imagine that, if intervortex distances and vortex radii of curvature are large compared to this core, things ought to simplify: this is a dilute gas approximation. Also, the singular behaviour of the phase at the core suggests that it is what happens in that region that dominates the vortex dynamics.

What we propose to do is as follows: find multivalued solutions of (3,4). They will depend, of course, on the position of the line singularity $\mathbf{X}(\sigma, t)$, where t is the time and σ is a Lagrangean parameter labeling the points along the line. Replace these solutions into the action (5), thereby rendering it a functional of the trajectory $\mathbf{X}(\sigma, t)$. Extremizing with respect to variations of this trajectory will yield dynamical equations governing the evolution of the line defects. This approach has been succesfully employed in determining the dynamics of defects in various areas of field theory[3]. The results of this approach are embodied in equations (16,17,19,20). In general, (16) and (19) are integrodifferential equations saying that the dynamics of vortex filaments is nonlocal in character: the behaviour of a given point along a filament is influenced by all other points, with an interaction that decreases like 1/(distance) , exactly as happens for vortex loops in classical hydrodynamics and current loops in magnetostatics. In contrast to the two dimensional case, there is vortex self-interaction that is not only non negligible but that dominates the dynamics when all distance scales are large by comparison with the core thickness: the leading order behaviour is given by (17), which says that the local vortex velocity is proportional to the local curvature. Since this curvature is zero in two dimensions, it is easy to see why there is no self interaction in that case. When vortex loops are infinitely long, straight, and parallel, one has in effect a two dimensional situation which allows for the recovery of point vortex dynamics. This is given by (20).

2. Computations and Results

Outside of the vortex core, the amplitude ρ is constant and Eqns. (3,4) reduce to Laplace's equation for the phase ϕ:

$$\nabla^2 \phi = 0. \tag{6}$$

The asymptotics implicit in this assertion has been recently worked out in detail by Neu[4]. Multivalued solutions to this equation are solid angles subtended by the vortex filament at the point of observation:

$$\phi(\mathbf{x}, t) = \frac{n}{2} \int_{S(t)} dS_i' \nabla_i' (|\mathbf{x} - \mathbf{x}'|^{-1}) \tag{7}$$

where n is the winding number and $S(t)$ is any surface bounded by the filament $\mathbf{X}(\sigma, t)$. Note that since ϕ is multivalued, the surface $S(t)$ must be introduced; it has no physical significance but it is needed in the mathematics to say exactly where the phase has a discontinuity. The gradient of the phase ϕ involves only the boundary $\mathbf{X}(\sigma,t)$ and not the surface $S(t)$:

$$\nabla_i \phi = \frac{n}{2} \oint d\sigma \varepsilon_{ijk} \nabla_j (|\mathbf{x} - \mathbf{X}|^{-1}) \mathbf{X}'_k. \tag{8}$$

It is easy to see that also $\dot{\phi}$ is also expressed in terms of a line integral: the time derivative of (7) involves only the changes at the boundary. Thus, since only derivatives of ϕ enter in the dynamical equations (3,4), the physics does not involve the surface $S(t)$, as promised.

The linearity of Laplace's equation insures that the contributions from many vortices just add up. The action (5) is now, up to unimportant additive constants,

$$S_\phi = \int dt dx (\partial_t \phi + (\nabla \phi)^2) \tag{9}$$

where, very importantly, the domain of integration must *exclude* the vortex core. It is this exclusion that turns the action into a functional of the vortex trajectories and makes the first term of the right hand side of (9) give a nontrivial contribution. Expressions (7,8) are now to be replaced into the action (9). It was already noted that a thin tube around the filament must be excluded from the volume of integration. Actually, since we are integrating multivalued functions, in addition to this tube a whole slab hugging the surface $S(t)$ must also be excluded. We have then, integrating by parts,

$$\begin{aligned} \int dt d^3x \dot{\phi} &= -\int dt \left(\int_{S+} + \int_{S-} + \int_{S_T} \right) dS_i V_i \phi + \int dt \frac{d}{dt} \int d^3 x \phi & (10) \\ &= 2\pi n \int dt \int_{S(t)} dS_i V_i, & (11) \end{aligned}$$

where V_i is the velocity of points on the surface $S(t)$, S^+ and S^- are the uper and lower lips of the slab sandwiching $S(t)$, and S_T is the surface of a thin tube surrounding the line defect. The last term on the right of (10), being the integral of a total time derivative, will not contribute to the equation of motion. The integral over S_T vanishes in the limit of zero tube thickness. We now show that the resulting integral over the surface $S(t)$ may be turned into an integral over its boundary $\mathbf{X}(\sigma,t)$. To see this, take $S(t)$ as a parametrized surface $\mathbf{X}(\xi_1, \xi_2, t)$with coordinates (ξ_1, ξ_2), and boundary ∂S given by the filament $\mathbf{X}(\sigma,t)$. Then,

$$\begin{aligned} \int dt \int dS_i V_i &= \int dt d\xi_1 d\xi_2 \varepsilon_{ijk} \frac{\partial X_j}{\partial \xi_1} \frac{\partial X_k}{\partial \xi_2} \frac{\partial X_i}{\partial t} \\ &= \frac{1}{3} \int dt d\xi_1 d\xi_2 \varepsilon_{ijk} \left(\frac{\partial}{\partial \xi_1} \left(X_j \frac{\partial X_k}{\partial \xi_2} \frac{\partial X_i}{\partial t} \right) \right. \\ & \left. - \frac{\partial}{\partial \xi_2} \left(X_j \frac{\partial X_k}{\partial \xi_1} \frac{\partial X_i}{\partial t} \right) + \frac{\partial}{\partial t} \left(X_i \frac{\partial X_j}{\partial \xi_1} \frac{\partial X_k}{\partial \xi_2} \right) \right) \end{aligned} \tag{12}$$

where the second equality is obtained integrating repeatedly by parts. The last term on the right of (12) is a total time derivative that can be dropped. Using Stoke's theorem in the form

$$\int_S d\xi_1 d\xi_2 \left(\frac{\partial}{\partial \xi_1} \left(f \frac{\partial g}{\partial \xi_2} \right) - \frac{\partial}{\partial \xi_2} \left(f \frac{\partial g}{\partial \xi_1} \right) \right) = \int_{\partial S} d\sigma f \frac{\partial g}{\partial \sigma}$$

we get

$$\int dt \int dS_i V_i = \frac{1}{3} \oint dt d\sigma \mathbf{X} \cdot \mathbf{X}' \wedge \dot{\mathbf{X}}, \tag{13}$$

which is a functional of the filament $\mathbf{X}(\sigma, t)$ only, as desired. An overdot means time derivative.

Similarly, the other term in (9) is, using $\nabla^2 \phi = 0$,

$$\int dt d^3x (\nabla\phi)^2 = \Gamma \int dt \int_{S(t)} dS_i \frac{\partial \phi}{\partial x_i} + \int dt \int_{S_T} dS_i \phi \frac{\partial \phi}{\partial x_i}.$$

The integral over S_T vanishes for a thin tube because of the behaviour of ϕ near the line defect. Using (8) and Stokes theorem, we have

$$\int dt \int_{S(t)} dS_i \frac{\partial \phi}{\partial x_i} = \frac{n}{2} \oint d\sigma \oint d\sigma' \frac{1}{|\mathbf{Y}(\sigma', t) - \mathbf{X}(\sigma, t)|} \frac{\partial \mathbf{X}}{\partial \sigma} \cdot \frac{\partial \mathbf{Y}}{\partial \sigma'} \tag{14}$$

where $\mathbf{Y}(\sigma', t)$ is a loop lying very close to the vortex filament. It is the boundary of $S(t)$ which, because of the thin tube sorrounding the filament, does not exactly coincide with $\mathbf{X}(\sigma, t)$; a cutoff is needed if divergent integrals in (13) are to be avoided. We choose

$$\mathbf{Y}(\sigma, t) = \mathbf{X}(\sigma, t) + \mathbf{h}$$

where $\mathbf{h}$ is much smaller than any radius of curvature. It does not have any dynamics of its own and its behaviour must be prescribed from the outside. This is a consequence of ignoring what is going on within the vortex in the present filamentary approximation.

Substituting (13) and (14) in (9) we get

$$\begin{aligned} S_\phi[\mathbf{X}(\sigma, t)] &= \frac{2\pi n}{3} \oint dt d\sigma \mathbf{X} \cdot \dot{\mathbf{X}} \wedge \mathbf{X}' \\ &+ \frac{\pi n^2}{2} \oint dt d\sigma d\sigma' \frac{1}{|\mathbf{X}(\sigma, t) - \mathbf{X}(\sigma', t) + \mathbf{h}|} \frac{\partial \mathbf{X}}{\partial \sigma} \cdot \frac{\partial \mathbf{X}}{\partial \sigma'} \end{aligned} \tag{15}$$

whose extrema $\delta S_\phi / \delta \mathbf{X} = 0$ yield

$$\dot{\mathbf{X}} \wedge \mathbf{X}' = \frac{n}{2} \oint d\sigma' \frac{((\mathbf{X}(\sigma) - \mathbf{X}(\sigma')) \wedge \mathbf{X}'(\sigma')) \wedge \mathbf{X}'(\sigma)}{|\mathbf{X}(\sigma) - \mathbf{X}(\sigma') + \mathbf{h}|^3} \tag{16}$$

which, in a hydrodynamical context, would be the well known Biot–Savart law, describing the motion of an isolated vortex filament in an infinite, inviscid, incompressible fluid. It is nonlocal, reflecting the fact that all points on the filament contribute to the dynamics, with an interaction that decreases as $1/(\text{distance})$. It is singular when $\sigma \sim \sigma'$ which means that the dynamics will be dominated by the local behaviour. To leading order this is given by a localized induction approximation, which is obtained by assuming that the integral on the right of (16) is dominated by those points $\mathbf{X}(\sigma')$ lying closest to $\mathbf{X}(\sigma)$, say within a distance δ, and it means keeping only the leading term of (16) when $h/\delta \to 0$. This approximation is good whenever the curvature is small and the vortex does not bend back on itself. If this is the case (16) yields

$$\dot{\mathbf{X}} \wedge \mathbf{X}' = \frac{\Gamma}{4\pi} ((\mathbf{X}' \wedge \mathbf{X}'') \wedge \mathbf{X}') |\mathbf{X}'|^{-3} \ln(\delta/h) + O(1) \tag{17}$$

in the limit $\delta \gg h$ and with $\mathbf{X}' \cdot \mathbf{h} = 0$. Thus, in this limit the defect velocity is given by its local curvature. The parameter (δ/h) is undetermined, and in practice it fixes the time scale.

If many vortex filaments are present, one has

$$\phi(\mathbf{x},t) = \sum_A \phi_A(\mathbf{x},t)$$

where ϕ_A is the potential of the Ath filament, with winding number n_A. Replacement into (9) gives

$$\begin{aligned} S_\phi[\mathbf{X}] &= \frac{2\pi}{3} \sum_A n_A \oint dt d\sigma \mathbf{X}_A \cdot \dot{\mathbf{X}}_A \wedge \mathbf{X}'_A \\ &+ \frac{\pi}{2} \sum_{A,B} n_A n_B \oint \frac{dt d\sigma d\sigma'}{|\mathbf{X}_A(\sigma,t) - \mathbf{X}_B(\sigma',t)|} \mathbf{X}'_A \cdot \mathbf{X}'_B \end{aligned} \tag{18}$$

whose extrema $\delta S_\phi = 0$ are given by

$$\dot{\mathbf{X}}_A \wedge \mathbf{X}'_A = \frac{1}{2} \sum_B n_B \oint d\sigma' \frac{((\mathbf{X}_A - \mathbf{X}_B) \wedge \mathbf{X}'_B) \wedge \mathbf{X}'_A}{|\mathbf{X}_A - \mathbf{X}_B|^3}. \tag{19}$$

This gives an integro-differential equation of motion for the velocity of the Ath filament. Only the component of velocity normal to the filament appears in (19). This makes physical sense, as motion of the filament along its tangent is no real motion but a relabeling of points. The localized induction approximation is unchanged for many filaments, provided they do not approach too closely (i.e., within a distance $\sim h$).

When the filaments are straight, very long, and parallel, they can be parametrized as

$$\mathbf{X}_A(\sigma,t) = (X^1_A(t), X^2_A(t), \sigma).$$

In this case $\mathbf{X}' = (0,0,1), \dot{\mathbf{X}} = (\dot{X}^1, \dot{X}^2, 0)$, and $\mathbf{X}'' = 0$. There is no self-induction. The integral along σ' in (19) can be done explicitely and the result is

$$\dot{\mathbf{X}}_A(t) = \sum_{B \neq A} n_B \frac{\hat{k} \wedge (\mathbf{X}_A - \mathbf{X}_B)}{|\mathbf{X}_A - \mathbf{X}_B|^2} \tag{20}$$

the well known equations of two-dimensional point vortex dynamics[5].

Acknowledgements

This work has been supported by DTI grant E-2854-8814 and FONDECYT grant 533-88.

References

[1] P. Coullet, L. Gil and J. Lega, *phys. Rev. Lett.* **62**, 1618 (1989); *Physica D* **37**, 91 (1989).

[2] S. Rica and E. Tirapegui, *Phys. Rev. Lett.* **64**, 878 (1990); S. Rica and E. Tirapegui, *Physica D*, to appear; C. Elphick and E. Meron, preprint; H. Sakaguchi, preprint; K. Kawasaki, *Prog. Theor. Phys. Suppl.* **79**, 161 (1984).

[3] F. Lund, *Phys. Fluids A* **1**, 606 (1989); F. Lund and N. J. Zabusky, *Phys. Fluids* **30**, 2306 (1987); F. Lund, *J. Mat. Res.* **3**, 280 (1988); F. Lund, *Phys. Rev.* **D33**, 3124 (1986); F. Lund, *Phys. Rev. Lett.* *54*, 14 (1985).

[4] J. C. Neu, *Physica D* **43**, 385 (1990).

[5] A. L. Fetter, *Phys. Rev.* **151**, 100 (1966).

Dynamics of spirals defects in two dimensional extended systems

Sergio Rica and Enrique Tirapegui
Laboratoire de Physique Théorique, Université de Nice and
Facultad de Ciencias Físicas y Matemáticas,
Universidad de Chile, Santiago.

Abstract.
A review of recent progress on the interaction of spiral defects in two dimensional extended systems is presented. The equations of motion of a diluted gas of spiral defects interacting with a new slowly varying global phase are derived in the frame of the Ginzburg-Landau equation with complex coefficients. We also derive the equation of motion of a single spiral when one adds a small constant term to the Ginzburg-Landau equation.

We present here some recent progress on the interaction of spiral defects arising in two dimensional extended systems. These defects appear in the Ginzburg-Landau equation with complex coefficients which is the normal form of reaction diffusion systems after going through a Hopf bifurcation to homogeneous oscillations. It has been shown by Walgraef, Dewel and Borckmans[1] that the presence of these defects can disorganize the system. Recently Coullet, Gil and Lega[2] have shown that spiral defects appear spontaneously in the complex Ginzburg-Landau equation (CGLE) due to phase instability and drive the system to a state of a *diluted gas* of spirals which leads to disorganization: *defects-mediated turbulence.* The CGLE for the complex field $A(\vec{r},t)$ reads

$$\frac{\partial A}{\partial t} = (\mu + i\Omega)A + (1+i\alpha)\nabla^2 A - (1+i\beta)|A|^2 A. \tag{1}$$

It admits spiral solutions[3,4,5] centered in any point $\vec{R}$ which are ($\vec{\rho} = \vec{r} - \vec{R}$, $\rho = |\vec{\rho}|$, $|m| = 1$, $\hat{\rho}$ is the unitary vector along $\vec{\rho}$)

$$A^{(o)} = D(\rho) exp\left\{ i\left(\omega t + m\varphi - S(\rho)\right)\right\}, \tag{2}$$

where for $\rho \to 0$ one has $D(\rho) = \lambda\rho$, $S(\rho) = \frac{\nu D_\infty^2}{8(1+\alpha^2)}\rho^2$, and for $\rho \to \infty$ one has $D(\rho) = D_\infty - \frac{q(1+\alpha^2)}{2\nu D_\infty}\frac{1}{\rho}$, $S(\rho) = q\rho + \frac{\gamma}{2\nu} ln(\rho)$, with $\gamma = 1 + \alpha\beta$, $\nu = \beta - \alpha$. The frequency $\omega = \Omega - \beta\mu + \nu q^2$, with $q = q(\alpha, \beta)$ chosen by the system, and

E. Tirapegui and W. Zeller (eds.), Instabilities and Nonequilibrium Structures III, 337–346.

$D_\infty = \sqrt{\mu - q^2}$. The functions D and S obey the equation (primes will denote derivatives, $\nabla_\rho^2 \equiv \frac{\partial^2}{\partial\rho^2} + \frac{1}{\rho}\frac{\partial}{\partial\rho}$)

$$\nabla_\rho^2 D - \left(\frac{1}{\rho^2} + S'^2 - q^2\right) D + \frac{\gamma}{1+\alpha^2}\left(D_\infty^2 - D^2\right) D = 0 \tag{3a}$$

$$2D'S' + D\nabla_\rho^2 S = \frac{\nu}{1+\alpha^2}\left(D_\infty^2 - D^2\right) D. \tag{3b}$$

In order to represent a *diluted gas* of spirals with charges $(m_1, \ldots, m_N)$ located at $(\vec{r}_1, \ldots, \vec{r}_N)$ we assume that (1) has a solution of the form ($\vec{\rho}_j = \vec{r} - \vec{r}_j$, $\rho_j = |\vec{\rho}_j|$, $cos\varphi_j = \frac{x - x_i}{\rho_j}$, $sin\varphi_j = \frac{y - y_i}{\rho_j}$)

$$A^{(o)} = (R^o + w)exp\left\{i\left(\omega t + \Theta^o\right)\right\}, \tag{4a}$$

where $R^o = D(\rho_k)$ in each region $D_k = \{\vec{r} : min(\rho_1, \ldots, \rho_N) = \rho_k\}$ (these are the regions around each spiral in the plane) and

$$\Theta^o = \sum_{i=1}^{N}\left(m_i\varphi_i - \tilde{S}(\rho_i)\right) - q\ min(\rho_1, \ldots, \rho_N) + \Psi(\vec{r}, t), \tag{4b}$$

where $\tilde{S}(\rho) \equiv S(\rho) - q\rho$. In (4a) w is a small correction term. We have also introduced a new global phase $\Psi(\vec{r}, t)$ of slow spatial variation in the scale of the core of the spirals which are of radius of order $\varepsilon = \frac{D_\infty}{\lambda}$ (see after eq.(2)) and allowed the spirals to move, i.e. $\{\vec{r}_j\} \to \{\vec{r}_j(t)\}$. The condition of a *diluted gas* of spirals on which we rely heavily in what follows is expressed as $|\vec{r}_i - \vec{r}_j| \gg \varepsilon$. The new variables are now $\{\vec{r}_j, j = 1, 2, \ldots, N\}$ related to translation invariance of (1) and $\Psi(\vec{r}, t)$ related to the invariance $A \to Aexp(i\delta)$ of (1)[6,7]. The justification of the *Ansatz* (4) is the requirement that the dominant behavior of $A^{(o)}$ near each point $\vec{r}_k$ must be locally that of an isolated spiral. From (4b) we see that in the region D_k one has

$$\Theta^o(\vec{r}, t) = m_k\varphi_k - S(\rho_k) + \Phi^{(k)} \tag{5a}$$

$$\Phi^{(k)} = \sum_{i \neq k}\left(m_i\varphi_i - \tilde{S}(\rho_i)\right) + \Psi(\vec{r}, t) \tag{5b}$$

where $\Phi^{(k)}$, which varies slowly through the core of the kth spiral according to our assumptions, is the total phase Θ^o near this spiral minus its self-phase $(m_k\varphi_k - S(\rho_k))$. Our *Ansatz* (4a) is nondifferentiable in the borders of the regions D_k in its present form and has to be modified accordingly. For example with 2 spirals we write

$$R^o = D(\rho_1)\tilde{\theta}(\rho_2 - \rho_1) + D(\rho_2)\tilde{\theta}(\rho_1 - \rho_2) \tag{6a}$$

$$\Theta^o = \sum_{i=1}^{2} \left(m_i \varphi_i - \tilde{S}(\rho_i) \right) - q \left(\rho_1 \tilde{\theta}(\rho_2 - \rho_1) + \rho_2 \tilde{\theta}(\rho_1 - \rho_2) \right) + \Psi(\vec{r}, t), \tag{6b}$$

where $\tilde{\theta}$ is a differentiable approximation to the step function $\theta(x) = 1$, $x > 0$, $\theta(x) = 0$, $x < 0$. The new equations of motion, i.e. the values of the derivatives $\partial_t \vec{r}_j(t)$ and $\partial_t \Psi(\vec{r}, t)$, will be determined by the solvability conditions for w which we assume depends on time only through $\{\vec{r}_j(t), \Psi(\vec{r}, t)\}$, i.e. $w = w(\vec{r}; \{\vec{r}_j(t), \Psi(\vec{r}, t)\})$. We assume also that $\left\{ \dot{\vec{r}}_j(t) \right\}$, $\partial_t \Psi(\vec{r}, t)$ and w are small quantities of the same order, then $\partial_t w = \mathcal{O}(w^2)$. Replacing the *Ansatz* (4a) in (1) one obtains to first order in w an equation $\mathcal{L} w = I$ with

$$\mathcal{L} w = [\mu + i(\Omega - \omega)]\, w + (1 + i\alpha) \Big[\nabla^2 w + 2i \vec{\nabla} \Theta^o \cdot \vec{\nabla} w + \\ + \left(i \nabla^2 \Theta^o - \left(\vec{\nabla} \Theta^o \right)^2 \right) w \Big] - (1 + i\beta) \left[2 R^{o2} \Re e\, w + R^{o2} w \right] \tag{7a}$$

$$I = \partial_t R^o - R^o \left[\mu + i(\Omega - \omega) - i \partial_t \Theta^o - (1 + i\beta) R^{o2} \right] + \\ - (1 + i\alpha) \left[\nabla^2 R^o + 2i \vec{\nabla} \Theta^o \cdot \vec{\nabla} R^o + R^o \left(i \nabla^2 \Theta^o - \left(\vec{\nabla} \Theta^o \right)^2 \right) \right]. \tag{7b}$$

Direct replacement of the isolated spiral solution $A^{(o)}$ for a spiral centered in $\vec{r}_k$ in (1) gives the equation

$$[\mu + i(\Omega - \omega)] D(\rho_k) + (1 + i\alpha) \Big[\nabla^2_{\rho_k} D + \\ - 2i S'(\rho_k) D'(\rho_k) - i D \nabla^2_{\rho_k} S - D \vec{L}_k^2 \Big] - (1 + i\beta) D^3 = 0, \tag{8}$$

where $\vec{L}_k \equiv \vec{\nabla}(m_k \varphi_k - S(\rho_k)) = \frac{m_k}{\rho_k} \hat{\varphi}_k - S'(\rho_k) \hat{\rho}_k$. Using (5) and (8) we obtain for $\mathcal{L} w$ and I in the region D_k the expressions

$$\mathcal{L} w \big|_{D_k} = (1 + i\alpha) \Big[\nabla^2 w + 2i \left(\vec{L}_k + \vec{\nabla} \Phi^{(k)} \right) \cdot \vec{\nabla} w + i \nabla^2 \Phi^{(k)} w + \\ - \left(2 \vec{L}_k \cdot \vec{\nabla} \Phi^{(k)} + \left(\vec{\nabla} \Phi^{(k)} \right)^2 \right) w + \\ + 2i S'(\rho_k) D'(\rho_k) w - \frac{\nabla^2_{\rho_k} D}{D} w \Big] - 2(1 + i\beta) D^2 \Re e\, w \tag{9a}$$

$$I \big|_{D_k} = - \left(\dot{\vec{r}}_k + 2i(1 + i\alpha) \vec{\nabla} \Phi^{(k)} \right) \cdot \left(\vec{\nabla} D(\rho_k) + i D \vec{L}_k \right) + \\ + i D \partial_t \Phi^{(k)} + (1 + i\alpha) D \left(-i \nabla^2 \Phi^{(k)} + \left(\vec{\nabla} \Phi^{(k)} \right)^2 \right). \tag{9b}$$

As we have explained we need to impose now the solvability conditions for w which are that the scalar product $\langle\chi, I\rangle = 0$ for $\chi \in ker\mathcal{L}^\dagger = \left\{\chi' : \mathcal{L}^\dagger\chi' = 0\right\}$, where $\mathcal{L}^\dagger$ is the adjoint of $\mathcal{L}$ with the same scalar product. The strategy to determine the elements of $Ker\mathcal{L}^\dagger$ is inspired in the study of the similar problem for the Ginzburg-Landau equation with real coefficients where Goldstone theorem is valid in its usual form ($\mathcal{L} = \mathcal{L}^\dagger$ there)[7]. The results there are that the equations of motion of the kth spiral, i.e. the value of $\dot{\vec{r}}_k$, are determined by an element $\chi^{(k)}$ of $Ker\mathcal{L}^\dagger$ which is different from zero only in the core of this spiral and vanishes outside this core. Concerning the equation of motion for the new global phase Ψ it is determined by a kernel element which is zero in the cores of the spirals and not vanishing only in the region outside these cores. We shall look then here for kernel elements $\chi^{(k)} \neq 0$ only in the core $V_\varepsilon(\vec{r}_k) = \{\vec{r} : |\vec{r} - \vec{r}_k| \leq \varepsilon\}$, $k = 1, 2, \ldots, N$, and an element $\chi \neq 0$ only outside all the cores. In order to determine $\chi^{(k)}$ we keep in (9a) only the dominant terms in $V_\varepsilon(\vec{r}_k)$ taking into account the known asymptotic behaviors of $D(\rho_k)$ and $S(\rho_k)$ in $V_\varepsilon(\vec{r}_k)$ which corresponds to $\rho_k \to 0$ (see after (2)). We have

$$\mathcal{L}w\big|_{V_\varepsilon(\vec{r}_k)} = (1 + i\alpha)\tilde{\mathcal{L}}w \equiv (1 + i\alpha)\left[\nabla^2 + 2i\frac{m_k}{\rho_k}\hat{\varphi}_k \cdot \vec{\nabla} - \frac{1}{\rho_k^2}\right] w. \tag{10}$$

We look for χ such that $\tilde{\mathcal{L}}^\dagger\chi = 0$ with $\tilde{\mathcal{L}}^\dagger = \tilde{\mathcal{L}}$ in the usual scalar product $\langle f_1, f_2\rangle = \int d\vec{r} f_1(\vec{r})^* f_2(\vec{r})$. The functions $\left\{\chi_n = \rho_k{}^{\pm|n+m_k|}e^{in\varphi_k},\ n \in Z\right\}$ are such that $\mathcal{L}^\dagger\chi_n = 0$ and for $n \neq -m_k$ they are singular when $\rho_k \to 0$ or vanish as $\rho_k{}^{|n+m_k|}$ and are then $\mathcal{O}(\varepsilon)$ in $V_\varepsilon(\vec{r}_k)$. The kernel element of interest to us is then $\chi^{(k)} = exp(-im_k\varphi_k)$ in $V_\varepsilon(\vec{r}_k)$ and zero outside. The expression in (9b) is still exact but we only need I in $V_\varepsilon(\vec{r}_k)$ where we can aproximate $\vec{\nabla}D + iD\vec{L}_k = \lambda e^{-im_k\varphi_k}\vec{j}_o + \mathcal{O}(\rho_k{}^2)$ with $\vec{j}_o = (1, im_k)$. Then ($\vec{u}\cdot\vec{v} = u_1^*v_1 + u_2^*v_2$)

$$I\big|_{V_\varepsilon(\vec{r}_k)} = -\vec{j}_o{}^* \cdot \left(\dot{\vec{r}}_k + 2i(1 + i\alpha)\vec{\nabla}\Phi^{(k)}\right)\lambda e^{-im_k\varphi_k} + iD\partial_t\Phi^{(k)} +$$
$$+ (1 + i\alpha)D\left(-i\nabla^2\Phi^{(k)} + \left(\vec{\nabla}\Phi^{(k)}\right)^2\right). \tag{11}$$

We express now the solvability condition $\langle\chi^{(k)}, I\rangle = 0$ and take into account that in the core $V_\varepsilon(\vec{r}_k)$ the phase $\Phi^{(k)}$ as well as its derivatives must be slowly varying functions in $V_\varepsilon(\vec{r}_k)$ and can be taken out of the integral over $d\varphi_k$ in the scalar product (the slow variation of $\Phi^{(k)}$ is a consequence of (5) and our assumption on Ψ). Then the scalar product of $\chi^{(k)}$ with the second term in (11) vanishes and one obtains

$$\vec{j}_o{}^* \cdot \left(\dot{\vec{r}}_k + 2i(1 + i\alpha)\left\langle\vec{\nabla}\Phi^{(k)}\right\rangle\right) = 0 \tag{12a}$$

$$\left\langle\vec{\nabla}\Phi^{(k)}\right\rangle \equiv \frac{1}{\pi\varepsilon^2}\int_{V_\varepsilon(\vec{r}_k)} d^2\vec{r}\ \vec{\nabla}\Phi^{(k)}. \tag{12b}$$

We see that $\left\langle\vec{\nabla}\Phi^{(k)}\right\rangle$ stands for the spatial average in the region $V_\epsilon(\vec{r}_k)$ (we use this notation from now on). From (12a) we obtain, separating real and imaginary parts, the equations of motion for the kth spiral[6,7] ($\hat{z}$ is the unitary vector perpendicular to the plane of the system)

$$\frac{d\vec{r}_k}{dt} = 2\alpha\left\langle\vec{\nabla}\Phi^{(k)}\right\rangle - 2m_k\hat{z}\wedge\left\langle\vec{\nabla}\Phi^{(k)}\right\rangle. \tag{13}$$

This central result shows that the motion of the kth spiral is determined by the phase $\Phi^{(k)}$ generated by the rest of the system on its location (more precisely the interpretation of $\Phi^{(k)}$ is given after eq. (5)). We recall that $\vec{\nabla}\Phi^{(k)}$ is the slowly varying part of $\vec{\nabla}\Theta^o$ in the core $V_\epsilon(\vec{r}_k)$. Two special cases of equation (13) has been discussed previously and correspond to the fact this equation expreses $\dot{\vec{r}}_k$ as the sum of two terms: the first term proportional to $\vec{\nabla}\Phi^{(k)}$ corresponds to the limit of the nonlinear Schrödinger equation ($\alpha = \beta \to \infty$) and was found by Fetter[8], the second term proportional to $\hat{z}\wedge\vec{\nabla}\Phi^{(k)}$ corresponds to the variational case of the real Ginzburg-Landau equation ($\alpha = \beta = 0$) and was found by Kawasaki[9]. Replacing $\Phi^{(k)}$ in (13) by expression (5b) one obtains the form

$$\begin{aligned}\dot{\vec{r}}_k =&2\left\{\sum_{i\neq k}\left(\frac{m_i m_k}{r_{ik}} - \alpha\tilde{S}'(r_{ik})\right)\hat{r}_{ik} + \alpha\vec{\nabla}\Psi(\vec{r}_k)\right\} + \\ &+ 2m_k\hat{z}\wedge\left\{\sum_{i\neq k}\left(\alpha\frac{m_i m_k}{r_{ik}} + \tilde{S}'(r_{ik})\right)\hat{r}_{ik} - \vec{\nabla}\Psi(\vec{r}_k)\right\},\end{aligned} \tag{14}$$

where $\vec{r}_{ik} = \vec{r}_k - \vec{r}_i$, $r_{ik} = |\vec{r}_{ik}|$, $\hat{r}_{ik} = \frac{\vec{r}_{ik}}{r_{ik}}$. We have based on (14) our discussion of some statistical properties of the *diluted gas of spirals* in Ref. 10. This form (14) has been found independently by Elphick and Meron[11]. One should remark that equations (5a) and (5b) which have been used in the derivation of (14) can be considered as the definition of the new global phase Ψ as it is clear from the form (4b) of the *Ansatz*, but it is $\Phi^{(k)}$ which is the total phase Θ^o minus the proper phase $(m_k\varphi_k - S(\rho_k))$ of the kth spiral what appears in the equation of motion (13). In fact nothing can be concluded from (14) unless something can be shown or assumed about the function Ψ.

We proceed now to determine an element $\tilde{\chi} \in Ker\mathcal{L}^\dagger$ which vanishes outside the cores and which will determine $\partial_t\Psi$. We rewrite first (7a) and (7b) using $\omega = \Omega - \beta\mu + \nu q^2$ (see after(2))

$$\begin{aligned}\mathcal{L}w =& (1+i\alpha)\left[\nabla^2 w + 2i\vec{\nabla}\Theta^o\cdot\vec{\nabla}w + \left(i\nabla^2\Theta^o - \left(\left(\vec{\nabla}\Theta^o\right)^2 - q^2\right)\right)w\right] + \\ &+ (1+i\beta)\left[\left(D_\infty^2 - R^{o2}\right)w - 2R^{o2}\Re e\ w\right]\end{aligned} \tag{15}$$

$$I = (1+i\beta)R^o\left(R^{o2} - D_\infty^2\right) + \partial_t R^o + iR^o\partial_t\Theta^o - (1+i\alpha)\Bigg[\nabla^2 R^o +$$

$$+ 2i\vec{\nabla}R^o \cdot \vec{\nabla}\Theta^o + R^o\left(i\nabla^2\Theta^o - \left(\left(\vec{\nabla}\Theta^o\right)^2 - q^2\right)\right)\Bigg]. \tag{16}$$

Since $\tilde{\chi} \neq 0$ only in the region $M = \mathbf{R}^2 - \cup_i V_\epsilon(\vec{r}_i)$ outside the cores we need an aproximation of $\mathcal{L}w$ in this region. Using now the asymptotic behavior of functions $D(\rho)$ and $S(\rho)$ for $\rho \to \infty$ (the region outside the cores correspond to $\rho > \varepsilon$ where the functions $D(\rho)$ and $S(\rho)$ have almost attained their asymptotic values for big ρ) and neglecting gradients in a first approximation we have in M that $\mathcal{L}w\big|_M = -2(1+i\beta)R^{o2}\Re e\; w$. In matrix notation with $w_1 = \Re e\; w$, $w_2 = \Im m\; w$, and with the usual scalar product in $\mathbf{R}^2$ (notice this is not the same as before) one has

$$\mathcal{L} = -2D_\infty^2\begin{pmatrix} 1 & 0 \\ \beta & 0 \end{pmatrix}, \; \mathcal{L}^\dagger = -2D_\infty^2\begin{pmatrix} 1 & \beta \\ 0 & 0 \end{pmatrix}. \tag{17}$$

Then $\tilde{\chi} = (\beta, -1)$ and the solvability condition is $\beta\Re e I - \Im m I = 0$ which using (16) becomes

$$\partial_t\Theta^o = \gamma\nabla^2\Theta^o + \nu\left(\left(\vec{\nabla}\Theta^o\right)^2 - q^2\right) + \frac{2\gamma}{R^o}\vec{\nabla}R^o \cdot \vec{\nabla}\Theta^o - \frac{\nu}{R^o}\nabla^2 R^o + \frac{\beta}{R^o}\partial_t R^o. \tag{18}$$

This equation is valid outside the cores and consequently can only determine the part of $\vec{\nabla}\Theta^o$ which is slowly varying in the cores, but this is all that we need since only $\vec{\nabla}\Phi^{(k)}$ appears in (13). Keeping in (18) only the dominant terms outside the cores one obtains

$$\partial_t\Theta^o = \gamma\nabla^2\Theta^o + \nu\left(\left(\vec{\nabla}\Theta^o\right)^2 - q^2\right). \tag{19}$$

On the other hand we remark at this point that in (13) we can replace $\left\langle\vec{\nabla}\Phi^{(k)}\right\rangle$ by $\left\langle\vec{\nabla}\Theta^o\right\rangle$ since $\left\langle\vec{L}_k \equiv \vec{\nabla}(\Theta^o - \Phi^{(k)})\right\rangle = 0$. We can regard the coupled equations (13) and (19) as the basic equations of our problem which must be solved taking into account the appropriate boundary conditions

$$\frac{1}{2\pi}\oint_{\Sigma_j^+} \vec{\nabla}\Theta^o \cdot d\vec{\ell}_j = m_j, \tag{20}$$

where Σ_j^+ is a closed curve encircling the jth spiral. Using equations (5) we can obtain from (18) the phase equation in the region D_k

$$\partial_t\Phi^{(k)} - \dot{\vec{r}}_k \cdot \left(\vec{L}_k - \frac{\beta}{D}\vec{\nabla}D(\rho_k)\right) = \gamma\nabla^2\Phi^{(k)} + \nu\left(\vec{\nabla}\Phi^{(k)}\right)^2 +$$

$$+ 2\nu\vec{L}_k \cdot \vec{\nabla}\Phi^{(k)} + \frac{2\gamma}{D}\vec{\nabla}D \cdot \vec{\nabla}\Phi^{(k)} +$$
$$+ \frac{1}{D}\left\{\nu D\left(\vec{L}_k^2 - q^2\right) - \nu\nabla^2_{\rho_k} D - \gamma D\nabla^2_{\rho_k} S - 2\gamma D'S'\right\}. \tag{21}$$

Using equations (3) we see that the last term in brackets $\{\cdots\}$ vanishes. Neglecting terms $\mathcal{O}\left(\rho_k^{-2}\dot{\vec{r}}_k, \rho_k^{-2}\vec{\nabla}\Phi^{(k)}\right)$ we obtain

$$\partial_t\Phi^{(k)} - \dot{\vec{r}}_k \cdot \vec{L}_k = \gamma\nabla^2\Phi^{(k)} + \nu\left(\vec{\nabla}\Phi^{(k)}\right)^2 + 2\nu\vec{L}_k \cdot \vec{\nabla}\Phi^{(k)}. \tag{22}$$

Replacing $\Phi^{(k)}$ using (5b) we can obtain an equation for Ψ. However we point out that (5) is an *Ansatz* for the relevant solution of (19) satisfying the boundary conditions (20) and the justification of this *Ansatz* depends essentially on what can be said about the phase Ψ. We can consider the coupled equations (13) and (22) as describing the state of a ***diluted gas of spiral defects***. We shall write down the explicit equation for Ψ in the case of two spirals using the differentiable *Ansatz* given by (6). Neglecting terms of order $\mathcal{O}\left(\rho_i^{-3}, \rho_i^{-2}\hat{\rho}_i \cdot \vec{\nabla}\Psi, \rho_i^{-2}\hat{\rho}_i \cdot \dot{\vec{r}}_i, q\rho_i^{-2}\right)$ we obtain $(\vec{\ell}_i \equiv \vec{\nabla}(m_i\varphi_i - \tilde{S}(\rho_i)) = \frac{m_i}{\rho_i}\hat{\varphi}_i - \tilde{S}'(\rho_i)\hat{\rho}_i)$

$$\partial_t\Psi - \sum_i \vec{\ell}_i \cdot \dot{\vec{r}}_i + q\left(\hat{\rho}_1 \cdot \dot{\vec{r}}_1\tilde{\theta}(\rho_2 - \rho_1) + \hat{\rho}_2 \cdot \dot{\vec{r}}_2\tilde{\theta}(\rho_1 - \rho_2)\right) = \gamma\nabla^2\Psi +$$
$$+ \nu\left(\vec{\nabla}\Psi\right)^2 + \nu\vec{\ell}_1^2\tilde{\theta}(\rho_1 - \rho_2) + \nu\vec{\ell}_2^2\tilde{\theta}(\rho_2 - \rho_1) + 2\nu\vec{\ell}_1 \cdot \vec{\ell}_2 +$$
$$+ 2\nu\sum_i \vec{\ell}_i \cdot \vec{\nabla}\Psi - 2q\nu\left(\hat{\rho}_1 \cdot \vec{\ell}_2\tilde{\theta}(\rho_2 - \rho_1) + \hat{\rho}_2 \cdot \vec{\ell}_1\tilde{\theta}(\rho_1 - \rho_2)\right) +$$
$$- 2q\nu\vec{\nabla}\Psi \cdot \left(\hat{\rho}_1\tilde{\theta}(\rho_2 - \rho_1) + \hat{\rho}_2\tilde{\theta}(\rho_1 - \rho_2)\right) +$$
$$+ 2\tilde{\delta}(\rho_1 - \rho_2)(1 - \hat{\rho}_1 \cdot \hat{\rho}_2)\left(q\gamma + \frac{\nu}{2D_\infty}(D'(\rho_1) + D'(\rho_2))\right), \tag{23}$$

which reduces to (19) in the regions $D_1(\rho_2 > \rho_1)$ and $D_2(\rho_1 > \rho_2)$ (in (23) $\tilde{\delta}(x) = \tilde{\theta}'(x)$).

We calculate now the correction w. It is easy to show that $w = 0$ inside the cores $V_\epsilon(\vec{r}_j)$ while outside one obtains from $\mathcal{L}w = I$ using (17) that $\Re e\ w = -(2D_\infty^2)^{-1}\Re e\ I$, $\Im m\ w = 0$ [7]. $\Re e\ I$ is obtained from (16) and using equations (6) we can see that its most important part will come from the term $\nabla^2\Theta^o$ which contains a δ-function type contribution

$$w = -\frac{\alpha q}{D_\infty}(1 - \hat{\rho}_1 \cdot \hat{\rho}_2)\tilde{\delta}(\rho_1 - \rho_2). \tag{24}$$

In order to estimate the width of de $\tilde{\delta}$ fuction we use the fact that numerical experiences[12] indicate that $\frac{\Re e\ w}{D_\infty} \approx 0.1$ in the region where it is maximal, i.e. in the line of action between the two spirals where $\hat{\rho}_1 \cdot \hat{\rho}_2 = -1$. If the height of $\tilde{\delta}$ is h its width is of order h^{-1} and from (24) we have $h^{-1} = -\frac{2\alpha q}{0.1 D_\infty^2}$. When $\gamma > 0$ we can determine q using the results of Hagan[3] as we have done in Ref. 7 and we obtain

$$h^{-1} = -\frac{2\alpha\gamma}{0.1}\frac{q_H \lambda_H}{(1+\alpha^2)(1-q_H^2)^{3/2}}\varepsilon, \tag{25}$$

where ε is the radius of the core, q_H is given numerically in Ref. 3 as a function $q_H(s)$ of $s = \frac{\nu}{\gamma}$ and λ_H is known at $s = 0$ to be such that $D(\rho) = \lambda_H(s = 0)\rho$, $\rho \to 0$, in the case $\alpha = \beta = 0$ in (1) (the usual vortex solution of the real Ginzburg-Landau equation). One knows $\lambda_H(s = 0) \approx 0.5$ and we use this value. If we take $\alpha = -0.5$, $\beta = 0.5$, $\gamma = 0.75$, we obtain $h^{-1} \approx 0.8\varepsilon$, i.e the width of the region where the term (20) is important is of the order of the diameter of the core which agrees with numerical observations[13].

Following a proposal of P. Coullet we have considered equation (1) when one adds to it a small constant real term g. Around the approximate solution $\sqrt{\mu} exp(i\omega' t)$, $\omega' = \Omega - \beta\mu$, the phase equation for constant perturbations ϕ, $A = (\sqrt{\mu} + \rho)\, exp(i\phi)$, is $\partial_t \phi = \omega' - g' sin(\phi + \delta)$, $g' = g\sqrt{\frac{1+\beta^2}{\mu}}$, $sin\delta = \frac{\beta}{\sqrt{1+\beta^2}}$, $cos\delta = \frac{1}{\sqrt{1+\beta^2}}$, and has two fixed points for $|g'| > |\omega'|$, one stable and one unstable, a situation arising in excitable media. We are interested in the motion of one single spiral when (1) is modified by the constant g, and accordingly we write instead of (4a) for a spiral with topological charge m the *Ansatz*

$$A = (D(\rho) + w)\, exp\left\{i\left(\omega t + m\varphi - S(\rho) + \Psi\right)\right\}, \tag{26}$$

where $\vec{\rho} = \vec{r} - \vec{r}_o(t)$, $\rho = |\vec{\rho}|$, $\vec{r}_o(t)$ is the position of the spiral and $\Psi(\vec{r}, t)$ is the new global phase. Proceeding in the same way as before we obtain now the equations of motion (with δ as above and $\vec{\nabla}\Psi(\vec{r}_o)$ has to be interpreted in the average sense indicated in (12))

$$\dot{\vec{r}}_o = 2\alpha\vec{\nabla}\Psi(\vec{r}_o) - 2m\hat{z}\wedge\vec{\nabla}\Psi(\vec{r}_o) - \frac{g}{\lambda}\left(cos(\omega t + \Psi)\hat{e}_1 - m\, sin(\omega t + \Psi)\hat{e}_2\right) \tag{27a}$$

$$\begin{aligned}\partial_t \Psi - \vec{L}_o \cdot \dot{\vec{r}}_o =& \gamma\nabla^2\Psi + \nu\left(\vec{\nabla}\Psi\right)^2 + 2\nu\vec{L}_o\cdot\vec{\nabla}\Psi \\ & - g\frac{\sqrt{1+\beta^2}}{D_\infty} sin\left(\omega t + m\varphi - S(\rho) + \delta + \Psi\right),\end{aligned} \tag{27b}$$

where $(\hat{e}_1,\hat{e}_2)$ are the unitary vectors of the cartesian coordinates in the plane and $\vec{L}_o \equiv \vec{\nabla}(m\varphi - S(\rho)) = \frac{m}{\rho}\hat{\varphi} - S'(\rho)\hat{\rho}$. Equation (27b) has to be solved outside the core where ρ is big, then up to first order in the small perturbation g and neglecting terms $\mathcal{O}(\tilde{g}q, \tilde{g}\rho^{-1})$ since q is small, $\tilde{g} = \frac{g\sqrt{1+\beta^2}}{\omega D_\infty}$, one has the solution

$$\Psi_o = \frac{g\sqrt{1+\beta^2}}{\omega D_\infty} \cos(\omega t + m\varphi - S(\rho) + \delta). \tag{28}$$

Replacement of this solution in (27a) shows that the center of the spiral $\vec{r}_o(t)$ performs a circular motion with the frequency ω of the spiral. We can also calculate the correction w as before. One finds $\Im m\ w = 0$ and

$$\Re e\ w = -\frac{1}{2D_\infty}\left(\left(\vec{\nabla}\Psi\right)^2 + 2\vec{L}_o \cdot \vec{\nabla}\Psi + \alpha\nabla^2\Psi\right) - \frac{g}{2D_\infty}\cos(\omega t + m\varphi - S(\rho) + \Psi). \tag{29}$$

Using the solution (28) and replacing $S(\rho) = q\rho + (smaller\ terms)$ one has to leading order

$$w = -\frac{g}{2D_\infty}\cos(\omega t + m\varphi - q\rho), \tag{30}$$

which implies according to (26) a spatially periodic modulation of the amplitude which has been observed[13]. One should notice that (28) tells us that this approximate solution can only be trusted for $\tilde{g} = \frac{g\sqrt{1+\beta^2}}{\omega D_\infty} < 1$. This condition can be written

$$\frac{g'}{\omega'} \frac{1}{\left(1 + \frac{\nu q^2}{\omega'}\right)\sqrt{1 - \frac{q^2}{\mu}}} < 1$$

and is compatible with $\frac{g'}{\omega'} > 1$ for $\omega' = \Omega - \beta\mu$ sufficiently small.

Acknowledgements
The authors are grateful to P. Coullet, K. Emilsson, L. Kramer, J. Lega and D. Walgraef for many interesting and stimulating discussions. They also thank FONDECYT and DTI for financial support. E.T. aknowledges the support of a J.S. Guggenheim fellowship.

References

[1]D. Walgraef, G. Dewel, and P. Borckmans, J. Chem. Phys. **78**, 3043 (1983).
[2]P. Coullet, L. Gil, and J. Lega, Phys. Rev. Lett. **62**, 1619 (1989).
[3]P.S. Hagan, SIAM J. Appl. Math. **42**, 762 (1982).

[4]Y. Kuramoto, *Chemical Oscillations, Waves and Turbulence*, edited by H. Haken (Springer-Verlag, New York, 1984).
[5]S. Koga, Prog. Theor. Phys. **67**, 164 and 454 (1982).
[6]S. Rica and E. Tirapegui, Phys. Rev. Lett. **64**, 878 (1990).
[7]S. Rica and E. Tirapegui, *Analytical description of a state dominated by spiral defects in two dimensional extended systems*, to appear in Physica **D**.
[8]A.L Fetter, Phys. Rev. **151**, 100 (1966).
[9]K. Kawasaki, Prog. Theor. Phys. Supp. **79**, 161 (1984).
[10]H. Calisto, S. Rica and E. Tirapegui, *Statistical properties of a diluted gas of spiral defects.* Preprint U. de Chile, F.C.F.M. submitted to J.Stat.Phys. (March 1990).
[11]C. Elphick and E. Meron, *Spiral vortex interaction.* Preprint U. Tec. Federico Santa María (Valparaíso) to appear in Phys. Rev. Lett.
[12]J. Lega, *Defects and Defect-Mediated Turbulence*, in Patterns, Defects and Materials Instabilities, D. Walgraef and N.M. Ghoniem Eds., Kluwer 1990.
[13]P. Coullet and K. Emilsson, private communication.

FRAGMENTATION

Etienne Guyon*
Laboratoire de Physique et Mécanique Hétérogene,
U.R.A. CNRS 857, Ecole Supérieure de Physique et
Chimie Industrielles de Paris,
10 rue Vauquelin, F-75231 PARIS Cédex 05, France

ABSTRACT. The statistical physics of disordered systems has been widely applied to a large number of classes of material (porous media, composites, suspensions, gels, granular media). We choose here to discuss the problem of compression of grains and of subsequent fragmentation. We show first that the heterogeneity in the contact distribution leads to a highly non linear response to deformation. The breaking characteristics should be influenced by the distribution of the subset of lines of stressed bonds. We show that the general features of the family or percolation problems apply to such systems. Finally we present some results of the breaking of a simple object and show how even weak heterogeneities control the large evolution of the system.

1. INTRODUCTION

1.1. Physics of disorder

The development of the field of "ill condensed matter" in the twenty last years (1), with the introduction of such concepts as fractals and multifractals, percolation, localisation, has had and appreciable impact on material science. In particular, whereas most models dealing with heterogeneous and disordered materials used homogeneisation technics where various averaging procedures tend to average the local fluctuations, the new concepts took into account explicitely, and emphasized, the role of disorder. Let us quote a few applications to which we have been associated and where review articles are available:

---

* Also at Palais de la Découverte, Education Nationale, Avenue Franklin Roosevelt, 75008 Paris.

E. Tirapegui and W. Zeller (eds.), Instabilities and Nonequilibrium Structures III, 347–355.

- electrical properties of composite materials (2)
- agregation of particles (3)
- concentrated suspensions (4)
- miscible and immiscible two phase flows in porous materials
- mechanics of granular systems (6)
- fracture of solid materials (7).

The four last examples involve the role of non linear local processes in the evolution of the global disorder as a function of the external solicitations. In the present article we develop the particular application of the fragmentation of grains which is a fairly open subject for statistical physicists and whose study combines some of the results obtained in the different examples listed above.

1.2. Fragmentation

Fragmentation occurs in nature at very different scales:

- in nuclear physics, the bombardment of heavy ions by projectiles (protons; other nuclei) very energetic (1 GEV) leads to distribution of fragments whose mass distribution has been analysed in scaling terms(8);
- at the other extreme astronomers are interested by fragmentation in the mechanims of formation of asteroids(9);
- there exists indeed a large range of intermediate scales to which we will be particularly insterested:

• reactions of degradation (or depolymerisation) of polymers play the study of their stability and applications(10). Various causes are involved (UV irradiation, heat, heavy particle bombardment, hydrolysis). Here again fragment size distribution are easily obtained.

• rupture of aggregates occur from atomic scale up to colloidal size particles and has various causes. We can add to the list the fragmentation of droplets: emulsification, in chemical engineering leads to larges increases of interfacial and, thus, of reactivity or adsorption.

• finally fragmentation of solid matter ranges from the fragile rupture of objects to the full range of processes of reduction of sizes of granular matter into finer grains. It is involved in flotation mechanisms of rocks to collect minerals present in divided form. In the cement industry it takes place both in the crushing of minerals involved before oven treatment and in the crushing of clinker produced which lead finally to cement powder. The processes are heavy energy consumers due to the very small efficiency (less than 10%) of the process (a unit rate would correspond to using the full energy provided as energy surface for the creation of fragments).

2. MODELISATION OF FRAGMENTATION

Beyond the variety of problems and of systems studied a number of basic common results can be looked for:

- geometrical characterisation of cluster shapes and sizes; establishment of static geometrical models using statistical tools such Poissonian schemes developped by Matheron, tesselations or simply combinatory analysis (8). A simple question is how to relate geometrically a distribution of fragments to the original intact structure? However this is more easily approached by treating the kinetics of the process which we will only consider below.
- formulation of scaling relations for the distribution of fragments as a function of time evolution.
- use of kinetic models and incorporation of geometrical characteristics in the model.
- consideration of the geometrical characteristics of the packing before and during crushing.
- fragmentation of a simple grain.

2.1. Scaling law for fragment distribution

Many physical examples exist where, at a given time, exists a disordered ensemble of particles of various sizes. Following M. Fisher's droplet model in statistical physics, such distributions have been characterized by general scaling laws with a control external variable (11). It can be the probability p of existence of a bond in a random lattice. Beyond a percolation threshold p_c there exist only finite clusters and the number density of clusters of mass s (number of sites of a cluster) can be written under the form (12)

$$r_s(P) \propto s^{-\tau} f(s/(p - p_c)^{\sigma}) \quad (1)$$

The exponent τ characterizes the decrease of the number of large clusters as a function of their size (mass). The function f describes the effect of upper cut off in size (when p reaches p_c, the power law form extends to infinity, which is characteristic of a fractal distribution). The exponents σ and τ as well as the form of f are independent of the local characteristics. Form (1) was used in particular in the analysis of Campi of nuclear fragments. He considered that fragmentation by impact amounts to breaking at random fraction of bonds. He found a reasonable agreement with percolation for an intermediate range of particle energies (large ones lead to a complete destruction into small fragments whereas small ones lead to the equivalent of surface erosion).

Similar laws can be expressed if the distribution evolves in time t using the time variable t insted of p:

$$n_s(t) \alpha s^{-\tau} f(s/t^{\gamma}) \quad (2)$$

This law has been used extensively in the study of cluster aggregation with two different limit forms observed in nature. A first type of variation of $n_s(t)$ corresponds to a continuous decrease of the number of clusters as time (and size) grows. A second one possesses a maximun for an intermediate range of sizes which itself increases with t. This is obtained in particular if the mobility of small clusters is larger that of large ones as they disappear faster by aggregation. The results are well illustrated by the review article of Jullien(3) and the film of Kolb(13).

Percolation and fractal aggregation models are two possible approaches for colloidal growth phenomena and for gel formation(14).

The extension to the problem of fragmentation is at hand and should lead to a better description of the statistical laws of distribution of fragments:

- in percolation one can increase continuously by the number of bonds broken with possible local correlations: numerical analysis by Aharony and Englman are along this line.
- in a kinetic description, positive time scale corresponds to decreasing fragment sizes, and the transposition of the aggregation models to fragmentation is not so direct. In particular the formation on an -intensive- gel phase which corresponds to the fact that a finite fraction of the total mass is in a given cluster which extends throughout the volume corresponds, in fragmentation, to the existence of a finite fraction of elements in grains of zero mass (this is indeed an unphysical limit and there exist a lower cut off of granules which depends on the crushing process as well as on the material properties).

2.2. Kinetic models

The classical model of Smoluchovski describes discuss the aggregation of particles of size m_i using a set of first order differential equations:

$$\frac{dmi}{dt} = \sum_{j+k=i} k_{jk}\, m_j m_k \;-\; m_i \sum_k k m_k \tag{3}$$

The first term to the right expresses the collision of two particles of size j and k to give one of size i=j+k, the second one the disapearence of particles of size i as they stick to particles of various sizes k.

Similarly there exists a widely used kinetic equation for crushing developed in particulary by Austin and his group (16):

$$\frac{dmi}{dt} = - S_i\, m_i(t) + \sum_{j>i} S_j b_{ij} m_j(t) \tag{4}$$

The first term expresses the decrease of particles of mass m_i by fragmentation. The coefficient Si can be evaluated by considering the initial rate of decrease of a monodisperse population of particles of given mass m_i; one generally finds that S_i remains constant along the process (which justifies the form of the term) if the grains are sufficiently small compared to the scale of the crushing elements. The second term shows how grains of size i form out of larger size grains j.

This approach is complementary from that of 2.1. and the knowledge of the variation law of S_i and b_{ij} with size can lead to the knowledge of the scaling laws for fragment size distribution. A significative step along this line has been made by Cheng and Redner (17) and calls for an experimental program with a well controled time evolution study.

Let us note the correspondances with systemic approaches in chemical engineering (18). The behavior of a crushing system like that of a chemical reactor involves a series of inputs and outputs and can be analysed using such concepts as distribution of residence times; at a fine level the notions of micro and macro mixing should be replaced by the concepts of time needed for breaking and average transit time in the machine. However for the more modern crushing machines where grain pass between two counter rotating horizontal cylinders under high pressure the meaning of a rate equation can be questionned as the full breaking operation takes place in the very short time of passage of the grains between the role. More generally in all the tools presented above, one neglects the geometrical organization of the individual particles. In this sense these models are "mean field". The question of the role of the local environment versus that of the averaged macroscopic properties is still open.

2.3. Fragmentation of a packing of grains

Breaking dense assemblies of unconsolidated grains is involved in many industrial processes as indicated in 1.2. The classical ball mill crusher where the breaking is due to heavy spheres put in rotation in a large rotating cylinder with the grains to be broken are being supplanted presently by roller mills where the grains are fed between two horizontal axis counter rotating cylinders. This last technique raises the problem of compaction and breaking of grains submitted to a quasi uniaxial compression. It may look, at first view, as if such a material can be treated using an homogeneous equivalent medium but a rapid consideration of the problem (just think of the breaking of one or of two walnuts in one hand) emphasizes the importance of the heterogeneties. We will first consider the effect of a statistical distribution of cylinders and, in the next paragraph, we will make some remarks on the opening of a simple grain.

Schnebili samples are made of packings of parallel axis cylinders and have been widely used to model the mechanical properties of granular arrays. In particular, if the cylinders are made of photoelastic material, they reveal the lines of dominant stresses when illuminated between crossed polaroids (21). One sees a loose lattice of bifurcating lines which progressively enriches as the pressure applied to the system increases. In particular a subset of the cylinders are completely loose and can be eliminated without any change of the characteristics thanks to arching by the surrounding stressed cylinders.

An appreciable and systematic basic experimental effort has been carried in particular around Bideau in Rennes to study the geometry and the mechanical properties of these systems by varying parameters such as size distribution, mixing of soft and hard cylinders, dilution and has led to a renewed inderstanding of classical problems of civil engineering. Quite generally it is found that, due to the progressive enrichment of the geometry with applied force (F), the displacement (d) characteristics is highly non linear

$$F \propto (d - d_o)^{\Delta} \qquad (5)$$

with a Δ exponent as high as 5 whereas the local characteristics for the contact between two grains (Herz contact) is only weakly non linear ($\Delta \sim 1.5$).

This problem has received detailed theoretical analysis using the concepts of central force and diode percolation, minimal path analysis and we refer to the review of reference (6) and to the thesis of Roux (22) for a detailed description.

Let us just mention a simple table top experiment which illustrates a basic feature of the problem (23). A regular intact lattice is made by soldering Zener diodes biased along the horizontal direction whose individual characteristics are given approximately by

$$i = 0 \qquad \theta<\theta_{cj}$$
$$i = c(\theta-\theta_c) \qquad \theta\geq\theta_{cj}$$

The voltage θ_{ci} is distributed randomly from diode to diode.

For an applied voltage V to the lattice of diodes such that for V > Vs the characteristics is linear as all the diodes are in the active states. For $V_c<V<V_s$ the characteristics is roughly parabolic unlike that of the individual elements. This comes from the fact that, when the voltage varies between V and V+dV, a fraction of the diodes passes from a passive to an active state. In a mean field like treatment this number dn, or the conductance dΣ, should increase as dV. And the total current should increase quedratically with the applied voltage.

The correspondance with the granular problem arises from the fact that a force carrying contact exists between two neighboring grains only if they are in compression (that is for one sign of the inter grain displacement). The progressive enrichment of contacts and the non linear stress-strain relation, which has also been directly studied numerically (24) is the analog of the non linear electrical characteristics.

Let us note, en passant, the analogy between V_c and a percolation threshold. V_c is the smallest voltage for which a continuous path of active diodes is formed across the system. It is given by

$$V_c = \min_{\text{all patterns } k} \left[\sum_{j\in k} \theta_{cj} \right] \qquad (6)$$

In a percolation problem characterized by a full lattice where each bond would be labelled by a random number p_j ($0<p_j<1$) the threshold is got by considering all continuous paths k, on the largest index met on each path and on the lowest of these last indices.

$$p_c = \min_{\text{all patterns } k} \left[\max_{j\in k} p_j\right] \qquad (7)$$

In fact the two relations can be seen as extrme forms of a minimal path analysis corresponding to extreme weightings (equal weight in first case, infinite moment of the distribution in the second one). This result also shows that percolation like concepts can apply beyond classical problems with dilute systems near a rigidity threshold (an extreme case for most mechanical applications) provided the local laws are non linear (25).

This analysis is being used for an experimental research program in our laboratory as it obvious that the dominant lines of stresses will be involved in the first stage of breaking before any reorganization of the fragment takes place. We consider in particular:

- The role of size distribution: large cylinders in Schnebeli packings are found to "localize" the stresses. On the other hand, a continuous distribution of fine grains can be expected to give a more even force distribution and oppose the needed localization of stress.
- The role of the kinetics in grain displacement. It has been observed that additional motions transverse to the force field leads to fast changing distribution of lattices stresses and, thus provides better opportunity for breakings (26).
- Mixtures of grains of different strengths, use of vibration are additional parameters which should be considered. It is to be emphasized again that this is a field where even small quantitative uncreases in efficiency mean large savings, and basic and applied researchers should combine their efforts!

2.4. Breaking of a simple object

This subject by itself could deserve a full treatment and recent approaches of statistical physics have strongly emphasized the role of the statistic of extreme in the understanding of breaking processes. In an otherwise perfectly uniform foil of size LxL a single crack transverse to applied tension force applied will control the breaking properties of the foil.

The general philosophy is also present in the results of numerical treatments of breaking lattice models of materials with various lattice structures and local breaking characteristics but with certain (even small) randomness introduced in each bar.

In the first stages of breaking the cracks can develop in a more or less uniform fashion depending on the distribution of local heterogeneities. In particular, for very large distribution of breaking strenghts and a uniform distribution of bars, the bars will break according to the increasing order of breaking strength like in a percolation problems where bonds are cut at random.

However, as fracture progresses, correlations develop due to the stress intensity factor which deals with the modification of stress field induced by a given crack, and multipolar type descriptions would be more appropriate.

The final stage is one where a dominant crack develop throughout the full system leading to the final disociation of the materials.

The intermediate stage could be considered as a ductile process whereas the latter one corresponds to localization of a crack. The simulation carried on two dimensional lattices of various LxL suggests a scaling form for the full force (F) and the displacement (d) characteristics as F/L^x and d/L^y, where $x \sim y \sim 0.7; 0.8$ for a number of local geometries and forms of rupture laws. This indicates that:

- the larger the size of a system, the more fragile it is. In the case of a contact macroscopic rupture stress, x and y would be 1. The larger the system, the more statistically probable it will be to contain a large fluctuation of disorder which will pilot the system.
- the relative "universality" of the result remains to be tested on 3D simulations and indeed on experimental systems of the same material of various scales.

Acknowledgments. The ideas developped in this article are largely due to work carried in common with Daniel Bideau and his group and with Stephane Roux. A support from Lafarge Coppée is also acknowledged.

REFERENCES

1) Chance and Matter, edited by J.Souletie, J.Vannimenus & R.Stora, North Holland (1987).

2) ETOPIM 2, édit. J.Lafait & D.B.Tanner, North Holland (1989).

3) R.Jullien, Ann. télécom. 41 343 (1986).

4) G.Bossis, R.Blanc, C.Camoin & E.Guyón, J. de Méca., N° spécial,p.141 (1989).

5) E.Guyon, C.D.Mitescu & S.Roux, Physica D 38 172 (1989).

6) E.Guyon, S.Roux, A.Hansen, D.Bideau, J.P.Troadec & H.Crapo, Rep. Progr. in Phys. 53 373 (1990).

7) H.Herrmann & S.Roux, Statistical models for the fracture of disordered solids, North Holland (1990).

8) X.Campi,Phys.Lett. B208,351 (1988); Nucl. Phys. A495, 2590 (1989).

9) J.S.Dohnayi, J. of Geophys. res. 75 3468 (1970).

10) R.Ziff & E.D.Mc Grady, J. of Phys. A 18 3027 (1985).

11) M.Fisher, Physics 3 255 (1967).

12) D.Stauffer, Introduction to percolation theory, Taylor & Francis London (1985).

13) M.Kolb, Agregation film 16 mm, 30 minutes.

14) B.Joulier, C.Allain, B.Gauthier & E.Guyon in "percolation and structures" Ann. Isr. Phys. Soc. 5 167 (1983).

15) A.Aharony, A.Levi, R.Englmen & Z.Jaeger, Ann. Isr. Phys. Soc. 8 112 (1986).

16) Austin L. & coworkers, Ind. Eng. Chem. Proc. 15 187 (1976).

17) Cheng Z. & Redner, Phys. Rev. Lett. 62 2321 (1989).

18) Villermaux J., Génie de la réaction chimique, Lavoisier Paris (1985).

19) In the ball mill, heavy and hard spheres are hitting the grain in a long and inclined cylinder containing the grains. In this continuous process the grains become progressively smaller along the axis of the cylinder.

20) In the roller mill the grains are brought between two counterrotatin cylinders. The operation takes place in one step where the grains are broken and form a weak powder aggregate.

21) T.Travers, M.Ammi, D.Bideau, A.Gervois, J.C.Messager & J.P.Troadec. Europhys. Lett. 4 329 (1987).

22) S.Roux, "Désordre et structures", These E.N.P.C. Paris (1990).

23) A.Gilabert, E.Guyon, S.Roux & Ben Ayad, Jour. de Phys. 49 1629 (1988).

24) D.Stauffer, H.J.Herrmann & S.Roux, J. de Phys. 48 347 (1987).

25) S.Roux, A.Hansen & E.Guyon, J.Phys. 48 2125 (1987).

26) M.Van der Born, Mechanical strength of porous catalyst grains (preprint).

QUALITATIVE THEORY OF DEFECTS IN NON-EQUILIBRIUM SYSTEMS

Pierre Coullet, Kjartan Pierre Emilsson, Frédéric Plaza

Institut Non Linéaire de Nice
Université de Nice-Sophia Antipolis
UMR CNRS 129
Parc Valrose
06034 NICE
FRANCE

1. Introduction

Defects play an essential role in the study of complex spatio-temporal dynamics in systems driven far from equilibrium. Evidence comes both from experimental and numerical work [1]. Defects arise naturally in physical or chemical systems which undergo symmetry breaking transitions. Special solutions of Partial Differential Equations (PDE) which model these singular objects have received a great deal of interest during the last decade. Actually, there are only few examples for which analytic expressions are available. In the context of equilibrium physics, a powerful qualitative theory of defects has been introduced in the seventies [2]. It uses the fact that the non-symmetric states of a system which are invariant under some group of symmetry have a non-trivial topology. The homotopy group of the manifold of these states allows to classify defects [3]. To some extent, these ideas have been generalized to non-equilibrium defects [4]. Such a pure topological approach, where defects are considered as singularities, gives no information about the inner structure of the defects nor of their dynamics.

In this paper we will sketch a simple approach to this problem which uses the theory of dynamical systems. General ideas will be discussed in the framework of reaction-diffusion equations. These allows us to introduce the elementary structures of kinks and vortices. These ideas will then be used to suggest a qualitative theory for the bifurcation of defects.

E. Tirapegui and W. Zeller (eds.), Instabilities and Nonequilibrium Structures III, 357–363.

2. Reaction-diffusion equations

Let $U(t;\vec{r}) = U_1, U_2, \ldots, U_N$ be a set of N fields which obey the equations:

$$\partial_t U = f(U) + D\nabla^2 U \tag{1}$$

where D is assumed, for simplicity, to be a positive diagonal matrix, and the reaction part, i.e. $\partial_t U = f(U)$ is a smooth dynamical system [5]. Let $U(0;\vec{r})$ be a set of smooth initial conditions. The question we want to answer is the following: Under what conditions do the dynamics of Eq.(1) naturally lead smooth initial conditions to evolve towards singular (defect-like) states? Although this question seems to be of a very general nature, it is possible to identify one of the possible causes of the formation of spatial inhomogeneities. Assuming that Eq.(1) is invariant under some symmetry group, and that non-symmetric asymptotic solutions exist (broken symmetry), these inhomogeneities will eventually evolve into defects. The basic reason for the appearance of singular solutions in solving Eq.(1) with smooth initial conditions is the existence of multiple attractors, which is again a direct consequence of the broken symmetry. The boundaries of the basins of attraction of the attractors generalize the notion of separatrix in dimensions higher than 2. These manifolds are actually the stable manifold of some unstable invariant set. Singularities will develop when the set corresponding to initial conditions tranversally cut one of these manifolds.

The structure of Eq.(1) (Reaction-Diffusion equation) allows a decoupling between space and time. In the singular limit $D \to 0$, Eq.(1) becomes a finite dimensional dynamical system:

$$\partial_t U = f(U) \tag{2}$$

Let C be one of the invariant set of the flow associated with Eq.(2). For our purpose of describing defects, C is either a fixed point, a circle or a torus. Let us introduce $W_S(C)$ and $W_U(C)$, as the stable and unstable manifolds of C, respectively. Since C belongs both to $W_S(C)$ and $W_U(C)$, one has:

$$\dim W_S(C) + \dim W_U(C) = N - \dim(C) \tag{3}$$

The set of smooth initial conditions $U(0;\vec{r})$ induces a mapping G from the physical space $\Re^d$ into the phase space of Eq.(2) (see Fig.1).

Figure 1. Mapping from physical space to phase space.

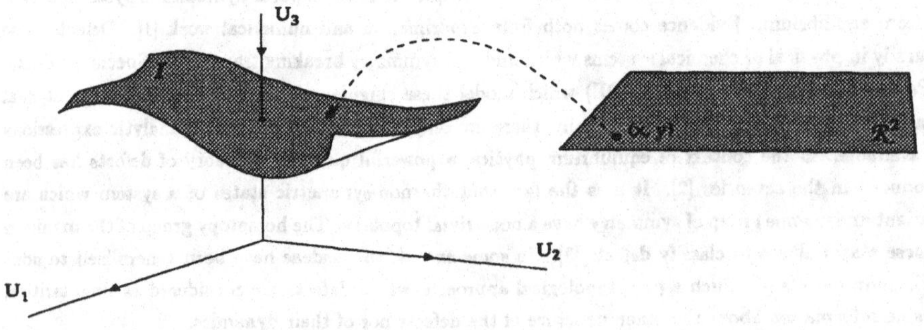

This mapping defines the manifold of initial states $\mathcal{I}$ ($\dim \mathcal{I} = d$). Whenever the intersection between $\mathcal{I}$ and $W_S(\mathcal{C})$, $\Sigma = \mathcal{I} \cap W_S(\mathcal{C})$ is non-trivial, spatial inhomogeneities are expected to appear in the early stage of the dynamics. Points in the phase space belonging to Σ flow towards $\mathcal{C}$, while nearby points flow elsewhere. In the physical space the corresponding points lie on a set $\mathcal{D}$ ($\mathcal{D} = G^{-1}(\Sigma)$) which dimension is the same as the dimension of Σ. In N-dimensional space, two manifolds $\mathcal{A}$ and $\mathcal{B}$ generically intersect whenever:

$$\dim(\mathcal{A} \cap \mathcal{B}) = \dim(\mathcal{A}) + \dim(\mathcal{B}) - N \tag{4}$$

is a positive number. If $\dim(\mathcal{A} \cap \mathcal{B})$ computed from this formula turns out to be negative, as for example in the case of two lines in three dimensions, the intersection between $\mathcal{A}$ and $\mathcal{B}$ is not generic, i.e. one or several parameters are necessary, in general, to insure that an intersection between $\mathcal{A}$ and $\mathcal{B}$ exists.

Let us apply these simple geometrical considerations to the problem of finding the nature of the spatial inhomogeneities associated with Eq.(1). Let us compute the dimension of Σ, i.e. the intersection between $\mathcal{I}$ and $W_S(\mathcal{C})$:

$$\dim(\Sigma) = \dim W_S(\mathcal{C}) + \dim(\mathcal{I}) - N \tag{5}$$

We are only interested in cases where $\dim(\Sigma) = \dim(\mathcal{D}) > 0$. Using the fact that $\dim(\mathcal{I}) = d$ one has:

$$\text{Codim}(\mathcal{D}) = \text{Codim } W_S(\mathcal{C}) \equiv n_U(\mathcal{C}) \tag{6}$$

where $\text{Codim}(\mathcal{D}) \equiv d - \dim(\mathcal{D})$, $\text{Codim } W_S(\mathcal{C}) \equiv N - \dim W_S(\mathcal{C})$ and $n_U(\mathcal{C})$ stands for the number of unstable directions of the invariant set $\mathcal{C}$. With these new definitions, the condition for a generic intersection becomes $\text{Codim}(\mathcal{D}) \leq d$. The classification of the singularities which emerges from this simple geometrical analysis is the following. Since the codimension of the set in the physical space where spatial inhomogeneities are expected to occur is given by n_U, only $n_U = 1, 2, 3$ are of interest. Furthermore, having in mind the practical difficulties to study experimentally 3-dimensional physical or chemical systems, we will restrict our attention to the cases $n_U = 1$ and $n_U = 2$.

3. Examples

For the sake of simplicity, let us choose $N = 3$ and let us consider the case of an unstable fixed point. Actually the dimension N of the phase space does not really matter, as the nature of the singularity will depend only on $\mathcal{C}$. As already discussed above, defects are natural consequences of the broken symmetry phenomenon. In the following, symmetry considerations are used to restrict the phase portrait associated with Eq.(2).

Let us first consider the case $n_U = 1$. In Fig.2, a typical phase space portrait has been sketched. Symmetry considerations enter into the problem in the following way: One assumes the origin O to be symmetric. This implies that its stable manifold $W_S(O)$ is also symmetric. Consequently the symmetry which is of interest in this case is the reflection with respect to $W_S(O)$ and thus $W_S(O)$ is a plane ($U_1 - U_3$ in the Fig.2).

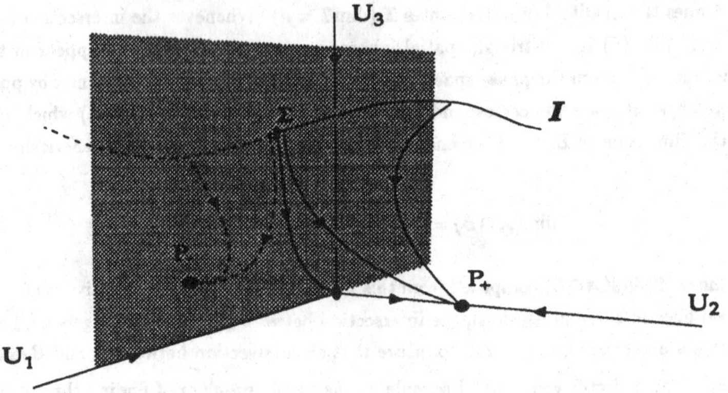

Figure 2. Phase space with two stable fixed points.

The existence of two stable fixed points $P_{\pm}$ has been assumed, which U_2 components are $\pm U^*$. Such stable solutions obviously break the reflectional symmetry. Whenever the set of initial conditions cuts $W_S(O)$, the corresponding points flow toward the stable fixed points. In the real space this singular behaviour is thus expected to occur on a point in one-dimensional systems, lines in two-dimensions, and surfaces in three-dimensions, i.e, on a set of codimension 1, as predicted.

In one-dimension, the plot of U^* as a function of the space variable x looks like a step function (see Fig.3a). A typical effect of the diffusion ($D \neq 0$) is to smooth this interface (see Fig.3b). More subtle effects of the diffusion can be expected when Eq.(2) is not a gradient flow. This simple geometrical picture allows one to understand the formation of kinks from smooth initial conditions. One also learns from this example that the core of the defect "sits", in some sense, on the unstable symmetric state.

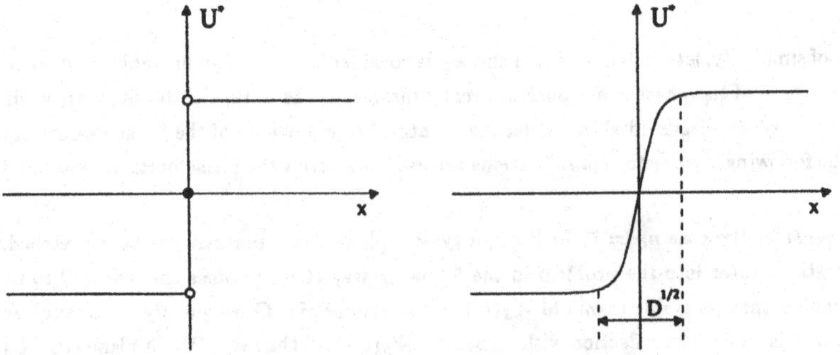

Figure 3a and 3b. Kink without and with diffusion

This lesson is far more general and applies for all defects charaterized by a core. This remark is at the root of a possible theory for defect bifurcation, since any bifurcation of the corresponding unstable state is likely to turn into a bifurcation for the defect itself. Let us now consider the case $n_U = 2$, which typical phase portrait has been sketched on Fig.(4). The stable manifold of the origin is now the U_3 axis and the natural symmetry is the rotational symmetry around this axis.

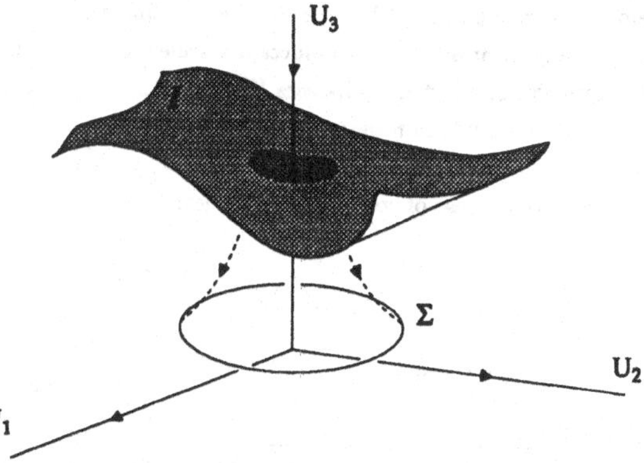

Figure 4. Phase space with one limit cycle

The existence of a stable circle P of radius U^* has been assumed. The unstable eigenvalue of O can be either real or complex. This difference will be discussed later. A point on the circle obviously breaks the rotational (phase) symmetry. Whenever the manifold of initial conditions $\mathcal{I}$ cuts transversally $W_S(O)$, i.e. the U_3 axis, the corresponding points flow toward the unstable (O) solution, while all other points flow toward the stable circle. Let us introduce $R = \sqrt{U_1^2 + U_2^2}$ and $\varphi = \arctan(\frac{U_1}{U_2})$.

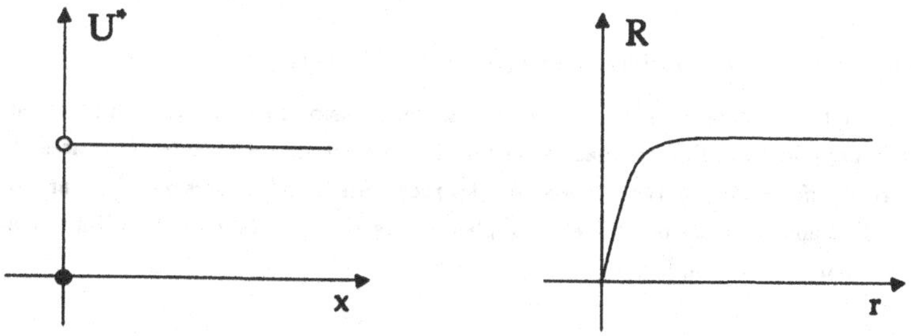

Figure 5.a and 5.b Modulus of vortex without and with diffusion

In the case $d = 2$, we have plotted in the Fig.5(a,b) the asymptotic value of R as a function of r (the radial coordinate from the singularity) without diffusion ($D = 0$) and with diffusion ($D \neq 0$), respectively. Fig.6 shows sketches of the equiphases ($\varphi = C^{st}$) in the x-y plane centered at the singularity. In the case of real eigenvalues, the stable circle is expected to be a circle of fixed points, the equiphases are radial (see Fig.6a) and clearly indicate the presence of a phase singularity. The effect of diffusion (Fig.6b) is to make this picture rotationally invariant. In the case of complex eigenvalues (Hopf bifurcation) the stable circle is a limit cycle, and the equiphases still exhibit a phase singularity. They take the form of spirals which rotate at a constant velocity (Fig.6c and Fig.6d). The effect of the diffusion is more subtle in this case. It makes the picture rotationally invariant, and it induces a wavelength selection from the center of the defect, which within the core exhibits radial equiphases [6]. In both cases, the defect associated with this symmetry is called a vortex .For obvious reason, such a defect does not exist in one dimension. Vortices of superfluid systems [7] correspond to the case of real eigenvalues, while spiral waves of chemical systems [8] and lasers [9] correspond to the case of complex eigenvalues.

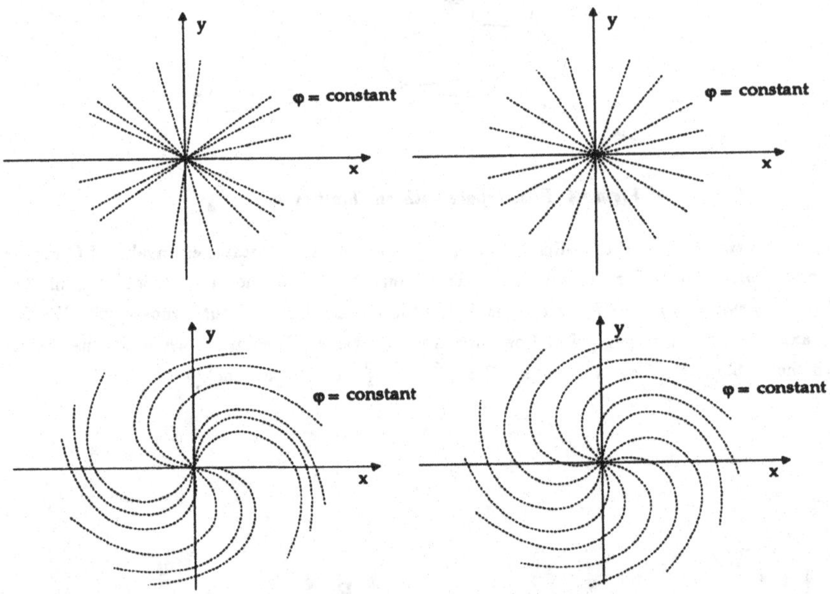

Figure 6.a-d Equiphases without and with diffusion for real (a,b) and complex eigenvalue (c,d)

A qualitative description of defects in the context of reaction-diffusion models has been sketched. We have illustrated the basic ideas of this approach with the simple examples of kinks and vortices. It can be extended to partial differential equations which includes propagation and dispersion effects and to more complex defects. This approach has been recently applied to describe defects near the roll-hexagon transition in Rayleigh-Bénard convection [10].

Acknowledgement:

This work has been supported by CEE twinning contract number 88100089FRPUJU1 and number SC1 0325-C (SMA). One of us (K. Emilsson) acknowledge the DRET for a financial support.

References

[1] "Pattern, Defects and Materials Instabilities", edited by D.Walgraef and N.M. Ghoniem, Kluwer Academic Publishers Group, 1990.

"Cellular Structures in Instabilities", edited by J.E. Wesfried and S.Zaleski, Springer Verlag, Berlin, 1984.

[2] D. Rogula, *Large Deformation of crystals, homotopy and defects* in "Trends in Application of Pure Mathematics to Mechanics", edited by G. Fichera, Pitman, New York, 1976.

G. Toulouse and M. Kléman, J. Phys. Lett. Paris **37** (1976) L 149

G. E. Volovik and V. P. Mineev, Zh. Eksp. Teor. Fiz. Pisma**24** (1976) 605 or JETP Lett. **24** (1976) 561.

[3] D. Mermin, Rev. Modern Phys. **51** (1979) 591

[4] "Défauts topologiques associés à la brisure de l'invariance de translation dans le temps", J. Lega, thesis, Université de Nice, 1989

[5] "Non-linear Oscillations, Dynamical Systems, and Bifurcations of vector fields, J. Guckenheimer, P. Holmes, Springer Verlag, Berlin, 1983.

[6] "Theory of Superconductivity", C.G. Kupper, Clarendon Press, Oxford, 1968

[7] "Spiral Waves in Reaction-Diffusion Equations", P. S. Hagan, S.I.A.M. J. Appl. Math. vol.42, **4**, 1982.

[8] "Chemical Oscillations, Waves and Turbulence", Y. Kuramoto, Springer Verlag, Berlin, 1984.

[9] P. Coullet, L. Gil, F. Rocca. Opt. Comm, *73*, 403, 1989.

[10] S. Ciliberto, P. Coullet, J. Lega, E. Pampaloni and C. Perez-Garcia. Phys. Rev. Lett. vol 65, 19, 2370, (1990)

SUBJECT INDEX

K

L

M

N

O

P

Q

R

of related interest

Instabilities and Nonequilibrium Structures
(Volume I)

Enrique Tirapegui and Danilo Villarroel (Eds.)

ISBN 90-277-2420-2, MAIA 33
1987, x + 337 pp.

TABLE OF CONTENTS

of related interest

Instabilities and Nonequilibrium Structures II

Dynamical Systems and Instabilities

Enrique Tirapegui and Danilo Villarroel (Eds.)

ISBN 0-7923-0144-7, MAIA 50
1989, x + 314 pp.

TABLE OF CONTENTS